Lecture Notes in Computer Science 8214

Commenced Publication in 1973
Founding and Former Series Editors:
Gerhard Goos, Juris Hartmanis, and Jan van Leeu'

Oren Kurland   Moshe Lewenstein
Ely Porat (Eds.)

String Processing
and Information Retrieval

20th International Symposium, SPIRE 2013
Jerusalem, Israel, October 7-9, 2013
Proceedings

 Springer

Volume Editors

Oren Kurland
Technion Institute of Technology
Faculty of Industrial Engineering and Management Technion
Haifa 32000, Israel
E-mail: kurland@ie.technion.ac.il

Moshe Lewenstein
Bar-Ilan University
Department of Computer Science
Ramat-Gan 52900, Israel
E-mail: moshe@cs.biu.ac.il

Ely Porat
Bar-Ilan University
Department of Computer Science
Ramat-Gan 52900, Israel
E-mail: porately@cs.biu.ac.il

ISSN 0302-9743   e-ISSN 1611-3349
ISBN 978-3-319-02431-8   e-ISBN 978-3-319-02432-5
DOI 10.1007/978-3-319-02432-5
Springer Cham Heidelberg New York Dordrecht London

Library of Congress Control Number: 2013948098

CR Subject Classification (1998): H.3, H.2.8, I.5, I.2.7, F.2, J.3

LNCS Sublibrary: SL 1 – Theoretical Computer Science and General Issues

Typesetting: Camera-ready by author, data conversion by Scientific Publishing Services, Chennai, India

Printed on acid-free paper

Springer is part of Springer Science+Business Media (www.springer.com)

Preface

In the 20 years since its inception in 1993 the International Symposium on String Processing and Information Retrieval (SPIRE) has become the reference meeting for the interdisciplinary community of researchers whose activity lies at the crossroads of string processing and information retrieval. This volume contains the proceedings of SPIRE 2013, the 20th symposium in the series.

The first four events concentrated mainly on string processing, and were held in South America under the title South American Workshop on String Processing (WSP) in 1993 (Belo Horizonte, Brazil), 1995 (Valparaiso, Chile), 1996 (Recife, Brazil), and 1997 (Valparaiso, Chile). WSP was renamed SPIRE in 1998 (Santa Cruz, Bolivia) when the scope of the event was broadened to include information retrieval. The change was motivated by the increasing relevance of information retrieval and its close interrelationship with the general area of string processing. From 1999 to 2007, the venue of SPIRE alternated between South / Latin America (odd years) and Europe (even years), with Cancun, Mexico in 1999; A Coruna, Spain in 2000; Laguna de San Rafael, Chile in 2001; Lisbon, Portugal in 2002; Manaus, Brazil in 2003; Padova, Italy in 2004; Buenos Aires, Argentina in 2005; Glasgow, UK in 2006; and Santiago, Chile in 2007. This pattern was broken when SPIRE 2008 was held in Melbourne, Australia, but it was restarted in 2009 when the venue was in Saariselkä, Finland, followed by Los Cabos, Mexico in 2010, Pisa, Italy in 2011, and in Cartagena de Indias, Colombia in 2012.

SPIRE 2013 was held in Jerusalem, Israel. The call for papers resulted in the submission of 60 papers. Each submitted paper was reviewed by at least three of the 40 members of the Program Committee, who eventually engaged in discussions coordinated by the three PC chairmen in cases of lack of consensus. We believe this resulted in a very accurate selection of the truly best submitted papers. As a result, 18 long papers and 10 short papers were accepted and have been published in these proceedings.

The main conference featured keynote speeches by Ido Dagan, Roberto Grossi, Robert Krauthgamer, and Yossi Matias, plus the presentations of the 18 full papers and 10 short papers. Following the main conference, on October 10, SPIRE 2013 hosted the Workshop on Compression, Text, and Algorithms (WCTA 2013).

We would like to take the opportunity to thank Yahoo!, Google, Bar-Ilan Univerity, and i-Core (Center of Excellence in Algorithms). All of them provided generous sponsorship. Thanks also to all the members of the Program Committee and to the additional reviewers, who went to great lengths to ensure the high quality of this conference, and to the coordinator of the SPIRE Steering

Committee, Ricardo Baeza-Yates, who provided assistance and guidance in the organization. We would like to thank the Local Organization Committee consisting of Amihood Amir, Tomi Klein, and Tsvi Kopelowitz (as well as ourselves). It is due to them that the organization of SPIRE 2013 was not just hard work, but also a pleasure.

October 2013   Oren Kurland
                                                          Moshe Lewenstein
                                                                 Ely Porat

Organization

Program Committee

Giambattista Amati	Fondazione Ugo Bordoni
Amihood Amir	Bar-Ilan University and Johns Hopkins University
Alberto Apostolico	Univ. of Padova and Georgia Tech
Ricardo Baeza-Yates	Yahoo! Research
Ayelet Butman	Holon Institute of Technology
Edgar Chavez	Universidad Michoacana
Raphael Clifford	University of Bristol
Carsten Eickhoff	Delft University of Technology
Johannes Fischer	Karlsruhe Institute of Technology
Inge Li Gørtz	Technical University of Denmark
Shunsuke Inenaga	Kyushu University
Markus Jalsenius	University of Bristol
Gareth Jones	Dublin City University
Jaap Kamps	University of Amsterdam
Tsvi Kopelowitz	Weizmann Institute of Science
Oren Kurland	Technion
Gad M. Landau	Haifa University
Avivit Levy	Shenkar College
Moshe Lewenstein	Bar Ilan University
Stefano Lonardi	UC Riverside
Andrew McGregor	University of Massachusetts, Amherst
Alistair Moffat	The University of Melbourne
Ian Munro	University of Waterloo
Gonzalo Navarro	University of Chile
Yakov Nekrich	University of Chile
Krzysztof Onak	IBM Research
Ely Porat	Bar-Ilan University
Berthier Ribeiro-Neto	Google Research
Benjamin Sach	University of Warwick
Rodrygo L.T. Santos	University of Glasgow
Srinivasa Rao Satti	University of Aarhus
Rahul Shah	Louisiana State Univeristy
Chris Thachuk	University of Oxford
Paul Thomas	CSIRO
Dekel Tsur	Ben Gurion University
Esko Ukkonen	University of Helsinki

Oren Weimann   University of Haifa
David Woodruff   IBM Almaden
Nivio Ziviani   Federal University of Minas Gerais
Guido Zuccon   CSIRO

Additional Reviewers

Atserias, Jordi   Ku, Tsung-Han
Bachrach, Yoram   Lecroq, Thierry
Bessa, Aline   Nakashima, Yuto
Bingmann, Timo   Patil, Manish
Biswas, Sudip   Petri, Matthias
Claude, Francisco   Rozenberg, Liat
Davoodi, Pooya   Shiftan, Ariel
Ferrada, Héctor   Sirén, Jouni
Flouri, Tomas   Tanaseichuk, Olga
Gagie, Travis   Tatti, Nikolaj
Gog, Simon   Thankachan, Sharma V.
Gupta, Ankur   Veloso, Adriano
Hata, Itamar   Vind, Soren
Hernandez, Cecilia   Wootters, Mary
Konow, Roberto

Table of Contents

Consolidating and Exploring Information via Textual Inference

Ido Dagan

Computer Science Department
Bar-Ilan University
dagan@cs.biu.ac.il

Effectively consuming information from large amounts of texts, which are often largely redundant in their content, is an old but increasingly pressing challenge. It is well illustrated by the perpetual attempts to move away from the flat result lists of search engines towards more structured fact-based presentations. Some recent attempts at this challenge are based on presenting structured information that was formulated according to pre-defined knowledge schemes, such as Freebase and Google's knowledge graph. We propose an alternative, as well as complementary, approach that attempts to consolidate and structure all textual statements in a document collection based on the inference relations between them. Generic textual inference techniques, formulated under the Textual Entailment paradigm, are used to consolidate redundant information into unique "core" statements, and then present them in an intuitive general-to-specific hierarchy. The talk will review some of the underlying concepts and algorithms behind our approach and present an initial demo.

O. Kurland, M. Lewenstein, and E. Porat (Eds.): SPIRE 2013, LNCS 8214, p. 1, 2013.

Pattern Discovery and Listing in Graphs

Roberto Grossi

Dipartimento di Informatica
Università di Pisa
grossi@di.unipi.it

Graphs are gaining increasing popularity in many application domains as they have the potential of modeling binary relations among entities. Along with textual and multimedia data, they are the main sources for producing large data sets. It is natural to ask how it is easy to extend the notion of patterns typically found in string matching and sequence analysis, to graphs and real-life networks. Unfortunately, even the basic problem of finding a simple path in a graph is NP-hard since this can establish if the graph is Hamiltonian. Also, the number of patterns can be exponentially large in the size of the graph, thus listing them is a challenge. We will discuss some output-sensitive and parameterized algorithms for listings patterns that are paths, cycles and trees, and provide a notion of "certificate" to attain this goal. This is joint work with Rui Ferreira.

O. Kurland, M. Lewenstein, and E. Porat (Eds.): SPIRE 2013, LNCS 8214, p. 2, 2013.
© Springer International Publishing Switzerland 2013

Efficient Approximation of Edit Distance

Robert Krauthgamer[*]

Faculty of Mathematics and Computer Science
Weizmann Institute of Science, Rehovot, Israel
robert.krauthgamer@weizmann.ac.il

Abstract. The similarity between two strings is often measured by some variant of edit distance, depending on the intended application. But even the basic version of this distance, which is simply the minimum number of character insertions, deletions and substitutions needed to transform one string to the other, presents remarkable algorithmic challenges.

This talk will examine the task of approximating the basic edit distance between two strings, starting with the classical RAM model and moving on to computational models which impose further constraints, such as the query complexity model and the sketching model. Beyond their concrete applications, these investigations provide a wealth of information about the problem, teaching us state of the art techniques and uncovering the limitations of certain methodologies. We will then come full circle with improved algorithms for the classical RAM model. During this journey, we may encounter special cases like permutation strings, whose patterns are all distinct, and smoothed instances, which are a mixture of worst-case and average-case inputs.

Finally, we shall discuss known gaps and open problems in the area, including variants of the basic edit distance, such as allowing block moves, or edit distance between trees, and hopefully touch upon related computational problems like nearest-neighbor search.

[*] Work was supported in part by the Israel Science Foundation (grant #897/13), the US-Israel BSF (grant #2010418), and by the Citi Foundation.

Nowcasting with Google Trends

Yossi Matias

Google

Since launching Google Trends we have seen extensive interest in what can be learned from search trends. A plethora of studies have shown how to use search trends data for effective nowcasting in diverse areas such as health, finance, economics, politics and more.

We give an overview of Google Trends and Nowcasting, highlighting some exciting Big Data challenges, including large scale engineering, effective data analysis, and domain specific considerations.

An extended summary will be available at `http://goo.gl/FbYh9`

O. Kurland, M. Lewenstein, and E. Porat (Eds.): SPIRE 2013, LNCS 8214, p. 4, 2013.
© Springer International Publishing Switzerland 2013

Space-Efficient Construction
of the Burrows-Wheeler Transform

Timo Beller[1], Maike Zwerger[1], Simon Gog[2], and Enno Ohlebusch[1]

[1] Institute of Theoretical Computer Science, University of Ulm, 89069 Ulm, Germany
{Timo.Beller,Maike.Zwerger,Enno.Ohlebusch}@uni-ulm.de
[2] Department of Computing and Information Systems, The University of Melbourne,
VIC, 3010, Melbourne, Australia
Simon.Gog@unimelb.edu.au

Abstract. The Burrows-Wheeler transform (BWT), originally invented
for data compression, is nowadays also the core of many self-indexes,
which can be used to solve many problems in bioinformatics. However,
the memory requirement during the construction of the BWT is often
the bottleneck in applications in the bioinformatics domain.

In this paper, we present a linear-time semi-external algorithm whose
memory requirement is only about one byte per input symbol. Our experiments show that this algorithm provides a new time-memory trade-off
between external and in-memory construction algorithms.

1 Introduction

In 1994 Burrows and Wheeler [5] presented the Burrows-Wheeler transform
(BWT). This reversible transformation produces a permutation of the input
string, in which symbols tend to occur in clusters. Because of this clustering,
in virtually all cases the BWT compresses much easier than the original string,
and Burrows and Wheeler suggested their transformation as a preprocessing step
in data compression. Data compression has become a major application for the
Burrows-Wheeler transform, e.g. it is the basis of the bzip2 algorithm.

Interestingly, the BWT has become the core of self-indexes [7, 14] which have
applications in bioinformatics and information retrieval. In the data compression
scenario it is possible to split a large input and construct the BWT for small
blocks, since decoding and encoding are done sequentially. However, this is not
possible for self-indexes because the optimal search routine requires the BWT
of the whole text. In this case, both the runtime and the memory requirement
of the construction of the BWT are critical. In the past, there were impressive
improvements in algorithms constructing the suffix array. Theoretical worst-case
time complexity, practical runtime and memory footprint have been improved.
As the BWT can easily (fast and space efficiently) be obtained from the suffix
array, the construction of the BWT profited indirectly from these improvements.
However, $n \log n$ bits seems to be a lower memory bound for fast suffix array
construction. On the other hand, this memory bound seems not to be valid
for BWT construction, as there are algorithms that directly construct the BWT

O. Kurland, M. Lewenstein, and E. Porat (Eds.): SPIRE 2013, LNCS 8214, pp. 5–16, 2013.
© Springer International Publishing Switzerland 2013

using less that $n \log n$ bits, but still depend on the input size. External algorithms take only a given amount of memory, which is independent of the input size and normally user defined. In the past, external algorithms were presented that compute the suffix array or the BWT, e.g. [4, 6, 8, 11]. While this approach finally solves the memory problem (the algorithm needs only as much memory as available), it is commonly known that external algorithms have a significant slow down.

Thus, external algorithms are only used when the input does not fit in RAM. Currently, this happens already for quite small files: In our experiments on a machine equipped with 8 GB RAM, the suffix array construction algorithm divsufsort[1] already suffered from swapping effects for inputs larger than 1.5 GB. A direct computation of the BWT may allow bigger inputs: An implementation of Sadakane[2] can construct the BWT for inputs up to 3 GB on that machine. But this implementation is limited to inputs of 4 GB (even if much more RAM would be available). We show in this paper that the space requirements can further be improved: We present a new semi-external algorithm to compute the BWT. Semi-external algorithms are in between internal algorithms and external algorithms. To be more precise, *semi-external algorithms* are—at least in this paper—algorithms that are allowed to use an input dependent amount of memory (like internal algorithms), but also use disk memory (like external algorithms). In practice, semi-external algorithms store all data on disk that is accessed sequentially, while data with random access pattern is kept in main memory. Our implementation has no limitation on the input size and can construct the BWT of a 6 GB file with only 8 GB of RAM. In contrast, internal suffix array construction algorithms would need over 54 GB of RAM (or 31 GB if bit compression would be used) to compute the suffix array of a 6 GB file, because they must keep at least the input and the output in memory.

2 Preliminaries

Let Σ be an ordered alphabet of size σ whose smallest element is the so-called sentinel character \$. In the following, S is a string of length n on Σ having the sentinel character at the end (and nowhere else). For $1 \leq i \leq n$, $S[i]$ denotes the *character at position* i in S. For $i \leq j$, $S[i..j]$ denotes the *substring* of S starting with the character at position i and ending with the character at position j. Furthermore, S_i denotes the i-th suffix $S[i..n]$ of S. The *suffix array* SA of the string S is an array of integers in the range 1 to n specifying the lexicographic ordering of the n suffixes of S, that is, it satisfies $S_{SA[1]} < S_{SA[2]} < \cdots < S_{SA[n]}$.

The suffix array SA is often enhanced with the so-called LCP-array containing the lengths of longest common prefixes between consecutive suffixes in SA. Formally, the LCP-array is an array so that $\mathsf{LCP}[1] = -1 = \mathsf{LCP}[n+1]$ and $\mathsf{LCP}[i] = |\mathsf{lcp}(S_{SA[i-1]}, S_{SA[i]})|$ for $2 \leq i \leq n$, where $\mathsf{lcp}(u, v)$ denotes the longest common prefix between two strings u and v. The Burrows-Wheeler

---

[1] http://code.google.com/p/libdivsufsort/
[2] http://researchmap.jp/muuw41s7s-1587/#_1587

transform [5] converts a string S into the permuted string BWT$[1..n]$ defined by BWT$[i] = S[SA[i] - 1]$ for all i with SA$[i] \neq 1$ and BWT$[i] = \$$ otherwise.

As in [15–18], we distinguish between S-type, L-type and LMS-type suffixes: S_i is called S-type if $i = n$ or $S_i < S_{i+1}$. Analogously, we call S_i an L-type suffix if $S_i > S_{i+1}$. An S-type suffix is (also) an LMS-type suffix provided that S_{i-1} is an L-type suffix. Note that S_1 is never an LMS-type suffix, but S_n is always an LMS-type suffix. We call $S[i..j]$ an LMS-substring if S_i and S_j are LMS-type suffixes and for every $k, i < k < j$, S_k is not of type LMS. Additionally, $\$S[1..k]$ is also an LMS-substring, where S_k is the first LMS-type suffix in S.

A rank query $rank_b(B, i)$ on a bit-vector B counts the number of occurrences of bit b in $B[1..i]$. Similarly a select query $select_b(B, i)$ on a bit-vector B returns the position of the i-th occurrence of bit b in B. By pre-processing B one can answer both queries in constant time [10].

3   Related Work

There are many suffix array construction algorithms with different time and space complexities. We refer to the overview article [19] for details. It is widely agreed that in practice Yuta Mori's divsufsort is one of the fastest algorithms to compute the SA. For $n < 2^{31}$, it uses $5n$ bytes and $9n$ bytes otherwise.

In contrast to suffix array construction algorithms, the direct computation of the BWT has received much less attention. In [13], it is shown how to compute the BWT for biological data in $\mathcal{O}(n \log n)$ time. In [18], a linear-time algorithm for computing the Burrows-Wheeler transform was presented. This algorithm uses $\mathcal{O}(n \log \sigma \log \log_\sigma n)$ working space.

External algorithms for computing the BWT are described in [8, 11]. They construct the BWT by splitting the input into blocks of fixed length and computing the BWTs of these blocks. Afterwards, one has to merge the BWTs of the blocks to obtain the BWT of the input. In contrast, [1] presented an external algorithm for computing the BWT of a collection of short strings. However, this task is conceptually easier and can not easily be adapted to the case of arbitrary strings.

Algorithms also exist for computing the suffix array in external memory, see e.g. [6]. Very recently, [4] presented an external algorithm, not only for suffix array construction, but also for the computation of the LCP array. This algorithm is also based on the induced sorting algorithm and it is reported to be faster than the previous external suffix array construction algorithms.

4   The Induced Sorting Algorithm

As our new algorithm is based on the induced sorting algorithm, we briefly revisit this elegant algorithm here. For more details and correctness, we refer to [15].

The suffix array can be divided into σ buckets, where all suffixes in a bucket start with the same character. Within a bucket, L-type suffixes are smaller than

S-type suffixes. So every bucket can further be divided in two ranges, an L-type range and an S-type range. In the following, assume that A is an array of size n, which is divided into buckets and ranges as described before. Fig. 1 illustrates the induced sorting algorithm by an example.

Step 1. Input S is scanned from right to left in order to detect all indexes j in S at which an LMS-type suffix starts. All these indexes are written consecutively to the rightmost free position in the S-type range of the corresponding $S[j]$ bucket in A.

Step 2. Array A is scanned from left to right. Assume we are at position i in A. If $A[i]$ is empty, we go to the next position $i + 1$. Otherwise let $j = A[i]$; we check if $S[j - 1] \geq S[j]$. If so, we delete $A[i]$ and write $j - 1$ to the leftmost free position in the L-type range of the corresponding $S[j - 1]$ bucket.

Step 3. After we finished the left-to-right scan, we scan A from right to left. Assume again that we are at position i of the array A. If $A[i]$ is empty, we go to the next position $i - 1$. Otherwise for $j = A[i]$ we check if $S[j - 1] \leq S[j]$. If so, we delete $A[i]$ and write $j - 1$ to the rightmost free position in the S-type range of the corresponding $S[j - 1]$ bucket.

Step 4. The two scans in steps 2 and 3 sort the LMS-substrings (but not the LMS-type suffixes). In this step, the induced sorting algorithm replaces each LMS-substring by its lexicographical name and concatenates them in text order. First, all LMS-type positions are moved to the second half of A. This is possible because there are at most $\frac{n}{2}$ LMS-substrings. Then the second half of A is scanned from left to right. Assume that we are at a non-empty position i in A and $j = A[i]$. We compare the LMS-substring starting at $S[j]$ with the LMS-substring starting at $S[A[i-1]]$. If the substrings are identical, j gets the same lexicographical name, otherwise, j gets the next larger lexicographical name. The name is moved to $A[\lfloor \frac{i}{2} \rfloor]$. Finally, all names are placed into the second half of A, overwriting the LMS-type positions there. We now interpret these values as a new string S'. Note that S' usually has a different alphabet size than S.

Step 5. The order of the LMS-type suffixes is now obtained from the suffix array of S'. If every symbol in S' is unique, then one can easily create the suffix array. Otherwise, the induced sorting algorithm recursively computes the suffix array of the string S'. In either case, the suffix array of S' is written to the first half of A.

Step 6. The inverse suffix array of S' is now calculated and stored in the second half of A (overwriting S'). Then, a right-to-left scan of S is executed to find all LMS-type positions (again). Each LMS-type position is written (with the help of the inverse suffix array of S') in the correct lexicographical order to the first half of A. Afterwards the induced sorting algorithm removes the inverse suffix array of S' and places the LMS-type positions stably into the S-type ranges of their corresponding buckets in A.

Step 7. Array A is scanned from left to right and the indexes are moved as described in step 2 . However, this time indexes placed into an L-type range are not erased.

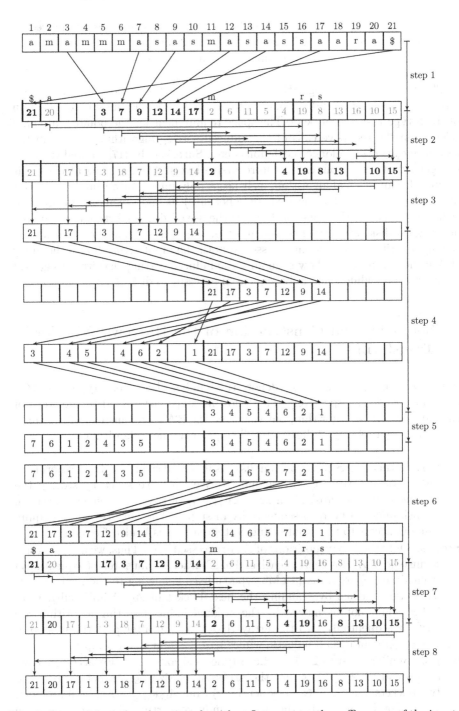

Fig. 1. Steps of the induced sorting algorithm: It computes the suffix array of the input string *amammmasasmasassaara*$. The movements of the indexes are illustrated with arrows, temporary results are shown in gray.

Step 8. Array A is scanned from right to left and the indexes are moved as described in step 3, but again indexes placed into an S-type range are not erased. After this step, A contains the suffix array of S.

The induced sorting algorithm, as described in this section, uses the input string S, the array A and σ pointers to the rightmost (leftmost) free position of the S-type (L-type) buckets. The space requirement for the pointers are only relevant in the recursive calls of the induced sorting algorithm because in the recursive calls σ is no longer negligible small. Surprisingly, [17] showed that one can get rid of these pointers in the recursive levels. The resulting algorithm is optimal for an internal algorithm, as it keeps only input, output and a constant number of variables (for constant alphabet size) in main memory. In order to reduce the space further, one has to allow the use of disk. Unfortunately, random accesses on disk are very slow and most of the accesses done by the induced sorting algorithm are random accesses to both the input string S and the array A. We show in Section 5 how to modify the induced sorting algorithm to get rid of the A array, while using only sequential accesses to disk.

5   Semi-external Construction of the Burrows-Wheeler Transform

In this section n_S, n_L, and n_{LMS} denote the number of S-type, L-type and LMS-type suffixes of S, respectively. The following steps correspond to the steps of the induced sorting algorithm, but this time the BWT of S instead of the suffix array is calculated. Fig. 2 illustrates all steps of the new algorithm.

Step 1. In this step S can reside on disk, as it is read sequentially. Furthermore, only n_{LMS} indexes are written into A. We can save space by storing the indexes (without gaps) in an array $A_{LMS,left}$ of size n_{LMS}, which is written to disk and will be read sequentially in step 2. The next two steps require random access to the input string S, therefore S is loaded from disk.

Step 2. Only the L-type positions of A are accessed here. Thus, we use A_L of size n_L instead of A. However, at each end of a bucket, we must read (sequentially) from $A_{LMS,left}$ to place all LMS-type suffixes belonging to the next bucket. Then we continue by scanning the next bucket of A_L. Additionally, if an index would not be moved by the original induced sorting algorithm, we write it sequentially into the array $A_{LMS,right}$. So after performing step 2, $A_{LMS,right}$ contains all indexes, while $A_{LMS,left}$ and A_L are empty.

Step 3. Similar to step 2, only the S-type positions of A are accessed now. So instead of A, we now use the array A_S of size n_S. As before, between two buckets one must read (sequentially) from $A_{LMS,right}$. Again, if an index would not be moved by the original induced sorting algorithm, we write it sequentially into $A_{LMS,left}$. At the end of step 3, $A_{LMS,left}$ contains the LMS-type indexes sorted according to their corresponding LMS-substrings.

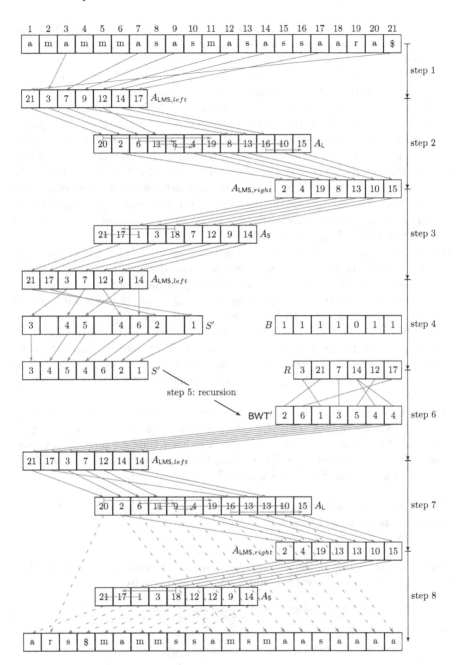

Fig. 2. Steps of the semi-external construction of the Burrows-Wheeler Transform for the input string *amammmasasmasassaara*$

Step 4. For the creation of the renamed string S' a bit-vector B of size n_{LMS} is computed, which indicates whether two consecutive entries in $A_{\mathsf{LMS},left}$ correspond to identical LMS-substrings or not. During this computation an array R is constructed, which contains the mapping of the lexicographical names for the LMS-substrings to their (end)positions in S. For identical LMS-substrings only one position has to be stored. Then $A_{\mathsf{LMS},left}$ is read sequentially again. The corresponding lexicographical name of $j = A_{\mathsf{LMS},left}[i]$ (determined with the help of B) is written to $S'[\lfloor \frac{i}{2} \rfloor]$. Afterwards, the gaps in S' are removed (preserving the order of the entries), and $A_{\mathsf{LMS},left}$ is deleted.

Step 5. The Burrows-Wheeler transform of S' (called BWT$'$) is calculated directly if the characters of S' are pairwise distinct, and recursively otherwise.

Step 6. The array R contains the mapping of the lexicographical names to the corresponding positions in S. So $A_{\mathsf{LMS},left}$ can be filled sequentially based on the equation $A_{\mathsf{LMS},left}[i] = R[\mathsf{BWT}'[i]]$.

Step 7. Array A_{L} is created again and scanned as in step 2 from left to right. Between two buckets, one must read (sequentially) from $A_{\mathsf{LMS},left}$. Again, if an index would not be moved by the original induced sorting algorithm, we write it sequentially to the array $A_{\mathsf{LMS},right}$. Additionally, we begin to produce the BWT of S. To be precise, every time an index i is placed into A_{L}, the character $S[i-1]$ is written to the correct position of the BWT. So after performing this step, every entry in the BWT that corresponds to an L-type suffix is set correctly, while there are gaps corresponding to S-type suffixes. Furthermore, $A_{\mathsf{LMS},right}$ contains all indexes, while $A_{\mathsf{LMS},left}$ and A_{L} are empty.

Step 8. Array A_{S} is created again and scanned as described in step 3 from right to left (but we do not need $A_{\mathsf{LMS},left}$). During this computation, the gaps of the BWT of S are filled. To be precise, each time an index i is placed into A_{S}, the character $S[i-1]$ is written to the correct position of the BWT. After this step, the BWT of S is completely calculated.

The correctness and linear runtime of this algorithm follows directly from the correctness and runtime of the induced sorting algorithm.

Table 1. Access pattern to the data structures during the different steps of the algorithm. In step 4, random access is needed first to S and then to S'.

step	random access	sequential access
1	$A_{\mathsf{LMS},left}$	S
2	S, A_{L}	$A_{\mathsf{LMS},left}$, $A_{\mathsf{LMS},right}$
3	S, A_{S}	$A_{\mathsf{LMS},left}$, $A_{\mathsf{LMS},right}$
4	S/S', B	$A_{\mathsf{LMS},left}$, R
5	S'	
6	R	BWT$'$
7	S, A_{L}	BWT$'$, $A_{\mathsf{LMS},right}$, BWT
8	S, A_{S}	$A_{\mathsf{LMS},right}$, BWT

Table 1 summarizes which data structures are needed in memory, and which can reside on disk because only sequential access is needed. The memory peak is now in steps 2, 3, 7, and 8 because in these steps the text and a relatively large array (A_L or A_S) is accessed randomly. However, one can reduce the space for A_L and A_S because of the special access pattern: These arrays are read sequentially, while the write access occurs only at positions that were not already read. We describe now how to replace A_L of size n_L with A'_L of size $k < n_L$. The idea is to split A_L in $\lceil \frac{n_L}{k} \rceil$ parts of size k. A'_L covers only one part of A_L, while for all other parts arrays P_i are created. Assume that we have to write value v to position p, where p does not belong to the part of A_L that corresponds to A'_L. In this case, both values v and p are written to the corresponding array P_i. When our reading position reaches the end of A'_L, we read the array P_i that covers the next part and write the values with an appropriate offset to A'_L. Because read and write accesses on P_i are sequentially, it can reside on disk. We deal analogously with A_S.

6   Practical Optimization for Very Small Alphabets

The BWT has important applications in bioinformatics. In this field, the alphabet size is very small, e.g. 4 or 5 in case of DNA data. Thus, it is worthwhile to optimize the algorithm for inputs with very small alphabet.

Let ℓ be a fixed natural number. We call an LMS-substring s *short* if $|s| \leq \ell$ and *long* otherwise. For a short LMS-substring s, we define its number as:

$$number(s) = \sum_{i=1}^{|s|} ord(s[i]) \cdot \sigma^{\ell-i} + \sum_{i=|s|+1}^{\ell} (\sigma - 1) \cdot \sigma^{\ell-i}$$

where $ord(a) = |\{a' \in \Sigma : a' < a\}|$ for every $a \in \Sigma$. For two short LMS-substrings s_1 and s_2, $number(s_1) < number(s_2)$ if and only if s_1 has a smaller lexicographical name than s_2. Now, we can obtain S' by another approach: We create a bit-vector B_{short} of size σ^ℓ to mark the numbers of all short LMS-substrings. By scanning S once from right to left, all LMS-substrings can be found. If the current LMS-substring s is short, we calculate its number $i = number(s)$ and set $B_{short}[i] = 1$. Otherwise, we store its starting position together with its position in S' (which is the number of LMS-type suffixes before the current one in S). Afterwards we (naively) sort the long LMS-substrings according to their lexicographical order. Then we create another bit-vector B_{LMS}, where $B_{LMS}[i] = 0$ if the i-th smallest LMS-substring is longer than ℓ and $B_{LMS}[i] = 1$ otherwise. B_{LMS} can be calculated by scanning V and B_{short} in parallel. During this scan, the lexicographical names of the long LMS-substrings can be written to S'. At last, the lexicographical names of the short LMS-substrings are inserted into S': S is scanned again from right to left. When we find a short LMS-substring s, we calculate the number of short LMS-substrings that are smaller than s by $r = rank_1(B_{short}, number(s))$, and obtain the lexicographical name with $select_1(B_{LMS}, r)$.

Sorting the long LMS-substrings can be done in $\mathcal{O}(n \log n)$ using multikey quicksort [3], so this optimization does not have a linear runtime. However, it is in practice faster than the linear method described in Section 5 because we can exploit that LMS-substrings are usually very short and thus (for $\ell = 8$) there are not so many long LMS-substrings. Unfortunately, this optimization does not work in the recursive steps because in the recursive calls the alphabet size is not small enough.

7   Experimental Results

We implemented the algorithm using Simon Gog's [9] library sdsl (http://github.com/simongog/sdsl). In particular, we used bit-compressed integers, which causes a slow down but avoids problems with inputs larger than 2^{32}.

The experiments were conducted on a machine with a Intel(R) Core i5-3570 processor (3.40 GHz; L1 Cache=256 KB, L2 Cache=1 MB, and L3 Cache=6 MB) and 8 GB RAM. The operating system was Ubuntu 12.04.2 LTS. All programs were compiled with g++ (version 4.6.3) using the provided makefile.

As test files we used DNA data of different size because this is the main application. We concatenated the genomes[3] from Human (hg19), Mouse (mm10) and Gorilla (gorGor3) and deleted all characters other than A, C, G, T and N. Then we took prefixes of size 1 GB (genome1), 3 GB (genome2), and 6 GB (genome3).

For a comparison with internal memory algorithms, we used Yuta Mori's divsufsort. It needs $5n$ bytes for inputs smaller than 2^{31} and $9n$ bytes otherwise. Additionally, an implementation from Sadakane (called dbwt in the following) was used. This implementation is based on [18] (but has some simplifications compared to the algorithm described in [18]) and usually uses less than $2.5n$ bytes. Unfortunately, dbwt is limited to inputs smaller than 2^{32} bytes and it is unclear if it can be modified so that it can handle bigger inputs without increasing the memory footprint or runtime.

For a comparison with external memory algorithms, we took the following three implementations: bwtdisk 0.9.0 from Giovanni Manzini based on the algorithms described in [8]. This program can handle compressed inputs and can produce compressed outputs, but we did not make use of that option. LS from Kunihiko Sadakane. It is an external memory variant of the Larsson-Sadakane algorithm presented in [12]. This implementation can use multiple processors and we tested it with all 4 available processors. eSAIS 0.5.2 [4] does not compute the BWT but the suffix array and (optional) the LCP array. We turned the LCP construction off to construct only the suffix array.

For a fair comparison with our new algorithm, we allowed each external implementation to take n bytes of RAM. However, LS can only take a power of 2, so we allowed it the usage of 2^{32} byte for the 3 GB input and 2^{33} byte for the 6 GB input.

---

[3] Downloaded from http://genome.ucsc.edu

Table 2. Each column shows the runtime in seconds and in parentheses the maximum memory usage in byte per input character. The files genome2 and genome3 were too large for divsufsort on the machine equipped with 8 GB of RAM. Because dbwt is limited to files smaller than 4 GB, genome3 (6 GB) could not be calculated with dbwt.

algorithm	genome1	genome2	genome3
divsufsort	204 (5.00)	-	-
dbwt	229 (1.95)	705 (2.00)	-
this paper	412 (1.00)	1 475 (1.00)	3 387 (1.00)
bwtdisk	1 751 (1.05)	5 693 (1.05)	12 342 (1.05)
eSAIS	4 042 (1.08)	14 225 (1.02)	28 324 (1.06)
LS	9 382 (0.82)	34 200 (1.07)	94 728 (1.07)

Table 2 shows the experimental results. On the small genome1 file, divsufsort is the fastest algorithm, followed by dbwt. Compared to dbwt our algorithm is about 2 times slower, but uses only about half of the space. The same is true for the genome2 file. The 6 GB file (genome3) was far too big for the internal memory algorithms divsufsort and dbwt on the machine with 8 GB of RAM. The suffix array construction algorithm divsufsort would require about 54 GB of RAM and dbwt is limited to inputs of at most 4 GB. That is why our algorithm is important. Of course, one can always resort to an external algorithm if internal memory algorithms need too much RAM. But as our experiments show, our algorithm is the faster alternative (provided that there is enough RAM for it): The implementation described in this paper is over 3 times faster than the fastest external algorithm bwtdisk. Compared to eSAIS it is nearly one order of magnitude faster. However, one should keep in mind that the comparison with eSAIS is not fair because eSAIS constructs the suffix array and not the BWT.

8   Conclusion and Future Work

In this paper we presented a new method to construct the BWT space efficiently. It is a semi-external algorithm, which is based on the induced sorting algorithm. The implementation is not limited to inputs smaller than 4 GB and experiments show that it needs only about n bytes to compute the BWT of a length n DNA sequence. Thus, it needs about half of the space dbwt uses and over 5 times less space than suffix array construction algorithms. Furthermore, it is faster than external algorithms when they are allowed to use n bytes of memory. So only in cases when the input does not fit in RAM, external algorithms must be used. In all other cases, one can construct the BWT with a non-external algorithm. Note that n bytes are enough to compute the LCP-array from the BWT as shown in [2] and also to construct the suffix array (semi-externally) from the BWT. So it is now possible to construct SA, BWT and LCP with about n bytes without using an external algorithm. These arrays are components of several full-text indexes.

In the full paper, we will show how the presented algorithm can be modified so that it directly computes the suffix array.

References

1. Bauer, M.J., Cox, A.J., Rosone, G.: Lightweight algorithms for constructing and inverting the BWT of string collections. Theoretical Computer Science 483, 134–148 (2013)
2. Beller, T., Gog, S., Ohlebusch, E., Schnattinger, T.: Computing the longest common prefix array based on the Burrows-Wheeler transform. Journal of Discrete Algorithms 18, 22–31 (2013)
3. Bentley, J.L., Sedgewick, R.: Fast algorithms for sorting and searching strings. In: Proc. 8th Annual ACM-SIAM Symposium on Discrete Algorithms, pp. 360–369 (1997)
4. Bingmann, T., Fischer, J., Osipov, V.: Inducing suffix and lcp arrays in external memory. In: Proc. Wkshp. Algorithm Engineering and Experiments (2013)
5. Burrows, M., Wheeler, D.J.: A block-sorting lossless data compression algorithm. Research Report 124, Digital Systems Research Center (1994)
6. Dementiev, R., Kärkkäinen, J., Mehnert, J., Sanders, P.: Better external memory suffix array construction. Journal of Experimental Algorithmics 12, Article No. 3.4 (2008)
7. Ferragina, P., Manzini, G.: Opportunistic data structures with applications. In: Proc. IEEE Symposium on Foundations of Computer Science, pp. 390–398 (2000)
8. Ferragina, P., Gagie, T., Manzini, G.: Lightweight data indexing and compression in external memory. Algorithmica 63(3), 707–730 (2012)
9. Gog, S.: Compressed Suffix Trees: Design, Construction, and Applications. PhD thesis, University of Ulm, Germany (2011)
10. Jacobson, G.: Space-efficient static trees and graphs. In: Proc. 30th Annual Symposium on Foundations of Computer Science, pp. 549–554. IEEE (1989)
11. Kärkkäinen, J.: Fast BWT in small space by blockwise suffix sorting. Theoretical Computer Science 387(3), 249–257 (2007)
12. Larsson, J., Sadakane, K.: Faster suffix sorting. Theoretical Computer Science 387(3), 258–272 (2007)
13. Lippert, R.A., Mobarry, C.M., Walenz, B.P.: A space-efficient construction of the Burrows-Wheeler transform for genomic data. Journal of Computational Biology 12(7), 943–951 (2005)
14. Navarro, G., Mäkinen, V.: Compressed full-text indexes. ACM Computing Surveys 39(1), Article No. 2 (2007)
15. Nong, G., Zhang, S., Chan, W.: Linear suffix array construction by almost pure induced-sorting. In: Proc. Data Compression Conference, pp. 193–202 (2009)
16. Nong, G., Zhang, S., Chan, W.: Two efficient algorithms for linear time suffix array construction. IEEE Transactions on Computers 60(10), 1471–1484 (2011)
17. G. Nong Practical Linear-Time O(1)-Workspace Suffix Sorting for Constant Alphabets. ACM Transactions on Information Systems (to appear, July 2013)
18. Okanohara, D., Sadakane, K.: A linear-time Burrows-Wheeler transform using induced sorting. In: Karlgren, J., Tarhio, J., Hyyrö, H. (eds.) SPIRE 2009. LNCS, vol. 5721, pp. 90–101. Springer, Heidelberg (2009)
19. Puglisi, S.J., Smyth, W.F., Turpin, A.: A taxonomy of suffix array construction algorithms. ACM Computing Surveys 39(2), Article No. 4 (2007)

Using Mutual Influence
to Improve Recommendations

Aline Bessa, Adriano Veloso, and Nivio Ziviani

Universidade Federal de Minas Gerais
Department of Computer Science, Belo Horizonte, Brazil
{alinebessa,adrianov,nivio}@dcc.ufmg.br

Abstract. In this work we show how items in recommender systems mutually influence each other's utility and how it can be explored to improve recommendations. The way we model mutual influence is cheap and can be computed without requiring any source of content information about either items or users. We propose an algorithm that considers mutual influence to generate recommendations and analyse it over different recommendation datasets. We compare our algorithm with the $Top - N$ selection algorithm and obtain gains up to 17% in the utility of recommendations without affecting their diversity. We also analyse the scalability of our algorithm and show that it is as applicable for real-world recommender systems as $Top - N$.

Keywords: Recommender systems, theory of choice, mutual influence, collaborative filtering.

1   Introduction

Consumers from widely varying backgrounds are inundated with options that lead to a situation known as "information overload", where the presence of too much information interferes with decision-making processes [1]. To circumvent it, content providers and electronic retailers have to identify a small yet effective amount of information that matches users expectations. In this scenario, Recommender Systems have become tools of paramount importance, providing a few personalized recommendations that intend to suit user needs in a satisfactory way. One type of such systems, known as Collaborative Filtering [2], generally works as follows: (i) prediction step - keeps track of consumers known preferences to predict items that may be interesting to other consumers; (ii) recommendation step - selects predictions, ranks and recommends them to consumers.

Traditionally, predictions are scores assigned to items with respect to a certain consumer. The higher the score the higher the compatibility between the items in question and consumer's known preferences. It is therefore intuitive to think that the N items with highest scores should be the ones chosen in the recommendation step. This approach though, known as $Top - N$ recommendation, does not consider the utility of the recommended list as a whole, focusing exclusively on individual scores. As we show in this work, items exert a mutual influence on

O. Kurland, M. Lewenstein, and E. Porat (Eds.): SPIRE 2013, LNCS 8214, pp. 17–28, 2013.

their utilities, i.e. the quality of an item depends not only on its own score but also on which other items are presented in the $Top - N$ recommendations.

This work is motivated by the theory of choice of Amos Tversky [3], which indicates that preference among items depends not only on the items' specific features, but also on the presented alternatives. In the context of movies, for instance, it is equivalent to state that an action fan may prefer a mediocre karate movie over romance titles, but would not interact with this same karate movie when presented with better action films. We investigate the influence alternatives exert on each other, and how this information can be used to improve recommendations utility. It could be either embedded on the predictions computation or weighed right after they are generated in the recommendation step. We here focus on the recommendation step.

We propose a novel algorithm that incorporates mutual influence to perform the selection of a set of N items, which we call *GSMI - Greedy Selection based on Mutual Influence*. We conducted a systematic evaluation of *GSMI* involving different recommendation scenarios and distinct datasets. In order to evaluate *GSMI*, we used the utility metric proposed in [4] and measured diversity using the framework proposed in [5].

In summary, the main contributions of this work are (i) a cheap way of modelling and computing the mutual influence items exert on each other's utility, (ii) a new algorithm that considers mutual influence to select items in the recommendation step of a collaborative filtering, (iii) a thorough evaluation of its benefits in recommendation tasks – we compare *GSMI* with $Top - N$ and obtain significant gains in the utility of recommendations without affecting their diversity –, (iv) an analysis of the scalability of *GSMI*, which indicates that the algorithm is applicable for real-world recommender systems.

This paper is structured as follows. Section 2 discusses previous related work and connects it to our study. Section 3 outlines some basic concepts that are the foundations of this work. Section 4 details *GSMI* and our evaluation methodology. Section 5 presents experiments that demonstrate the efficiency and efficacy of *GSMI*, taking different datasets into account. Finally, Section 6 details our conclusions and future work.

2   Related Work

In this section, we present related work in $Top - N$ recommendations, dependencies among items, and learning to rank, as described next.

Top − N Recommendation. The state of the art prediction algorithms for $Top - N$ recommendations, when explicit feedback is available, are PureSVD and NNCosNgbr (Non-normalized Cosine Neighborhood) [6]. A $Top - N$ recommendation step when predictions are generated by these algorithms performs better than some sophisticated Learning to Rank methods. PureSVD is based on latent factors, i.e. users and items are modelled as vectors in a same vector space and the score of user u for item i is predicted via the inner-product between their corresponding vectors. NNCosNgbr works upon the concept of neighborhood, computing predictions according to feedback of similar users/items. [6] is related to our work because we use both PureSVD and NNCosNgbr as our prediction algorithms and compare *GSMI* with *Top − N* for the recommendation step.

Dependencies among Items. In 1972, Amos Tversky proposed a model acording to which a user chooses among options by sets of item aspects – an example would be {*price* < $100.00} [3]. We do not assume that items' features are available and therefore do not model aspects in our approach, but we do rely on the idea that user choice depends on all presented alternatives – i.e. such alternatives interfere with each other's utility. Another work that relies on Economics principles to model dependencies among items is that of Wang [7]. Inspired by the Modern Portfolio Theory in finance, Wang derives a document ranking algorithm that extends the Probability Ranking Principle by considering both the uncertainty of relevance predictions and correlations between retrieved items. This work is the closest to ours.

Learning to Rank. LTR (*Learning to Rank*) are supervised methods to automatically build ranking models for items [8]. Although we do not generate an ordering among the items selected in the recommendation step, we do use supervised learning to compute mutual influence and perform selections. An LTR work that is somewhat close to ours is that of Xiong et al [9]. In an advertisement scenario, they observed that the CTR (*Click-Through Rate*) of an ad is often influenced by the other ads shown alongside. Based on it, they designed a Continuous Conditional Random Field for click prediction focusing on how ads influence each other. Another work that models influence among items and explores it to perform an LTR is [10]. In this paper, influences are modelled as similarities among items and embedded in a latent structured ranking method afterwards.

3   Basic Concepts

In this work, there are two fundamental sources of evidence that are used to select which items should be recommended to a certain user: (i) individual scores ϕ generated in the prediction step by either PureSVD or NNCosNgbr, and (ii) pairwise scores θ that quantify mutual influence among items. Both ϕ and θ are real values in the interval $[0, 1]$.

The pairwise scores θ work in a positive way: the higher they are, the higher the utility of the items in question when selected to a same recommendation list. Given items a and b, $\theta(a, b)$ should ideally be computed considering only cases where they are simultaneously selected to recommendation. Unfortunately, it is not possible to reconstruct the recommendation lists in none of the studied datasets. As a consequence, it is not possible to know which items were presented to users together and therefore we compute $\theta(a, b)$ considering users historical data as a whole, as detailed in Section 3.2.

In this work, we assume that predictions are generated to K items, and then N, $N \leq K$, items must be selected to compose a recommendation list. Typical values for N are 5 and 10, and depending on the prediction algorithm K can be equivalent to the total number of items in the dataset [8]. We also assume that users explicitly give feedback to items, and depending on the system it can be a rating, a purchase signal (0/1), a click (0/1) etc. Next, we detail how we compute individual scores ϕ, pairwise scores θ, and how they are combined.

3.1   Individual Scores ϕ

Individual scores are generated by prediction algorithms that are divided into two categories: neighborhood-based and model-based [8]. The latter have recently enjoyed much interest due to related outstanding results in the Netflix competition, a popular event in the recommender systems field that took place between 2006 and 2009[1]. Nonetheless, neighborhood-based prediction algorithms usually provide a more concise and intuitive justification for the computed predictions, and are more stable, being little affected by the addition of users, items, or ratings [8]. The predictors used in this paper are PureSVD (model-based) and NNCosNgbr (neighborhood-based), both state of the art methods for $Top - N$ recommendations when explicit feedback is available.

The input for PureSVD is a User $\times$ Item matrix M filled up as follows:

$$M_{ui} = \begin{cases} \text{numerical feedback, if consumer } u \text{ gave feedback about item } i, \\ 0, \text{if not.} \end{cases} \tag{1}$$

PureSVD consists in factorizing M via SVD as $M = U \times E \times Q$, where U is an orthonormal matrix, E is a diagonal matrix with the first γ singular values of M, and Q is also an orthonormal matrix. The prediction of an individual score $\phi(i)$ given a user u is thus given by:

$$\phi(i) = M_u \times Q^T \times Q_i \tag{2}$$

where M_u is the u-th row of M corresponding to user u latent factors, Q^T is the transpose of Q, and Q_i is the i-th row of Q corresponding to item i latent factors.

NNCosNgbr is a neighborhood model that bases its predictions on similarity relationships among either users or items. Working with items similarities usually lead to better accuracy rates and more scalability [11]. In this case, recommendations can be explained in terms of the items that users have already interacted with via ratings, purchases, likes etc [11]. Due to these reasons, we focus on item-based NNCosNgbr. The prediction of an individual score $\phi(i)$ given a user u is computed as follows:

$$\phi(i) = b_{ui} + \sum_{j \in D^k(u;i)} d_{ij}(r_{uj} - b_{uj}) \tag{3}$$

---

[1] http://en.wikipedia.org/wiki/Netflix_Prize

where b_{ui} is a combination of user and item biases as in [12], $D^k(u; i)$ is the set of k items rated by u that are the most similar to i, d_{ij} is the similarity between items i and j, r_{uj} is an actual feedback given by u to j and b_{uj} is the bias related to u and j.

Biases are taken into consideration because they mask the fundamental relations between items. Item biases include the fact that certain items tend to receive better feedback than others. Similarly, user biases include the tendency of certain users to give better feedback than others. Finally, the similarity among items, used to compute both $D^k(u; i)$ and d_{ij}, is measured with the adjusted cosine similarity [6].

3.2   Pairwise Scores θ

The pairwise scores $\theta(i, j)$ capture the mutual influence items i and j have on their own utility. In other words, $\theta(i, j)$ quantifies to what extent the selection of i is correlated with the selection of j and vice-versa. Given that it is not possible to track at what times i and j were selected together in the studied datasets, we compute $\theta(i, j)$ considering all their co-occurences in the historical data, regardless of when they were presented to users. A straightforward way of computing $\theta(i, j)$ is via Maximum Likelihood Estimator (MLE), as follows:

$$\theta(i, j) = \frac{l_{ij}}{f_{ij}} \tag{4}$$

where l_{ij} is the number of consumers that liked i and j and f_{ij} is the number of consumers that gave feedback to i and j. It turns out that this MLE computation yields good results, as detailed in Section 5.

The notion of *"liked"* can be understood as *"clicked"*, *"received a high rating"*, *"purchased"*, etc. The problem with computing θ via Equation 4 is that most items do not receive much feedback – i.e., recommendation datasets. As a consequence, using MLE to approximate the value of θ can lead to arbitrarily bad approximations. To make more realistic approximations, one can penalize pairs of items with a poor support, shrinking the computation with a factor λ [6]:

$$\theta(i, j) = \frac{f_{ij}}{\lambda + f_{ij}} \times \frac{l_{ij}}{f_{ij}} \tag{5}$$

Note that Equation 5 converges to Equation 4 when $\lambda \to 0$. The main challenge in using Equation 5 is to find an adequate value for λ.

3.3   Combining Scores

In this work, we combine individual and pairwise scores to select N items out of K for recommendation. The problem is therefore posed as selecting a set of items $I = \{i_1, ..., i_N\}$ that maximizes the following utility function:

$$\sum_{i_a \in I} \frac{\phi(i_a)}{|I|} + \sum_{(i_l, i_m) \in I^2} \frac{\theta(i_l, i_m)}{|I|^2} \tag{6}$$

where the normalization in both summations is important to keep their contributions fair – i.e., both values will remain in the interval $[0, 1]$.

There are some different ways of obtaining an exact solution to this optimization problem. For instance, one can trivially enumerate all N-combinations of a set with K items and choose the one that sums up to the highest value. It is also possible to use integer programming to solve it (NP-Hard) [13]. To the best of our knowledge, all these techniques are costly and there is no polynomial algorithm that maximizes this function in an exact way.

4   The *GSMI* Algorithm

Combining scores to select items for recommendation leads to an intractable optimization, as discussed previously. To tackle with this problem under a practical viewpoint, we propose *GSMI*, a greedy algorithm that selects N items, one at a time, taking into account items that were selected previously.

The algorithm receives a set of items $I = \{i_1, \ldots, i_k\}$ and their individual scores $\{\phi(i_1), \ldots, \phi(i_k)\}$, and returns a set R with N selected items, where $N \leq K$. It is described as follows.

Algorithm 1. *GSMI* Algorithm

1: $i \leftarrow \underset{i \in I}{\operatorname{argmax}} \, \phi(i)$
2: $R \leftarrow \{i\}$
3: $I \leftarrow I \setminus \{i\}$
4: **while** $|R| < N$ **do**
5:     $j \leftarrow \underset{j \in I}{\operatorname{argmax}} \sum_{a \in R_j} \dfrac{\phi(a)}{|R_j|} + \sum_{(b,c) \in R_j^2} \dfrac{\theta(b, c)}{|R_j|^2}$, where $R_j = R \cup \{j\}$
6:     $R \leftarrow R_j$
7:     $I \leftarrow I \setminus \{j\}$
8: **end while**
9: **return** R

GSMI starts selecting the item that has the best individual score, i. All other $N - 1$ selected items are chosen in a way that maximizes the equation in line 5, where the maximized set is comprised by all items that were already chosen, R, and the new item itself. The crucial greedy choice of *GSMI* is selecting the item with best individual score first.

GSMI runs in polynomial time. The loop in line 4 will be executed exactly $N - 1$ times. In line 5, an item is chosen out of $K - 1$ in the worst case; in the best case, out of $K - N + 1$ ones. It means that $O(K)$ items need to be analysed at each time. In line 5, the first summation is performed in $O(|R|)$ time. The second summation is performed in $O(|R|^2 \times \beta)$ time, where $O(\beta)$ is the time complexity of θ. An upper bound for the time complexity of *GSMI* is therefore $O(K + |R| \times (K \times |R|^2 \times \beta)) = O(KN^3\beta)$, given that $|R| \leq N$.

The time complexity of computing θ, $O(\beta)$, depends on the size of the dataset, on the maximum number of feedback given by a certain user to the items in question, and on the used data structures. For each loop iteration, it is possible to reuse partial summations from the previous iteration, in a way that the total time complexity is reduced to $O(KN^2\beta)$. Besides that, it is possible to optimize function calls to compute θ by taking advantage of its symmetry and by using memoization. Therefore, as we discuss in Section 5, it is simple to speed up *GSMI* and make it scalable to big datasets.

It is worth pointing out that *GSMI* is compatible with any recommender system where it is possible to estimate I, its corresponding scores $\{\phi(i_1),\ldots,\phi(i_k)\}$, and approximations for pairwise scores θ. Therefore, the proposed algorithm is *a priori* compatible with systems that employ both matrix factorization techniques and sketching/fingerprinting methods for dealing with big data.

To validate *GSMI*, we use the explicit feedback users give over items as a utility measurement: the better it is, the more useful the recommendations are [14]. The hypothesis we started investigating can therefore be simply posed as *"Does GSMI select items that receive better feedback when compared to those selected by Top − N?"* For all studied datasets, feedback consists of ratings. To perform the comparison between *GSMI* and *Top − N* we thus measure the average rating users gave to recommended items, applying 5-fold cross-validation [4]. We generate individual scores for all $(user, item)$ pairs in the test set and then perform items selection using both *GSMI* and *Top − N*. The recommendation list containing items that receive higher ratings is the one that is considered more useful.

In this work we also compare *GSMI* and *Top − N* under a diversity perspective. It has recently become a consensus that a desirable feature for successful recommender systems is the ability of generating diverse, non-monotonous recommendations to users [15]. Diversity is usually defined as the opposite of similarity, and the most explored approach for measuring it uses content-based similarity between items [8]. The diversity metric we apply, intra-list distance (ILD), was proposed by Zhang and Hurley [16] and works as follows:

$$ILD = \frac{2}{|R|(|R|-1)} \sum_{i_k, i_l \in R, l < k} 1 - sim(i_k, i_l) \qquad (7)$$

where R is comprised by all selected items and $sim(i_k, i_l)$. More details of how we performed experiments with ILD are given in Section 5.

5   Experiments

In this work, we investigated mutual influence in three different datasets: MovieLens 100K[2], MovieLens 1M[3], and Jester 1[4]. Table 1 summarizes some of their characteristics.

---

[2] http://www.grouplens.org/system/files/ml-100k.zip
[3] http://www.grouplens.org/system/files/ml-1m.zip
[4] http://goldberg.berkeley.edu/jester-data/jester-data-1.zip

Table 1. Succint characterization of the studied datasets

Characteristic	MovieLens 100K	MovieLens 1M	Jester 1
Domain	Movies	Movies	Jokes
Feedback	Ratings (1 - 5)	Ratings (1 - 5)	Ratings (-10.00 - 10.00)
Number of users	943	6,040	24,983
Number of items	1,682	3,900	100
Number of feedback	100,000	1,000,209	1,810,455
Minimum ratings/user	20	20	36
Sparsity rate	0.937	0.958	0.275

Table 2. Average ratings given by users to items recommended by two different methods: $Top - N$ and $GSMI$. $GSMI - 5$, $GSMI - 10$, and $GSMI - 20$ correspond to $GSMI$ selecting $N = 5, 10, 20$ items respectively. The average ratings were computed for different values of N using PureSVD and NNCosNgbr as predictors. For each $GSMI/Top - N$ pair, we performed a t-test over each dataset, and with a 95% confidence level only the underlined results are not statistically different.

Predictor	Method	MovieLens 100K	MovieLens 1M	Jester 1
PureSVD	$Top - 5$	3.924	4.127	1.292
	$GSMI - 5$	3.974	4.175	1.518
	$Top - 10$	3.837	4.004	1.031
	$GSMI - 10$	3.878	4.057	1.188
	$Top - 20$	3.738	3.908	<u>0.892</u>
	$GSMI - 20$	3.765	3.939	<u>0.911</u>
NNCosNgbr	$Top - 5$	3.821	4.027	2.312
	$GSMI - 5$	3.875	4.096	2.363
	$Top - 10$	3.775	3.928	1.529
	$GSMI - 10$	3.824	3.993	1.589
	$Top - 20$	3.691	3.836	<u>0.939</u>
	$GSMI - 20$	3.726	3.890	<u>0.940</u>

The MovieLens datasets are significantly more sparse than Jester 1. While in the former users rated at least 20 movies, in the latter users gave feedback to at least 36% of the jokes. MovieLens 1M is comprised by many more users and items than MovieLens 100K, and its total amount of ratings is similar to Jester's. Finally, while ratings in the MovieLens dataset are discretized and vary from 1 to 5, users in Jester 1 can assign any real number from -10.00 to 10.00 to any rated joke.

We compared $GSMI$ with the $Top - N$ approach in order to evaluate how mutual influence alone can bring up gain to recommender systems. Table 2 presents results for experiments with $N = 5, 10, 20$. For the MovieLens datasets, we considered that movies were liked by users if their ratings were equal or higher than 4; in the case of Jester 1, if they were equal or higher than 5.00. For all experiments, PureSVD was executed with 50 latent factors, the number of neighbors in $D^k(u; i)$ in Equation 3 was fixed in 60, and the value of λ for pairwise scores θ in Equation 5 was fixed in 0.5.

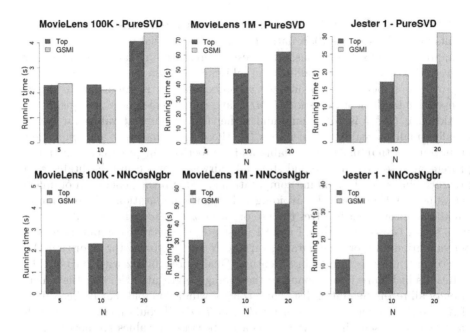

Fig. 1. Average running times per validation fold, in seconds, for different combinations of datasets and predictors, with $N = 5, 10, 20$

As shown in Table 2, predictor PureSVD generated better average ratings for both $Top - N$ and $GSMI$ methods with respect to the MovieLens datasets. Concerning the Jester 1 dataset, NNCosNgbr performed better. In all cases, either $GSMI$ produced superior average ratings or was statistically equivalent to the $Top - N$ results. The obtained gains were up to 17%. Although the difference between $Top - N$ and $GSMI$ approaches may seem small, it is known that such differences have a huge impact on recommender systems [17].

5.1 Efficiency and Scalability of $GSMI$

$GSMI$ is a greedy algorithm for the maximization problem posed by Equation 6. It is thus useful to compare it against exact solutions. We implemented such solutions for the MovieLens 100K dataset, with predictors PureSVD and NNCosNgbr and $N = 5$. The computations were carried out via the enumeration of all items combinations and posterior selection of the one that maximized Equation 6. Nonetheless, such computations took more than 3 hours to be completed, while $GSMI$ generates results per validation fold in around 70 seconds in the worst case, as shown in Figure 1.

Regarding the utility of results, the average ratings obtained with the exact solutions were 3.984 and 3.895 for PureSVD and NNCosNgbr respectively. Both results were not statistically different from the corresponding average ratings obtained with $GSMI$ for $N = 5$, 3.974 and 3.875, according to a t-test with a 95%

confidence level. The fact that $GSMI$ for $N = 5$ generated statistically equivalent results for the MovieLens 100K dataset is an indicative that it is a good heuristic to approximate exact solutions. Also, results presented in Table 2 consistently indicate that $GSMI$, by embedding mutual influence in its selection strategy, can improve recommendations utility. As a consequence, it is important to devise competitive implementations for $GSMI$ that scale in real-time situations.

Although $GSMI$ is polynomial and rather fast, given that values for N are usually small in real-world scenarios [8], there are some easy and important optimizations that makes it scalable and competitive in practice. A first improvement is to precompute and store all pairwise scores θ in a hashtable as a preprocessing step. This offline computation speeds up the generation of different values for Equation 6 by avoiding redundant computations of Equation 5. Another improvement involves the use of memoization to reuse partial summations in the $GSMI$ algorithm. Figure 1 illustrates the average computation time per validation fold for each dataset, varying N and the predictor algorithm. All experiments were performed in a Pentium Dual-Core 2.0GHz with 2GB RAM.

Results in Figure 1 correspond to the average aggregated time for the generation of all recommendation lists concerning a validation fold. For higher values of N, the time difference between $GSMI$ and $Top - N$ could increase, but such analysis is not useful in real-world scenarios because N values are never big in practice. Therefore, for realistic values of N, $GSMI$ scales well and its average running times per validation fold are only slightly bigger than those obtained with $Top - N$. In spite of that, the time difference for generating a single recommendation list with both methods is irrelevant. Given that in real-world systems recommendation lists are generated once at a time via the interaction with users, and $GSMI$ yields better utility results, it is thus a feasible alternative.

5.2 Relation between $GSMI$ and Recommendations Diversity

To investigate the relation between $GSMI$ and recommendations diversity, we computed the ILD metric, as in Equation 7, for the MovieLens datasets. Movie similarities were computed via the Jaccard's similarity over their corresponding genres, as properly indicated in the datasets. We did not perform such experiments over the Jester 1 dataset because it does not provide any content-based information. Results for both $Top - N$ and $GSMI$ with respect to predictors PureSVD and NNCosNgbr are summarized in Table 3.

According to our experiments, $GSMI$ and $Top - N$ do not generate statistically significant diversity differences in recommendations. This is an evidence that $GSMI$ is not likely to hurt recommendations diversity – at least when compared to $Top - N$. It also indicates that considering mutual influence via pairwise scores θ does not imply in either redundant or monotonous recommendations.

6   Conclusions and Future Work

In this work, we investigated how items can interfere with their own utility in recommendation scenarios. We stated that there is a mutual influence among

Table 3. *ILD* results for different values of N using PureSVD and NNCosNgbr as predictors. For each $GSMI/Top - N$ pair, we performed a t-test over each dataset. With a 95% confidence level, none of the results are statistically different.

Predictor	Method	MovieLens 100K	MovieLens 1M
PureSVD	$Top - 5$	0.8557	0.7459
	$GSMI - 5$	0.8544	0.7557
	$Top - 10$	0.8619	0.7591
	$GSMI - 10$	0.8615	0.7660
	$Top - 20$	0.8645	0.7666
	$GSMI - 20$	0.8643	0.7696
NNCosNgbr	$Top - 5$	0.8649	0.7617
	$GSMI - 5$	0.8652	0.7636
	$Top - 10$	0.8649	0.7694
	$GSMI - 10$	0.8652	0.7709
	$Top - 20$	0.8649	0.7712
	$GSMI - 20$	0.8652	0.7727

them that increases their utilities when simutaneously selected. It is thus possible to take advantage of these mutual influences to improve recommendation systems. The main intuition behind this project is that not only individual features matter in decision-making processes: the presented set of alternatives as a whole also plays an important role on it.

We proposed a means of computing such mutual influence, pairwise scores θ, and an algorithm that incorporates it to improve the recommendation of items, *GSMI*. To analyse mutual influence as an isolated evidence, we compared *GSMI* with $Top - N$, an item selection technique that does not rely upon any type of signal but sorted individual items scores. These individual scores, namely ϕ, were generated by two different state of the art predictors: PureSVD and NNCosNgbr [6].

We showed that for three distinct datasets – MovieLens 100K, MovieLens 1M, and Jester 1 – *GSMI* consistently generated recommendation lists with higher utility measures, i.e. higher average ratings [14], when compared to $Top - N$. We also present evidence that *GSMI* is easily scalable and therefore useful for real-world scenarios. Finally, we show that this algorithm does not seem to hurt recommendations diversity.

Given that we show that mutual influence is an important evidence for recommender systems, we intend to develop LTR algorithms that embed it in a near future. We also want to investigate exact solutions for our optimization problem (Equation 6) that can be feasible in practice, as well as ways of overcoming data sparsity as a means to compute scores θ in a more stable fashion. A thorough assessment of which of all studied algorithms leads to less performance variations is also planned as future work. Finally, we plan on implementing different baselines that also consider some type of influence or dependency among items, such as Latent Structured Ranking [10] or the mean-variance ranking model proposed by Wang [7].

Acknowledgements. This work was partially sponsored the Brazilian National Institute of Science and Technology for the Web (grant MCT/CNPq 573871/2008-6), and by the authors' individual grants and scholarships from CAPES and CNPq. The first author is also thankful for fruitful discussions with Google Software Engineer Davi M. J. Barbosa.

References

1. Toffler, A.: Future Shock. Random House (1970)
2. Adomavicius, G., Tuzhilin, A.: Towards the next generation of recommender systems: A survey of the state-of-the-art and possible extensions. IEEE Transactions on Knowledge and Data Engineering 17(6), 734–749 (2005)
3. Tversky, A.: Elimination by aspects: A theory of choice. Psychological Review 79(4), 281–299 (1972)
4. Passos, A., Gael, J.V., Herbrich, R., Paquet, U.: A penny for your thoughts? the value of information in recommendation systems. In: NIPS Workshop on Bayesian Optimization, Experimental Design, and Bandits, pp. 9–14 (2011)
5. Vargas, S., Castells, P.: Rank and relevance in novelty and diversity metrics for recommender systems. In: RecSys., pp. 109–116 (2011)
6. Cremonesi, P., Koren, Y., Turrin, R.: Performance of recommender algorithms on top-n recommendation tasks. In: RecSys., pp. 39–46 (2010)
7. Wang, J.: Mean-variance analysis: A new document ranking theory in information retrieval. In: Boughanem, M., Berrut, C., Mothe, J., Soule-Dupuy, C. (eds.) ECIR 2009. LNCS, vol. 5478, pp. 4–16. Springer, Heidelberg (2009)
8. Ricci, F., Rokach, L., Shapira, B., Kantor, P.B. (eds.): Recommender Systems Handbook. Springer (2011)
9. Xiong, C., Wang, T., Ding, W., Shen, Y., Liu, T.Y.: Relational click prediction for sponsored search. In: WSDM, pp. 493–502 (2012)
10. Weston, J., Blitzer, J.: Latent structured ranking. In: UAI, pp. 903–913 (2012)
11. Papagelis, M., Plexousakis, D.: Qualitative analysis of user-based and item-based prediction algorithms for recommendation agents. Engineering Applications of Artificial Intelligence 18(7), 781–789 (2005)
12. Koren, Y.: Factorization meets the neighborhood: a multifaceted collaborative filtering model. In: SIGKDD, pp. 426–434 (2008)
13. Nemhauser, G., Wolsey, L.: Integer and combinatorial optimization. Wiley (1988)
14. Breese, J., Heckerman, D., Kadie, C.: Empirical analysis of predictive algorithms for collaborative filtering. In: UAI, pp. 43–52 (1998)
15. McNee, S.M., Riedl, J., Konstan, J.A.: Being accurate is not enough: how accuracy metrics have hurt recommender systems. In: SIGCHI, pp. 1097–1101 (2006)
16. Zhang, M., Hurley, N.: Avoiding monotony: improving the diversity of recommendation lists. In: RecSys., pp. 123–130 (2008)
17. Bell, R., Koren, Y.: Lessons from the netflix prize challenge. ACM SIGKDD Explorations Newsletter 9(2) (2007)

Position-Restricted Substring Searching
over Small Alphabets*

Sudip Biswas, Tsung-Han Ku, Rahul Shah, and Sharma V. Thankachan

Louisiana State University, USA
National Tsing Hua University Hsinchu, Taiwan
{sudip,rahul,thanks}@csc.lsu.edu, thku@cs.nthu.edu.tw

Abstract. We consider the problem of indexing a given text $T[0...n-1]$ of n characters over an alphabet set Σ of size σ, in order to answer the position-restricted substring searching queries. The query input consists of a pattern P (of length p) and two indices ℓ and r and the output is the set of all $occ_{\ell,r}$ occurrences of P in $T[\ell...r]$. In this paper, we propose an $O(n \log \sigma)$-word space index with $O(p + occ_{\ell,r} \log \log n)$ query time. Our solution is interesting when the alphabet size is small. For example, when the alphabet set is of constant size, we achieve exponential time improvement over the previously best-known linear space index by Navarro and Nekrich [SWAT 2012] with $O(p + occ_{\ell,r} \log^\epsilon n)$ query time, where $\epsilon > 0$ is any positive constant. We also study the property matching problem and provide an improved index for handling semi-dynamic (only insertions) properties, where we use position-restricted substring queries as the main technique.

1 Introduction and Related Work

Let $T[0...n-1]$ be a text of size n over an alphabet set Σ of size σ. The fundamental problem in text indexing is to preprocess T and maintain an index for reporting all occ occurrences of a query pattern P within T. Linear space data structures such as suffix trees and suffix arrays can answer this query in $O(p + occ)$ and $O(p + \log n + occ)$ time respectively [18,16,15]. In this paper, we revisit the well studied *Position-restricted substring searching* (PRSS) problem as defined below:

> The query input consists of a pattern P (of length p) and two indices ℓ and r, and the task is to report all $occ_{\ell,r}$ occurrences of P in $T[\ell...r]$.

Many text searching applications, where the objective is to search only a part of the text collection can be modeled as PRSS problem. For example, restricting the search to a subset of dynamically chosen documents in a document database, restricting the search to only parts of a long DNA sequence, etc [14]. The problem also finds applications in the field of information retrieval as well.

---

* This work is supported in part by US NSF Grant CCF–1017623 (R. Shah and J. S. Vitter) and CCF–1218904 (R. Shah).

O. Kurland, M. Lewenstein, and E. Porat (Eds.): SPIRE 2013, LNCS 8214, pp. 29–36, 2013.
© Springer International Publishing Switzerland 2013

The PRSS problem was introduced by Mäkinen and Navarro [14], where they show an elegant reduction of PRSS problem into a two dimensional orthogonal range reporting problem. Their data structure consists of a suffix tree (for initial pattern matching) and an orthogonal range reporting structure in two dimension (RR2D). Their structure takes $O(n)$-word space and the queries can be answered in $O(p + \log n + occ_{\ell,r} \log n)$ query time. By using the most recent two-dimensional range reporting structure by Navarro and Nekrich [17], this query time can be improved to $O(p + \log^\epsilon n + occ_{\ell,r} \log^\epsilon n)$, where ϵ is any positive constant. Another trade-off given by Mäkinen and Navarro [14] is $O(n \log^\epsilon n)$-word space and near optimal $O(p + \log \log n + occ_{\ell,r})$ query time. This query time is further improved to $O(p + \log \log \sigma + occ_{\ell,r})$ by Kopelowitz et al. [13] and then to optimal $O(p + occ_{\ell,r})$ by Bille and Gortz [3] without changing the space bounds. Crochemore et al. [6] also have proposed an optimal time solution with a much higher space requirement of $O(n^{1+\epsilon})$ bits. Recently, Hon et al. [10] have studied the possibility of indexing the text in succinct space and answering PRSS queries efficiently. They proved that designing a succinct index with poly-logarithmic query bounds is at least as hard as designing a three dimensional range reporting structure in linear space with poly-logarithmic query bounds, using Geometric BWT techniques [5]. However, they provided optimal time and succinct space solutions for a special case where the pattern is sufficiently long.

The counting version of $PRSS$ is also an interesting problem. For this, the linear space index by Mäkinen and Navarro [14] takes $O(p + \log n)$ time. A solution by Bille and Gortz [3] can perform counting in faster $O(p + \log \log n)$ time, which is slightly improved to $O(p + \log \log \sigma)$ by Kopelowitz et al. [13]. However these indexes consumes $O(n \log n / \log^2 \log n)$ words of space. Finally Gagie and Gawrychowski proposed a space efficient solution of $O(n)$ words, where counting queries can be answered in $O(p + \log \log n)$ time for general alphabets and in optimal $O(p)$ time, when $\sigma = \log^{O(1)} n$ [7]. In this paper, we revisit the reporting version of PRSS problem and obtain the result summarized in the following theorem:

Theorem 1. *There exist an $O(n \log \sigma)$-word index supporting PRSS queries in $O(p + occ_{\ell,r} \log \log n)$ time.*

Using the existing techniques, one can easily design an $O(n \log \log n)$-word space index with $O(p + occ_{\ell,r} \log \log n)$ query time. Therefore, our result is interesting when the alphabet size is small (i.e., when $\sigma = 2^{o(\log \log n)}$). Note that when $\sigma = O(1)$, we achieve exponential time improvement over the previously best-known linear space index.

In property matching problem, in addition to the text T, a set $\pi = \{[s_1, e_1], [s_2, e_2], ...\}$ of intervals is also given. Our task is to preprocess T and π and maintain an index, such that when ever a pattern P (of length p) comes as a query, return those occurrences of P in T which are within (at least) one of the intervals in π. Efficient linear space [1,11] and compressed space [9] indexes are known for this problem. In [12], Kopelowitz have studied the dynamic case of this problem, where π can be updated (i.e., intervals can be inserted to or deleted from π), and provide a linear space index with $O(e - s + \log \log n)$ update time, where (s, e) is the interval inserted/deleted. In semi-dynamic case, (i.e., only insertions or deletions) the update time is $O(e - s)$. Note that $e - s$ can be even $\Theta(n)$ in the worst case. In this paper, we describe a semi-dynamic index (only insertions) with the result summarized in the following theorem. We use position-restricted substring queries as one of the main technique to achieve this result.

Theorem 2. *There exists an $O(n \log^\epsilon n)$ space index for semi-dynamic (only insertions) property matching with query time $O(p + \sqrt{n} \log \log n + occ_\pi)$ and amortized update time $O(\sqrt{n})$, where occ_π represents the output size.*

2   Preliminaries

2.1   Suffix Trees

For a text $T[0...n-1]$, a substring $T[i..n-1]$ with $i \in [0, n-1]$ is called a suffix of T and $T[0...i]$ called a prefix of T. The suffix tree [18,16] of T is a lexicographic arrangement of all these n suffixes in a compact trie structure of $O(n)$ words space, where the ith leftmost leaf represents the ith lexicographically smallest suffix of T. The suffix range of a pattern P (of length p) is given by the maximal range $[L, R]$ such that for $L \le j \le R$, P is a prefix of (lexicographically) jth suffix of T. Using suffix tree, the suffix range of P can be computed in $O(p)$ time and all the occurrences of P within T can be reported in optimal $O(p + occ)$ time, where occ is the number of occurrences of P within T. The following is a useful result from [2].

Lemma 1. *For a given pattern $P = P[0...p-1]$, a substring of the form $P[i... p-1]$, $0 \le i \le p-1$ is called a suffix of P, and the suffix range of all suffixes of P can be computed in $O(p)$ time.*

2.2   Orthogonal Range Reporting in Two Dimensions (RR2D)

Let S be a given set of n points of the form (x_i, y_i) in an $[0, n-1] \times [0, n-1]$ grid. An orthogonal range reporting query consists of two input ranges (x', x'') and (y', y''), and the task is to output all those k points (x_j, y_j) such that, $x' \le x_j \le x''$ and $y' \le y_j \le y''$. For our purpose, we used the data structure (in RAM model) by Chan et al. [4], where the space requirement is $O(n \log \log n)$ words and the query time is $O(\log \log n + k \log \log n)$.

3   The Index

Based on the alphabet size σ and the pattern length p, we consider 3 cases as follows:

3.1   Index for $\sigma = \log^{\Omega(1)} n$

The index consists of a suffix tree ST of the text T. Then for each suffix $T[x...n-1]$, we define a two dimensional point (x, y) such that y be the lexicographic rank of $T[x...n-1]$ among all suffixes of T, and maintain an RR2D structure over these n two-dimensional points using the data structure described in preliminaries. The index space can be bounded by $O(n \log \log n) = O(n \log \sigma)$ words. In order to answer a PRSS query, we first obtain the suffix range $[L, R]$ of P in $O(p)$ time via navigating in the suffix tree and report all those points within the box $[\ell, r - p] \times [L, R]$ by querying on the RR2D structure. The y coordinate of each output is an answer to the original PRSS problem. The query time can be bounded by $O(p + (occ_{\ell,r} + 1) \log \log n)$. The range emptiness problem states given a range in the text and a substring, check if the substring exists in the given range of the text or not. Bille and Gortz [3] showed how to solve the range emptiness problem in $O(p)$ time. Using this result, the query time for PRSS problem can be improved to $O(p + occ_{\ell,r} \log \log n)$.

Lemma 2. *There exist an $O(n \log \sigma)$ space index supporting PRSS queries in $O(p + occ_{\ell,r} \log \log n)$ time for $\sigma = \log^{\Omega(1)} n$.*

3.2   Index for $\sigma = \log^{O(1)} n$ and $p \geq \sqrt{\log n}$

In this section, we introduce suffix sampling techniques to achieve the desired space bound. Based on a sampling factor $\delta = \lfloor \frac{1}{3} \log_\sigma \log n \rfloor$, we define the followings:

- δ-*sampled suffix:* $T[x...n-1]$ is an δ-sampled suffix if $x \; mod(\delta) = 0$.
- δ-*sampled block (or simply block):* any substring $T[x, x+\delta-1]$ of T of length δ with $x \; mod(\delta) = 0$ is called a block.

The number of blocks in T is $\Theta(n/\delta)$, where the number of all possible distinct blocks is at most the number of distinct strings of length δ, and is bounded by $\sigma^\delta = O(\log^{1/3} n)$. Let B_i represent the lexicographically ith smallest string of length δ. We categorize the δ-sampled suffixes into σ^δ categories $C_1, C_2, ..., C_{\sigma^\delta}$ such that C_i contains the set of all δ-sampled suffixes whose previous block is B_i. For each category C_i, we maintain a two-dimensional range reporting structures $RR2D_i$'s (Section 2.2) on the set of points (x, y), where $T[x...n-1]$ is a δ-sampled suffix in C_i and y is its lexicographic rank among all suffixes of T (i.e., $SA[y] = x$). Note that each δ-sampled suffix belongs to exactly one category and the number of δ-sampled suffixes is $\Theta(n/\delta)$. Therefore, the overall space for all $RR2D_i$ structures can be bounded by $O((n/\delta) \log \log n) = O(n \log \sigma)$ words.

Query Answering: Since $p \geq \sqrt{\log n} > \delta$, the starting and ending positions of an occurrence of P will not be in the same block of T. Based on the block in which a match starts, we shall categorize the occurrences into $\sigma^\delta \delta$ types as follows:

We call an occurrence as a type-(i, j) occurrence, if the prefix of P and the suffix of block B_i matches for j characters.

Here $1 \leq i \leq \sigma^\delta$ and $0 \leq j \leq \delta - 1$. In other words, a type-(i, j) occurrence is an occurrence of P at an index y satisfying:

(1) The block of T containing y is equal to B_i.
(2) $y = \delta - j \pmod{\delta}$.

Then type-(i, j) occurrences for a fixed i and j can be retrieved as follows: firstly check if the prefix of P and the suffix of block B_i matches for j characters. This takes only $O(j) = O(\delta) = O(\log \log n)$ time. If it is not matching, $occ_{i,j} = 0$. Otherwise, corresponding to each type-(i, j) occurrence, there exist a δ-sampled suffix $T[x...n-1]$ of T which is (i) prefixed by $P[j...p-1]$, (ii) occurring after a block B_i, and (iii) $x - j \in [\ell, r]$. By issuing a query on the $RR2D_i$ structure with $[\ell + j, r - p + j] \times [L_j, R_j]$ as the input range, all such suffixes can be retrieved. Here $[L_j, R_j]$ represents the suffix range of $P[j, ..., p-1]$. Note that $[L_j, R_j]$ for $j = 0, 1, ...p - 1$ can be computed in total $O(p)$ time (Lemma 1). From every reported point (x, y), we shall output $x - j$ as an answer to the original PRSS query. This way, all type-(i, j) occurrences can be reported in $O(\log \log n)$ time plus $O(\log \log n)$ time per output. Hence the total time for reporting all possible type-$(., .)$ occurrences is bounded by $O(p + \sigma^\delta \delta \log \log n + occ_{\ell,r} \log \log n) = O(p + \sqrt{\log n} + occ_{\ell,r} \log \log n)$.

Lemma 3. *There exist an $O(n \log \sigma)$ space index supporting PRSS queries in $O(p + occ_{\ell,r} \log \log n)$ time, given $p \geq \sqrt{\log n}$.*

3.3 Index for $\sigma = \log^{O(1)} n$ and $p \leq \sqrt{\log n}$

Here we maintain $\sqrt{\log n}$ separate structure for answering PRSS queries for patterns of length $1, 2, 3...., \sqrt{\log n}$. Structure for a fixed pattern length (say $\alpha \leq \sqrt{\log n}$) is described as follows: the number of distinct patterns of length α is σ^α. Each such distinct pattern can be encoded using an integer from $\Sigma_\alpha = \{1, 2, ..., \sigma^\alpha\}$ in $\alpha \log \sigma$ bits. Next we transform the original text $T[0...n-1]$ into $T_\alpha[0...n - \alpha]$, such that

- Each character in T_α is drawn from an alphabet set $\Sigma_\alpha = \{1, 2, ..., \sigma^\alpha\}$ and can be represented in $\alpha \log \sigma$ bits.
- $T_\alpha[i]$: the ith character of T_α is the integer in Σ_α corresponding to the encoding of the string $T[i...i + \alpha - 1]$.

Hence, T_α can be seen as a sequence of $(n - \alpha + 1)$ integers drawn from an alphabet set Σ_α. We shall maintain T_α in $O(|T_\alpha| \log |\Sigma_\alpha|) = O(n\alpha \log \sigma)$ bits using the data structure described in [8], such that rank/select queries on any character within T_α can be supported in $O(\log \log |\Sigma_\alpha|) = O(\log \log n)$ time. Since we are maintaining T_α for $\alpha = 1, 2, 3, ..., \sqrt{\log n}$, the total space can be bounded by $O(n \log \sigma \sum_{\alpha=1}^{\sqrt{\log n}} \alpha) = O(n \log \sigma \log n)$ bits or $O(n \log \sigma)$ words.

Query Answering: A PRSS query for a pattern P of length $p \leq \sqrt{\log n}$ can be answered as follows: first we find the integer β in Σ_p corresponding to P in $O(p)$ time. An occurrence of β at position i in T_α corresponds to an occurrence of P at position i in T. Therefore, all occurrences of P in $T[\ell...r]$ can be answered by reporting all those occurrences of β in $T_\alpha[\ell...r]$. Using rank/select queries on T_α, this can be easily handled as follows: find the number of occurrences (say a) of β in T_α before the position ℓ and the number of occurrences (say b) of β until the position r in T_α. Using two rank queries on T_α, the values of a and b can be obtained in $O(\log \log n)$ time. Next we output the kth occurrence of β in T_α for $k = a + 1, a + 2, ..., b$ using $(b - a)$ select queries in total $O((b - a) \log \log n)$ time, and each output corresponds to an answer to the original PRSS problem. Hence the total query time can be bounded by $O(p + \log \log n + occ_{\ell,r} \log \log n)$. In order to remove the additive $\log \log n$ factor, we also maintain the linear space structure by Gagie and Gawrychowski [7] for counting the number of outputs in $O(p)$ time (for poly-logarithmic alphabet). And while querying, we first count the output size using this structure, and then we query on our structure only if the output size is non zero.

Lemma 4. *There exist an $O(n \log \sigma)$ space index supporting PRSS queries in $O(p + occ_{\ell,r} \log \log n)$ time, for $p \leq \sqrt{\log n}$.*

By combining Lemma 2, Lemma 3 and Lemma 4, we obtain Theorem 1.

4   Semi Dynamic Index for Property Matching

In this section we design an index for the Property Matching problem which supports insertion. We use the following two existing structures as the building blocks of our index for P.

Lemma 5. *[12] There exists a linear space index (which we call as I_{SDPM}) for semi-dynamic (only insertions) property matching with query time $O(p + occ_\pi)$ and update time $O(e - s)$, where (e, s) is the inserted interval.*

Lemma 6. *[3] There exists an $O(n \log^\epsilon n)$ space index (which we call as I_{PRSS}) supporting PRSS queries in optimal $O(p + occ_{\ell,r})$ time, where $\epsilon > 0$ is any constant.*

Our index consists of I_{SDPM}, and I_{PRSS}. An interval (e, s) be called *small* if $e - s \leq \sqrt{n}$, otherwise it is *large*. Note that a small interval can be inserted into π, and can update I_{SDPM} in $O(\sqrt{n})$ time. However, if the interval is *large*, we do not update I_{SDPM}. Actually we insert it into a set S of intervals. When ever the size of S grows up to $\sqrt{n}$, we do a batched insertions of all the intervals in S into I_{SDPM} using the following lemma, and then initialize S to a tree with zero nodes.

Lemma 7. *[12] A set I of intervals can be inserted into π, and we can update I_{SDPM} in time $O(|coversize(I)|)$, where $coversize(I) = \{i \in [1, n] : \exists [s, e] \in I,$ such that $i \in [s, e]\}$.*

Here $|cover.size(I)|$ can be at most size of the text, n. Batch insertion will be required after $\sqrt{n}$ number of insertions. Therefore, from Lemma 7, the time for batch insertion can be bounded by $O(n)$. Since each batch insertion can support $O(\sqrt{n})$ number of subsequent insertions, the time for insertion of a single interval can be bounded by $O(\sqrt{n})$ in amortized sense.

Query Answering: Our algorithm has the following 3 phases:

1. Issue a query on I_{SDPM} and retrieve the corresponding occurrences. How-ever, we cannot obtain the complete outputs just by this step alone, as we do not (always) update I_{SDPM} immediately after an insertion.
2. The task of finding those missing occurrences, which are from the region corresponding to those *large* intervals in S are obtained in this Step. Firstly we sort all intervals in S in $O(\sqrt{n})$ time as follows: if $|S| < \sqrt{n}/\log n$, then the time for sorting is $O(|S|\log|S|) = O(\sqrt{n})$, otherwise use radix sort and the required time is $O(|S|) = O(\sqrt{n})$. After initializing a variable h as 1 (this is an invariant storing the last occurrence retrieved so far), we perform the following steps for each interval $[s_i, e_i] \in S$ (where $s_i \leq s_{i+1}$) in the ascending order of i: if $h \in [s_i, e_i]$, then issue a PRSS query with $[h + 1, e_i - p + 1]$ as the input range, and if $h < s_i$, then issue another PRSS query with $[s_i, e_i - p + 1]$ as the input range, where as if $h > e_i$, skip this step. Based on the retrieved occurrences, we update h with the last occurrence. The overall time required for this step can be bounded by $O(|S|\log\log n + occ_\pi)$ $= O(\sqrt{n}\log\log n + occ_\pi)$.
3. There can be occurrences which are retrieved by both Step 1 and 2. In order to eliminate these duplicated, we sort the occurrences retrieved from Step 1 and Step 2 separately (in $O(\sqrt{n} + occ_\pi)$ time as described before), and then merge these results without duplicates by scanning both lists simultaneously.

By combining all, the query time can be bounded by $O(p + \sqrt{n}\log\log n + occ_\pi)$. This completes the proof of Theorem 2.   □

References

1. Amir, A., Chencinski, E., Iliopoulos, C.S., Kopelowitz, T., Zhang, H.: Property matching and weighted matching. Theoretical Computer Science 395, 298–310 (2008)
2. Amir, A., Farach, M., Idury, R.M., La Poutré, J.A., Schäffer, A.A.: Improved Dynamic Dictionary Matching. Information and Computation 119(2), 258–282 (1995)
3. Bille, P., Gørtz, I.L.: Substring Range Reporting. In: Giancarlo, R., Manzini, G. (eds.) CPM 2011. LNCS, vol. 6661, pp. 299–308. Springer, Heidelberg (2011)
4. Chan, T.M., Larsen, K.G., Patrascu, M.: Orthogonal range searching on the RAM, revisited. In: SoCG, pp. 1–10 (2011)
5. Chien, Y.-F., Hon, W.K., Shah, R., Thankachan, S.V., Vitter, J.S.: Geometric BWT: Compressed Text Indexing via Sparse Suffixes and Range Searching. Algorithmica, 1–21 (2013)

6. Crochemore, M., Iliopoulos, C.S., Kubica, M., Rahman, M.S., Walen, T.: Improved Algorithms for the Range Next Value Problem and Applications. In: STACS, pp. 205–216 (2008)
7. Gagie, T., Gawrychowski, P.: Linear-Space Substring Range Counting over Polylogarithmic Alphabets. CoRR, arXiv: 1202.3208 (2012)
8. Golynski, A., Munro, J.I., Rao, S.S.: Rank/Select Operations on Large Alphabets: A Tool for Text Indexing. In: SODA, pp. 368–373 (2006)
9. Hon, W.K., Patil, M., Shah, R., Thankachan, S.V.: Compressed Property Suffix Tree. In: IEEE Data Compression Conference, pp. 123–132 (2011)
10. Hon, W.K., Shah, R., Thankachan, S.V., Vitter, J.S.: On position restricted substring searching in succinct space. Journal of Discrete Algorithms (2012); Hon, W.-K., Ku, T.-H., Shah, R., Thankachan, S.V., Vitter, J.S.: Compressed text indexing with wildcards. In: Grossi, R., Sebastiani, F., Silvestri, F. (eds.) SPIRE 2011. LNCS, vol. 7024, pp. 267–277. Springer, Heidelberg (2011)
11. Juan, M.T., Liu, J.J., Wang, Y.L.: Errata for "Faster index for property matching". Information Processing Letter 109(18), 1027–1029 (2009)
12. Kopelowitz, T.: The Property Suffix Tree with Dynamic Properties. In: Amir, A., Parida, L. (eds.) CPM 2010. LNCS, vol. 6129, pp. 63–75. Springer, Heidelberg (2010)
13. Kopelowitz, T., Lewenstein, M., Porat, E.: Persistency in Suffix Trees with Applications to String Interval Problems. In: Grossi, R., Sebastiani, F., Silvestri, F. (eds.) SPIRE 2011. LNCS, vol. 7024, pp. 67–80. Springer, Heidelberg (2011)
14. Mäkinen, V., Navarro, G.: Position-Restricted Substring Searching. In: Correa, J.R., Hevia, A., Kiwi, M. (eds.) LATIN 2006. LNCS, vol. 3887, pp. 703–714. Springer, Heidelberg (2006)
15. Manber, U., Myers, G.: Suffix Arrays: A New Method for On-Line String Searches. SIAM Journal on Computing 22(5), 935–948 (1993)
16. McCreight, E.M.: A Space-Economical Suffix Tree Construction Algorithm. Journal of the ACM 23(2), 262–272 (1976)
17. Nekrich, Y., Navarro, G.: Sorted Range Reporting. In: Fomin, F.V., Kaski, P. (eds.) SWAT 2012. LNCS, vol. 7357, pp. 271–282. Springer, Heidelberg (2012)
18. Weiner, P.: Linear Pattern Matching Algorithms. In: SWAT (1973)

Simulation Study of Multi-threading in Web Search Engine Processors*

Carolina Bonacic** and Mauricio Marin

Universidad de Santiago, Chile
Yahoo! Labs Santiago, Chile

Abstract. Modern cluster processors have been steadily increasing the number of cores able to execute concurrent threads. Web search engines critically rely on multithreading to efficiently process user queries and document insertions to support real-time search. This requires synchronization of readers and writers which, for large number of threads, poses the question of what concurrency control strategies are capable of scaling to hundreds of cores and more. This paper presents a comparative study of a number of such strategies. To this end, we focus on the development of suitable simulation models for performance evaluation of search algorithms on dedicated single-purpose multi-threaded processors. We validate our model against actual implementations of the multi-threading strategies to then go further on studying performance on very large processors. We conclude that intra-query parallelism scales up more efficiently than inter-query parallelism.

1 Introduction

Web search engines are large-scale systems devised to achieve efficient performance using the least possible hardware resources from data centers. Efficiency is achieved throughout a composition of distributed indexing, partial-results caching and parallel query processing algorithms, all devised to be deployed on large clusters of processing nodes (processors). These clusters form a distributed memory systems among processors, wherein each processor is a shared memory multi-core node that allows concurrent execution of multiple threads. Queries processing is usually divided in a number of tasks and each processor is in charge of a single task. Communication latency among processors is small with respect to the relative cost of query processing.

In this paper we develop a process-oriented simulation model that enables performance evaluation studies of multi-threaded algorithms for search engine processors. We then use the model to investigate the scalability of search strategies involving R/W operations to hundreds of threads in multi-core processors. The aim is to realize what strategies are the most efficient ones to support real-time search. In this case, insertion and elimination of documents in the search index (inverted file) are allow to take place concurrently with normal user query

---

* Partially funded by research grant DICYT 061319BC and FONDEF CA12i10314.
** Corresponding author.

O. Kurland, M. Lewenstein, and E. Porat (Eds.): SPIRE 2013, LNCS 8214, pp. 37–48, 2013.

processing. This requires the use of thread synchronization strategies to prevent from R/W conflicts on the shared inverted index at processor level.

The simulation model is constructed on the following concepts that we believe make it of practical interest:

1. Web search engines are daily subjected to highly dynamic and intensive query traffic. Performance is totally dependent on user intention which is unpredictable. This means that any credible performance evaluation study must consider the execution of millions of actual user queries (large sets of queries are publicly available for research). This is because there are dependencies across terms in query streams that can be observed only after processing many queries (e.g., partial-results cache design). This implies that a practical simulation model program must be light enough to deliver results in a fairly short time, at most a few hours, hopefully minutes.

2. Point 1 rules out sophisticated simulators that model processor hardware details including aspects such as instruction/data pre-fetching and processor cache consistency protocols. To overcome this obstacle we resort to models of parallel computation intended to cost just relevant key features of hardware rather than low level details. This is possible thanks to the nature of our application since it can be considered as a coarse-grained application where cost dominating operations can be clearly differentiated each other. In particular, we use the so-called Multi-BSP model of parallel computation [8] proposed recently for multi-core processors. The well-defined structure of the model enables us to (a) run benchmark programs on the actual hardware to get the average value of parameters representing the effect of hardware in the cost of algorithms and (b) simplify debugging in the software development process required to produce correct simulation programs.

3. As modelling world-view we use process-oriented simulation since it provides a direct mapping between a multi-threaded algorithm and its simulation. Processes (programmed as co-routines that can be blocked and unblocked at will during simulation) represent threads and the cost of relevant operations is caused in the simulation time by means of a process blocking *hold(cost)* operation. Using this approach to simulation, we formulate a model of a generic multi-core processor which extends the Multi-BSP model with implementations of processor caches and special global variables such as exclusive locks. On top of this simulated multi-core processor, we perform the simulation of thread computations. The cost of relevant operations is determined with benchmark programs run on actual datasets.

To our knowledge, literature reports no work on performance evaluation methods located in the intersection of search engine multi-threaded processor computations, models of parallel computation for multi-core processors, and modelling and simulation of Web search problems. Certainly there are works on each separate topic but not in the intersection (cf. [2]). For our application setting (e.g., research on multi-threaded search algorithms, and partial or full results caching), the goal of this intersection is to compare different solutions for a given problem under the same conditions and understand reasons of efficient performance.

2   Background and Running Example

Web search engines use inverted indexes to determine the top K results for queries. Namely, the K documents that best match the query terms in accordance with a given document ranking method. The index is composed of a vocabulary table and a set of inverted or posting lists. The vocabulary contains the set of relevant terms found in the text collection. Each of these terms is associated with an inverted list which contains the document identifiers where the term appears in the collection along with the term frequency in the document. Frequencies are used for document ranking purposes. To solve a query, it is necessary to get from the inverted lists the set of documents associated with the query terms and then to perform a ranking of these documents in order to select the top K documents as the query answer. A query receptionist machine, called the broker, sends queries for solution to a set of search nodes (in our case *processors*) and blends their results to produce the global top K ones (cf. [4,5]).

Query processing efficiency is increased by using application caches (cf. [3,7]). Usually the size of the inverted index is huge and only the part that is most referenced by queries in a given period of time is maintained in main memory by means of a posting lists cache. Also caches holding pre-computed results can be kept in each search node such as a top-K results cache for frequent queries.

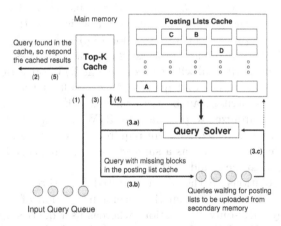

Fig. 1. Paths followed by queries in a single search node processor

Figure 1 illustrates a possible search node layout. The posting lists cache consists of a large number of blocks which are used to store inverted list items, namely pairs (doc_id, term_freq). Each inverted list usually contains several blocks. The second cache holds the top K results of the most frequent queries resolved by the search node. The whole inverted file is kept on local disk or in main memory but in highly compressed format.

A feasible road map for the threads is illustrated in Figure 1. In this example, queries arrive at the input queue of the search node (processor). A given number of threads are in charge of solving the queries. Every time a thread takes a new

Table 1. Solutions to thread synchronization for read/write transactions

	Serial	Type of Parallelism	Sync. Primitive
BP	Yes	PRT and PWT	Thread Barrier
CR	Yes	CRT and PWT	Thread Barrier
TLP1	Yes	CRT (list sharing) and CWT	Inverted List Locking
TLP2	Yes	CRT (no list sharing) and CWT	Inverted List Locking
RTLP	No	non-atomic: CRT and CWT	Inverted List Locking
RBLP	No	non-atomic: CRT and CWT	Inverted List Block Locking

CRT= concurrent read transactions   PRT= read transaction in parallel
CWT= concurrent write transactions   PWT= write transaction in parallel

query from the input queue it checks whether the same query is already stored in the top-K cache (1). If there is a hit on this cache, the thread responds with the K document IDs stored in the respective entry (2). Otherwise, the thread verifies whether all of the blocks holding inverted list items are already in the posting lists cache (3). If so, the thread uses those blocks to solve the query by applying a document ranking algorithm (3.a). Once the thread calculates the top K document IDs, it applies the top-K cache admission and eviction policy to store the new entry in the cache (4), and responds with the calculated top-K results (5). If there are missing inverted list blocks in the cache (3.b), the thread places the query in another queue of secondary memory requirements (a subordinate thread manages these transfers of new blocks) and verifies whether in this second queue there is another query for which the transferring of its blocks has finished to proceed with its solution (3.c), (4) and (5).

The above multi-threaded query processing approach assumes the existence of a strategy able to properly deal with the concurrent R/W operations demanded by the threads on the caches and queues. As a result of the eviction policy, concurrent reads and writes can be performed on the same cache entries so threads must be synchronized to prevent from R/W conflicts. The situation gets further involved when inverted lists are required to receive new pairs (doc_id, term_freq) in an on-line manner as a support for real-time search (recent work on real-time search can be found in [1]).

A number of solutions to the above inverted lists thread synchronization problem are listed in Table 1 from [1]. Insertion/deletion of new documents are represented by write-only transactions whereas user queries are represented by read-only transactions. Parallelism means all threads dedicated to solve one transaction at a time whereas concurrency means each thread completely processes a different transaction. A research question is which of those strategies is able to scale up efficiently to more threads than current multi-core processors are able to support efficiently. A simulation model as we describe in the following should be able to predict performance to answer that question. Another research question is what are the relevant reasons for a given strategy to outperform other alternatives. Executions with actual implementations on actual hardware would tend to show us the fact but not the reasons as it is difficult to repeatedly reproduce the same running conditions in each experiment, and it is difficult to properly trace computations for queries that last a few milliseconds.

3   Simulation Model

The simulation model uses a processes and resources approach. We describe the model for the running examples described in Section 2. Processes represent threads in charge of transaction processing. Resources are artifacts such as inverted lists, processor caches and global variables like exclusive access locks. The simulation program is implemented using a library like [6], where each process is implemented by a co-routine that can be locked and unlocked at will during simulation. The operations *passivate()* and *activate()* are used for this purpose. Co-routines execute the relevant operations of threads/transactions. The operation *hold(t)* blocks the co-routine during *t* units of simulation time. This is used to simulate the different costs of transactions and the cost of the different tasks performed by the simulated multi-core processor. For the running example, the dominant costs come from ranking of documents, intersection of inverted lists and inverted list updates. These costs are determined by benchmark programs implementing the same operations. Figure 2 illustrates the main ideas where we placed the *rank()* operation to emphasize that these simulations can be driven at each point by the outcome of the actual process being simulated.

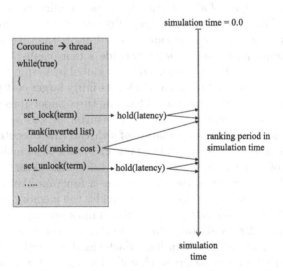

Fig. 2. Concurrent routines simulating the steps taken by a thread to process a query (read-only transaction) and generating cost in simulation time

   The processor cache entries (cache lines) are managed with the LRU replacement policy, where each entry is a memory block of 64 bytes like Intel processors, and we use a directory based coherence protocol for cache entries. In the Multi-BSP model [8], a multi-core processor is seen as a hierarchy of memories (caches and main memory) that form a tree with root being the main memory and leaves being pairs (core, cache L1). The model is general, but we apply it in a manner that suits our purposes. We believe that the resulting model is also generic and includes the key features that affect performance in our application setting.

To simply description let us assume a processor composed of eight cores, where four pairs (core, L1 cache) are connected to a single L2 cache, and two L2 caches are connected to the main memory. Cores (CPUs) can only work on local data stored in their respective L1 caches. When a core references data out of its L1 cache, the data is look at the respective cache L2. If found, a cache line transfer (or several lines if data pre-fetching is enabled) is performed between caches L2 and L1. The cost of this transfer is given by a parameter g_1. If not found, a cache line sized piece of the data is transfered from main memory to cache L2 at cost g_2, and then from L2 to L1 at cost g_1. The values of g_1 and g_2 are determined as explained below by using benchmark programs.

Computations in Multi-BSP are organized as sequences of three supersteps where each superstep is ended with a synchronization barrier. During the first superstep, all data transfers take place between main memory and the two L2 caches. During the second superstep, all data transfers take place between the L2 and L1 caches. During the third superstep, all cores are allowed to perform computations on the data stored in their respective L1 caches. The cost of barriers at levels 1 and 2 are given by the parameters ℓ_i and ℓ_2. Certainly this bulk-synchronous structure of computations is intended to make mathematically tractable the analysis of algorithms. We do not require this structure but it helps during simulation program debugging. Process oriented simulation enables us to simulate these actions asynchronously. In addition, the LRU policy and the read/write coherence protocol on caches provide a more realistic setting to our processor model. Lock variables are forced to be global by keeping them updated in main memory at all time. This accounts for its fairly larger cost $(g_1 + g_2)$ than standard local and global program variables. The three supersteps structure also helps to construct benchmark programs for measuring g_1 and g_2.

The steps followed during the simulation of a thread that executes a cost dominating transaction operation that takes `time_cpu` units of simulation time and works on a piece of data whose space address is `[base_address:offset_bytes]` are the following. The respective co-routine executes a function like `run( time_cpu, base_address, offset_bytes)`. This function divides the memory space in chunks of similar size (line caches) and retrieves the chunks one by one. The cost of chunk memory transfers and computation performed on it by cores is simulated by corresponding *hold(t)* operations. Each simulator cache object contains space to store a set of pairs (address, chunk) (line caches). Cache entries that get replaced by the LRU policy and have been modified by threads, are copied by the simulator to the lower memory in the hierarchy (L1 → L2, L2 → main memory). The coherence protocol maintains track of cache entries that contain copies of the same data so that when a thread modifies an instance of a replicated data, the remaining copies are invalidated. This takes place instantaneously with respect to simulation time as only one co-routine is active at any time instant. The implementation of exclusive locks associates a name string with a lock variable along with a queue for co-routines (threads) waiting to get access. Lock administration takes a latency $g_1 + g_2$ in simulation time, as they are always read and written in main memory. Synchronization barriers are implemented with locks.

Simulator Parameters. In line with the above L1-L2 processor model, we describe experiments to obtain g_1 and g_2 on a cluster node containing two Intel Xeon Quad-Core processors with caches L1 and L2 as described above. Similar method can be applied in three-level cache processors like Intel $i7$.

We first performed runs with actual implementations of the running examples to measure the cost of each relevant operation separately. Running times had less than 1% variation. From the total running time, we found that approximately 6% was due to the administration of data structures and the cost of the memory hierarchy. For example, one execution took 571 seconds for read-only transactions with ranking of documents set to enabled, while the same run but with document ranking disabled took 32 seconds, thus $32/(571 - 32) = 0.059$. The total running time is reduced to 218 seconds when introducing 40% write-only transactions, which increases to 17% the effect of data management and memory hierarchy. This is because write-only transactions are processed much faster than read-only transactions.

Since we know the size of the inverted lists, it is possible to determine the average cost of processing an inverted list for document ranking, excluding the cost of data management and memory hierarchy. The same can be determined for the other two relevant operations, namely updating an inverted list and determining the global top-K results for a query. If the cost of performing the ranking on an inverted list takes x units of simulation time, it was found that the average cost of the other two operations is $x/10$ and $x/5$ approximately.

The transfer rate of memory blocks through the hierarchy of caches is obtained by running benchmark programs that perform consecutive accesses within arrays of integers (this resembles the data access pattern performed by the running example strategies). We define arrays A_i and B_i for each thread i. Before start measuring running time, each thread i writes in B_i the consecutive positions in A_i that it will visit during the experiment. The idea is to reference both arrays simultaneously during measurement. Runs were performed by varying the sizes of the arrays A_i and B_i to determine g_2/g_1. After obtaining this value, the values g_1 and g_2 were adjusted to make the total cost of operations to access shared data structures to represent 6% of the total simulation time, this by using read-only transactions. This was achieved for $g_1 = 0.010$ and $g_2 = 0.044$ as we found that $g_2/g_1 = 4.4$ is a good estimate for our application.

The ratio $g_2/g_1 = 4.4$ was obtained as follows. We varied the size n of the array A_i and B_i for 8 cores and obtained $T_{g_2+g_1}/T_{g_1}$ for different values of n. Time $T_{g_2+g_1}$ represents the case where all threads begin their access to arrays A_i and B_i, ensuring that measurement is started when they are stored in main memory and not in any of the L1 and L2 caches. To ensure this, each thread previously executes random accesses to auxiliary arrays of a size much larger than the capacity of the L2 caches. After this, a total of m accesses to the arrays A_i and B_i are performed to measure running time. The effective size of each L1 cache is 16KB and each L2 cache is 4MB.

To obtain T_{g_1}, we first made a copy of the array B_0 in the two L2 caches, and ensured that each thread i has its array A_i stored in its L1 cache. Then we measured T_{g_1} by making that all threads i perform m accesses on adjacent positions of B_0 and assign the values to corresponding positions in their local arrays A_i. This caused data transfers between L2 and L1 caches in a semantics similar to that promoted by the Multi-BSP model. In this way, the ratio $T_{g_2+g_1}/T_{g_1}$ can be represented by the following equation:

$$\frac{T_{g_2+g_1}}{T_{g_1}} = \frac{m \cdot c + 2 \cdot n \cdot g_1 + 2 \cdot n \cdot g_2}{m \cdot c + n \cdot g_1}$$

where c is the average time demanded by each of the m accesses performed on the arrays A_i and B_i, both of size n. From the results for T_{g_1} with 8 cores, we measured the values of c and g_1. The values of c were very stable with an average of 3.6×10^{-9} seconds approximately. Since experimentally we determined $T_{g_2+g_1}$, T_{g_1}, c and g_1, it was possible to determine g_2. We observed that the average $g_2/g_1 \approx 4.4$ was kept with small variance when n grows. We emphasize that the inverted lists are usually stored in contiguous memory space that is similar to the space occupied by the arrays A_i and B_i for large n.

4   Average Case Analysis

The Multi-BSP model allows validation of intuition about performance of read only transactions (queries) as follows. We assume queries with only one term. We define the parameters α_t and β_t as estimates of the average hit rate that a term t achieves in the caches L1 and L2 respectively. The term t has an inverted list of size n_t. Each strategy must incur in a constant software latency γ to administer the state of queries being solved.

In the bulk-synchronous strategy, each of the p threads work in parallel to solve a single query at a time. Each thread computes the relevance of each document in its n_t/p sized piece of inverted list for term t and sorts the results to determine the best k local results at a cost $O(\gamma + n_t/p + (n_t/p) \cdot \log(n_t/p) + \ell_0)$. However, the threads can maintain a *heap* of size $O(k)$ to keep the top ranked documents and therefore the cost computation can be reduced to $O(\gamma + n_t/p + (n_t/p) \cdot \log k + \ell_0)$. To run that amount of work it was required to pay an average cost of $O((1 - \alpha_t) \cdot (n_t/p) \cdot c \cdot g_1 + \ell_1)$ for data transfers among the L2 and L1 caches. It was also necessary to pay $O((1 - \alpha_t) \cdot (1 - \beta_t) \cdot (c \cdot (n_t/p)) \cdot (p/c) \cdot g_2 + \ell_2) = O((1 - \alpha_t) \cdot (1 - \beta_t) \cdot n_t \cdot g_2 + \ell_2)$ for transfers among the L2 caches and main memory.

After each thread has determined the best k local documents, one of them is responsible for sorting the $k \cdot p$ results to determine the best k documents that are the answer for the query. Since each of the p sets of size k are already ordered, it is only necessary to perform a *merge* of the p sets at a cost $O(k \cdot p \cdot \log p)$. As cores are arranged so as to have c cores for each L2 cache, then the total cost of data transfers from the L2 cache to the L1 cache and managing $k \cdot p$ results is $O((c + p) \cdot k \cdot g_1 + \ell_1 + (c \cdot k) \cdot ((p/c) - 1) \cdot g_2 + \ell_2)$, which for constant c is

$O(p{\cdot}k{\cdot}(g_1{+}g_2){+}\ell_1{+}\ell_2)$. On the other hand, it is not necessary that all the cores send k results, they can do iterations by increasingly sending their next k/p best results. The worst case is to have p iterations. Most likely after a few iterations it will get the best k documents. That is, the cost can be reduced to $O(k \cdot \log p)$, which has a cost in data transfers among caches of $O(k \cdot (g_1 + g_2) + \ell_1 + \ell_2)$.

For fair comparison with the asynchronous strategy, it is necessary to consider a group of p, since the asynchronous strategy processes p queries in parallel, one per thread. For a large number of queries, it is assumed that in a Multi-BSP superstep p queries can be processed sequentially but each one in parallel. For any term the average case is $\alpha = \bar{\alpha}_t$, $\beta = \bar{\beta}_t$, $n = \bar{n}_t$. Then, the total cost of processing p queries in bulk-processing strategy is given by:

$$
\begin{aligned}
\text{Sync. Comp. Cost} \quad &\to p \cdot \gamma + n + n \cdot \log k + k \cdot \log p + \ell_0 + \\
\text{L1-L2 Cache Cost} \quad &\to (1 - \alpha) \cdot n \cdot g_1 + k \cdot p \cdot g_1 + \ell_1 + \\
\text{L2-Ram Cache Cost} \quad &\to (1 - \alpha) \cdot (1 - \beta) \cdot n \cdot p \cdot g_2 + k \cdot p \cdot g_2 + \ell_2 .
\end{aligned}
$$

Similar arguments can be applied to the asynchronous strategy concluding a similar average cost for processing p queries. In this case, the total cost is:

$$
\begin{aligned}
\text{Async. Comp. Cost} \quad &\to \gamma + n + n \cdot \log k + \ell_0 + \\
\text{L1-L2 Cache Cost} \quad &\to (1 - \alpha) \cdot n \cdot g_1 + k \cdot g_1 + \ell_1 + \\
\text{L2-Ram Cache Cost} \quad &\to (1 - \alpha) \cdot (1 - \beta) \cdot n \cdot p \cdot g_2 + k \cdot p \cdot g_2 + \ell_2 .
\end{aligned}
$$

Current cluster search engines, include main memories of very large sizes in their multi-core nodes. This implies that the inverted lists can be very large making the cost $O(n)$ of scoring documents very dominant. The k value is constant and small compared to the average n, and so is p. That is, in practice both strategies should have cost

$$
O(\ n \ + \ (1 - \alpha) \cdot n \cdot g_1 \ + \ (1 - \alpha) \cdot (1 - \beta) \cdot n \cdot p \cdot g_2 \ ).
$$

As a result, in the average case, the synchronous and asynchronous strategies have a similar cost, but as the following simulation study shows they do not scale up similarly when write transactions are included in the work-load.

5 Simulation Study

Using the above defined simulation model we experimented with the concurrency control strategies presented in Table 1. The inverted lists were constructed from a large sample of the UK Web, and read transactions were generated from a query log containing queries submitted by actual users of the Yahoo! UK search engine. To introduce write transactions, small documents were extracted from the Web sample and were randomly mixed with the stream of user queries processed during the experiments. This represent a case in which the search engine is able to include in the results of user queries, documents (blogs, facebook, twitter) that were generated by other users within a time window of a few seconds or minutes

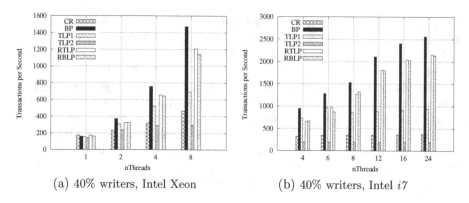

(a) 40% writers, Intel Xeon   (b) 40% writers, Intel *i7*

Fig. 3. Throughput (transactions per second) achieved by actual implementations of concurrency control strategies presented in Table 1

in the past. Overall, 40% of transactions were write-only, namely insertion of new documents in an on-line manner. We also experimented with 20% writers.

Figure 3.a presents comparative performance results for the actual implementations of the strategies of Table 1, all executed on our cluster node formed by two Intel Xeon Quad-Core processors. We observed a similar trend with an Intel *i7* processor that is able to efficiently support 24 threads. These results are shown in 3.b, which confirm that comparative performance is less dependent on the processor architecture than on the synchronization method used to prevent from read/write conflicts. Therefore, to understand the reasons of such differences in performance we resorted to simulation.

We focus next on the strategies labeled BP and RTLP as they achieve the best performance in Figure 3. Besides, both strategies are representative of two different forms of multi-threading. BP is a bulk-synchronous approach which uses all threads to process one transaction at a time. Threads are barrier synchronized between transactions. RTLP is a more classical approach where threads process different transactions concurrently (one thread per transaction). In this case, inverted lists are locked and unlocked by the thread as soon as it has finished processing on it. Notice, however, that RTLP is a more relaxed strategy than BP in terms of the serial and atomic properties of transactions. To force RTLP to be at least atomic, we included in our simulations a RTLP version, called RTLP-RB, that tries to preserve atomic consistency of transactions by performing rollbacks when the same inverted lists are modified by writers whilst read-only transactions are working with them.

Figure 4 shows results for the throughput (transactions per second) achieved by the simulator ranging from 1 to 128 threads. Each thread is assigned to a different core. Considering the range between 1 and 8 threads, the results show a very similar trend in performance to that observed in Figure 3. For more than 8 threads, it is observed that the strategies RTLP and RTLP-RB do not scale efficiently. Traces from simulator executions reveal that there are threads that have to wait too long for locks on inverted lists. This happens because user queries tend to share the same terms, and popular terms in queries are also popular

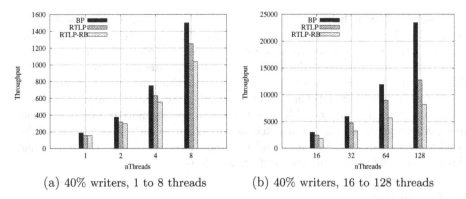

(a) 40% writers, 1 to 8 threads   (b) 40% writers, 16 to 128 threads

Fig. 4. Throughput obtained with the simulation model for two-level caching

Table 2. Metrics illustrating loss of performance in RTLP and RTLP-RB

	20%		40%	
NT	RB	LB	RB	LB
2	0.04	1	0.06	0.95
4	0.10	0.95	0.13	0.98
8	0.18	0.95	0.21	0.89
16	0.26	0.92	0.31	0.87
32	0.33	0.82	0.42	0.78
64	0.41	0.70	0.51	0.69
128	0.46	0.56	0.57	0.41

	0%		40%	
NT	P_w	LB	P_w	LB
2	0	1	0	1
4	0.01	0.99	0.01	1
8	0.02	0.99	0.03	1
16	0.04	0.97	0.09	1
32	0.06	0.94	0.14	0.99
64	0.10	0.88	0.17	0.97
128	0.21	0.84	0.37	0.90

(a) 20% and 40% writers (RTLP-RB)   (b) 0% and 40% writers (RTLP)

in new documents inserted in the inverted lists. This also produces read/write inconsistencies that the RTLP-RB must solve with an excessive number of rollbacks which degrades its performance.

Table 2.a show results that illustrate the loss of performance of the strategy RBLP-RB. The amount of rollbacks (RB columns for different percentage of writers) increases to the point where little more than half of the transactions must be re-executed. The table also shows the effect that these rollbacks have in load balance of computations performed by cores (LB columns). Load balance is measured considering the amount of computation performed by threads, and is defined as the average load observed in the threads divided by the maximum load in any of the threads. This metric is calculated at regular intervals of simulation time. Load balance is significantly degraded due to rollbacks.

Table 2.b shows results that explain the loss of performance in the RTLP strategy as the total number of threads increases. There is a non-negligible probably that for any given transaction a thread has to wait for a lock. This probability increases with the number of threads. Columns P_w show values for this probability. This causes imbalance as shown in columns LB. However, the main factor in performance degradation is the time spent waiting for locks. We believe this fact is difficult to observe in a real implementation.

These simulations help us to understand that as we increase the total number of threads (transactions), the probability of threads working with the same popular terms increases and thereby performance is degraded by the increment of threads waiting for locks. The bulk-synchronous strategy does not suffer from this problem which explains why it scales up better than the asynchronous strategy. Another fact, that is difficult to understand by examining traces from a real implementation, is that the oblivious barriers tend to amortize imbalance due to variations in the lengths of inverted lists assigned to each thread in the bulk-synchronous strategy (while a thread is blending top K results, the remaining threads can start processing the next transaction).

6 Conclusions

This paper has presented a performance evaluation study of the scalability to hundred core processors of different query processing strategies devised to support real-time search in search engines. To this end, we have proposed a simulation method that is suitable for Web search in multi-core processors. The method is constructed upon a model of processors based on concepts from the Multi-BSP model of computation, and process-oriented simulation. We believe that our simulation method is generic and useful to other IR applications. Model validation shows good agreement between what is observed with an actual implementation of Web search and what is observed with the respective simulator. The simulation results clearly show that synchronous multi-threading is able to maintain efficiency under large number of threads whereas asynchronous multi-threading loses performance significantly.

References

1. Bonacic, C., Garcia, C., Marin, M., Prieto-Matias, M., Tirado, F.: Building efficient multi-threaded search nodes. In: CIKM (2010)
2. Cacheda, F., Carneiro, V., Plachouras, V., Ounis, I.: Performance analysis of distributed information retrieval architectures using an improved network simulation model. Information Processing and Management 43, 204–224 (2007)
3. Fagni, T., Perego, R., Silvestri, F., Orlando, S.: Boosting the performance of Web search engines: caching and prefetching query results by exploiting historical usage data. ACM Transactions on Information Systems 24(1), 51–78 (2006)
4. Moffat, A., Webber, W., Zobel, J., Baeza-Yates, R.: A pipelined architecture for distributed text query evaluation. Information Retrieval 10(3) (2007)
5. Marin, M., Gil-Costa, V., Bonacic, C., Baeza-Yates, R., Scherson, I.D.: Sync/Async parallel search for the efficient design and construction of Web search engines. Parallel Computing 36(4), 153–168 (2010)
6. Marzolla, M.: LibCppSim: A SIMULA-like, portable process-oriented simulation library in C++. In: European Simulation Symposium, ESM (2004)
7. Gan, Q., Suel, T.: Improved techniques for result caching in Web search engines. In: WWW (2009)
8. Valiant, L.G.: A bridging model for multi-core computing. Journal of Computer and System Sciences 77(1) (2011)

Query Processing
in Highly-Loaded Search Engines

Daniele Broccolo[1,2], Craig Macdonald[3], Salvatore Orlando[1,2], Iadh Ounis[3],
Raffaele Perego[2], Fabrizio Silvestri[2], and Nicola Tonellotto[2]

[1] Università Ca'Foscari of Venice
[2] ISTI-CNR of Pisa
[3] University of Glasgow
`firstname.surname@unive.it, firstname.surname@isti.cnr.it,`
`firstname.surname@glasgow.ac.uk`

Abstract. While Web search engines are built to cope with a large
number of queries, query traffic can exceed the maximum query rate
supported by the underlying computing infrastructure. We study how
response times and results vary when, in presence of high loads, some
queries are either interrupted after a fixed time threshold elapses or
dropped completely. Moreover, we introduce a novel dropping strategy,
based on machine learned performance predictors to select the queries
to drop in order to sustain the largest possible query rate with a relative
degradation in effectiveness.

Keywords: Distributed Search Engines, Efficiency, Effectiveness,
Throughput.

1 Introduction

In this paper we study strategies for query processing in highly-loaded Web
Search Engines (SEs). We refer to a classical distributed SE architecture, adopt-
ing a Document Partitioning strategy [2], where each query server manages a
local index partition (shard), built on a non-overlapping subset of the whole document collection. Queries are processed on all the shards in parallel, and partial
results, ordered by their score, are returned to a broker for the final ranking.
Dynamic pruning strategies (e.g. `WAND` [3] or `MaxScore` [4]) have been proposed
to reduce query processing times, by avoiding to score a subset of documents
(possibly those that are likely to not be present in the final list of results). We
can thus use these techniques to improve the throughput when unsustainable
bursts of queries arrive to the SE, even if they potentially reduce the quality of the retrieved results. Another ranking strategy that trades effectiveness
for retrieval efficiency is based on impact-sorted indexes [1], but since Boolean
querying becomes harder to support and inclusion of new documents is also
complex, postings lists are commonly maintained in document-sorted order. Alternatively, we can choose to fully score arriving queries, and drop during peak
load the queries that cannot be processed within a fixed time threshold [5].

O. Kurland, M. Lewenstein, and E. Porat (Eds.): SPIRE 2013, LNCS 8214, pp. 49–55, 2013.
© Springer International Publishing Switzerland 2013

In this paper, we investigate the performances of different *dropping solutions* with the goal of maintaining the query response time under a user specified *time threshold*. We compare naïve solutions with a novel method, based on the prediction of query processing time which leverages a *machine learning* technique. We consider disjunctive query processing with full DAAT as the baseline strategy [4]. Since our reference architecture is distributed, we design a model to predict query processing times[1] to be deployed at each query server. We use the predictors to understand when a query has to be dropped in order to reserve the current capacity of the SE for processing the remaining query traffic. We test our solution while varying the query arrival rate, from 5 to 100 queries per second (q/s), and measuring the query response time and the effectiveness in terms of NDCG@20 for all the methods proposed. Our approach can remarkably decrease the total number of queries dropped and also improve the overall SE effectiveness, whilst attaining query response times within the time threshold. For instance, for a query arrival rate of 100 per second, our strategy is able to answer up to 40% of the queries without degrading effectiveness, while for our baseline strategies this happens for only 10% of queries.

2   Prediction-Based Dropping

We consider that each query server of our distributed SE receives a query stream from the query broker, and processes one query at a time. If a query server is processing a query, and other queries arrive, they are locally enqueued until they can be processed. Hence, the query response time for a query q is the sum of the time spent waiting in the queue $wt(q)$ and the processing time $pt(q)$. The length of the queue at each query server depends on the query arrival rate and the processing time of the previous queries. In general, for higher query arrival rates, the query response time increases, due to the longer waiting times. To ensure low query response times in a high load environment, we fix a maximum processing threshold T that queries must be answered within. We adopt two baseline strategies that define how a query server responds to a query q for which T has elapsed during processing. The first strategy (hereinafter, Drop), whenever $wt(q) + pt(q) \geq T$, interrupts the processing of q and returns an empty list of results. Similar to the Drop strategy, the second baseline (hereinafter, Partial-Drop) returns the partial results list that has been computed thus far (instead of dropping all results that have already been computed). Finally, we note that each query server acts independently from the other servers, in an autonomous fashion: each queue is managed locally, and any dropping strategy is enforced locally. Hence, even if a query is fully processed on one query server, it can be (partially-)dropped on another server, causing the final results returned by the query broker to the user to be partial in nature.

Unlike the previously described baselines, in this paper we aim to use the predicted response times at each query server to understand if a query q can be processed within the remaining time on that server before T has elapsed. Given

---

[1] Query efficiency predictors have been proposed in [6] for WAND and MaxScore.

the predicted response time $\widehat{pt}(q)$ of query q, if the inequality $\widehat{pt}(q) \leq T - wt(q)$ does not hold, then the query is dropped before processing starts and the next query is processed from the queue. In this way, the query server does not consume processing resources for queries that cannot be fully (and effectively) completed within the remaining time until the threshold T has elapsed. Query efficiency prediction for full DAAT can be achieved using a machine learned algorithm designed for a specific number of query terms and using the total number of postings to be scored as feature [6]. We adopt a different learned model, where the number of query terms is a feature, thus obtaining a single model instead of a model for each query length. To further improve the quality of estimations based on the total number of postings only, we use five additional features, which are listed in Table 1(a). All features can be easily computed during the processing time without affecting the query response time. The response times are predicted using a machine learning model, i.e., a *linear regression* of all these features. The coefficients of the regression model are computed by minimising the mean squared error on a set of training queries. In the following, we refer to our prediction-based dropping strategy as ML-Drop, and experiment to ascertain its properties in terms of efficiency and effectiveness.

Table 1. (a) Features used for predicting DAAT processing time. (b) prediction accuracy using different feature sets.

Query Efficiency Prediction Features
total no. of postings in the query's term lists
no. of terms in the query
variance of the length of the posting lists
mean of the length of the posting lists
length of the shortest posting list
length of the longest posting list

(a)

# features	RMSE	err $\leq$ 10 ms
1^a	$8.78 \cdot 10^{-3}$	87.83 %
6	$\mathbf{4.98 \cdot 10^{-3}}$	**95.53 %**

(b)

^{a} total no. of postings in the query's term lists

3   Experiments

The research questions addressed in this paper are: *(i)* What is the accuracy of our response time predictors? *(ii)* What are the benefits of our ML-Drop strategy with respect to the two baseline strategies, Drop and Partial-Drop?

First we define the setup for all the experiments. The SE is implemented in C++, exploiting multi-threading to handle multiple queries, and communications between query servers and the broker are implemented by low-level socket interfaces to reduce overheads. Each query is processed in disjunctive mode using a full DAAT strategy, where documents are ranked using BM25 with its default parameters. We use a cluster of twelve machines, where each machine has one Intel Xeon 2.40GHz X32230 CPU and 8GB of RAM, connected using Gigabit Ethernet. Ten machines are used for the query servers, one for the broker and the last one for the client that simulates. For the experiments, we use 40,000

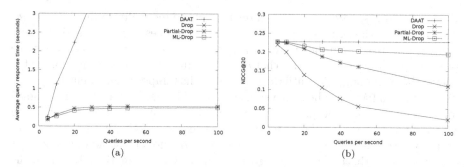

Fig. 1. (a) Average query response time (in seconds) for different dropping strategies; (b) Effectiveness (NDCG@20)

queries from the TREC Million Query Track 2009 [7], 678 of which have corresponding relevance assessments: 30,000 queries are used as the training set for learning regression models for response time prediction; the other 10,000, including the 687 with relevance assessments, are used for testing the accuracy of the predictors, and retrieval experiments. The corresponding document corpus is ClueWeb09 (cat. B), which comprises 50 million English Web documents. We index the document collection using the Terrier search engine[2], removing standard stopwords and applying Porter's English stemmer (our C++ retrieval system can read Terrier's indices). The resulting index is document-partitioned into ten separate index shards, while maintaining the original ordering of the collection. We retrieve 1,000 results for each query. Finally, we set $T = 0.5\ s$ as our time threshold. We choose this value because is a reasonable time from the user perspective. Indeed, in our architecture, 98% of queries can be answered within 0.5 seconds using the full DAAT strategy when the system is not heavy loaded.

Prediction Accuracy. Table 1(b) shows the average accuracy of our prediction models measured in terms of root mean square error (RMSE) respect to the actual query execution time. The first row shows the performance of the predictors using only one feature, namely the total number of postings for each query. The second row shows the performance obtained when all the six features of Table 1(a) are used. Our model with six features halves the RMSE over the 10,000 queries used for the test set. Given an average processing time for DAAT of 110 ms, we compute the percentage of test queries with a predicted processing time with a maximum absolute error of 10 ms, and our model performs markedly better ($\sim$96%) than the single feature model ($\sim$88%). Hence, in conclusion to our first research question, we find that the proposed additional features enhance remarkably the accuracy of the predicted response times.

Dropping Strategies. In order to analyse the performance of three different query dropping strategies, namely Drop, Partial-Drop and ML-Drop, we

---

[2] http://terrier.org/

Table 2. Effectiveness (NDCG@20) for the different methods. Statistically significant degradations vs. DAAT, as measured by the paired t-test, are denoted by $\triangledown$ ($p < 0.05$) and $\blacktriangledown$ ($p < 0.01$).

Method	5 q/s	10 q/s	20 q/s	30 q/s	40 q/s	50 q/s	100 q/s
DAAT	0.228	0.228	0.228	0.228	0.228	0.228	0.228
Drop	0.219 $\blacktriangledown$	0.200 $\blacktriangledown$	0.140 $\blacktriangledown$	0.105 $\blacktriangledown$	0.076 $\blacktriangledown$	0.056 $\blacktriangledown$	0.021 $\blacktriangledown$
Partial-Drop	0.227	0.224 $\triangledown$	0.210 $\blacktriangledown$	0.189 $\blacktriangledown$	0.173 $\blacktriangledown$	0.161 $\blacktriangledown$	0.110 $\blacktriangledown$
ML-Drop	0.227	0.227	0.217	0.207 $\triangledown$	0.205 $\triangledown$	0.203 $\triangledown$	0.195 $\triangledown$

compute the average query response time for the various strategies and we compare them to the full DAAT processing strategy without any dropping. Figure 1(a) shows the average query response time (measured on the broker) vs. the number of queries per second (denoted q/s). We observe that using the full DAAT processing for all the queries implies an increasing query response time that is caused by congestion at the queues. However, all the other strategies (Drop, Partial-Drop and ML-Drop) manage to answer, on average, within the time threshold $T = 0.5\ s$, as the superimposed curves show. As expected, the Drop and Partial-Drop strategies respect this threshold, as they are both defined such that query processing terminates within T. Our approach (ML-Drop), instead, can respect the threshold since our predictor are able to identify queries to drop that cannot be processed within T. Next, we examine the impact on effectiveness of the different processing methods. Figure 1(b) presents effectiveness in terms of NDCG@20, while Table 2 reports the same NDCG@20 values, in conjunction with statistical significance tests using the paired t-test. As expected, full processing (DAAT) always obtains the best effectiveness, at the price of a higher query response time. The other strategies obtain an effectiveness dependent on the system load, since the number of dropped queries is impacted by the remaining time for processing queries. This time is inversely proportional to the waiting time of the query itself. The least effective method is Drop: even though it achieves high effectiveness when the system is unloaded, NDCG@20 decreases quickly as the load increases, because the processing of many queries cannot be finished within the permitted time. Consequently, these queries are dropped by the query server and the time spent is wasted, as no results are returned to the broker. The other baseline, Partial-Drop, obtains a better effectiveness in comparison to Drop. This is expected, because by returning partial results that have been computed within the limited processing time, some relevant results for some queries can be retrieved on average. On the other hand, the effectiveness of ML-Drop is always higher than the two baselines. For instance, when queries arrive at a rate of 100 q/s, ML-Drop results in an effectiveness drop of 15% NDCG@20 (0.228 to 0.195, significant for $p < 0.05$), compared to Partial-Drop which would result in a 52% drop in effectiveness, significant for $p < 0.01$. Similarly, for query arrival rates up to 20 q/s, ML-Drop exhibits no significant degradation in effectiveness, which is in contrast with both Drop and Partial-Drop. The higher effectiveness of ML-Drop compared to the baselines is explained by the pro-active control over the query dropping behaviour: queries

Table 3. Percentage of globally dropped (G) and partially evaluated queries (P)

Methods	10 q/s		20 q/s		30 q/s		40 q/s		50 q/s		100 q/s	
	P+G	G	P+G	G	P+G	G	P+G	G	P+G	G	P+G	G
Drop	10%	1%	36%	9%	51%	25%	62%	34%	70%	41%	90%	57%
Partial-Drop	9%	-	32%	-	48%	-	60%	-	71%	-	91%	-
ML-Drop	**6%**	1%	**20%**	2%	**31%**	7%	**38%**	11%	**44%**	15%	**59%**	28%

which cannot satisfy threshold T are immediately discarded, thus leaving the potential for more queries to be fully processed. To illustrate this, we analyse the number of queries that are *globally* dropped for the different methods. A query is globally dropped when it is dropped by all query servers processing it. Indeed, as query servers are independent, a query can be dropped only in a subset of the query servers. It is therefore possible that some queries have partial results, even when the Drop strategy is used. Table 3 shows, for each strategy and query rate, the percentage of queries that are either *partially* evaluated or *globally* dropped (see columns P+G, where the best values are in bold). The various columns G show the percentages of queries that are *globally* dropped. In the case of Partial-Drop, since the expiry of the time threshold causes some local partial results to be sent back to the broker, no global drops are observed. For high query loads, i.e., 100 q/s, this impacts 90% of processed queries. For the same high arrival rate, the Drop strategy globally drops around 57% of queries while returning partial results for 33% of queries. However, in the case of the ML-Drop strategy, the number of queries globally dropped or with partial results markedly decreases in relation to the other strategies. Hence, in addressing our second research question, we find that the proposed ML-Drop strategy reduces the number of queries dropped under high load, resulting in improved effectiveness. Indeed, when 100 queries per second arrive, ML-Drop is able to answer up to 40% of the queries without effectiveness degradations, while for Drop and Partial-Drop strategies this happens for only 10% of queries.

4   Conclusions

In this paper, we analysed dropping and stopping methods for query processing in presence of an unsustainable workload. Our aim was to answer queries within a fixed time threshold, whilst maintaining overall effectiveness of the results. We proposed a novel dropping method based on the prediction of query execution time. We test the proposed method and the baseline on a distributed SE using 10, 000 queries and a collection of 50 million documents, varying the number of queries per second. Moreover, effectiveness measures use the relevance assessments from the TREC Million Query track. Our efficiency predictor models are able to predict the query response time for DAAT with an error less than 10 ms in more than 93% of the cases. We showed that by using these predictors to select the queries to drop, our proposal obtains up to 80% improvement in comparison to the most effective of the used baselines. Finally, we showed that our method decreases the number of dropped queries when the system is overloaded.

Acknowledgements. This work was partially supported by the EU projects InGeoCLOUDS (no. 297300), MIDAS (no. 318786), E-CLOUD (no. 325091), the Italian PRIN 2011 project "Algoritmica delle Reti Sociali Tecno-Mediate" (2013-2014) and the Regional (Tuscany) project SECURE! (FESR PorCreo 2007-2011).

References

1. Anh, V.N., de Kretser, O., Moffat, A.: Vector-space ranking with effective early termination. In: Proceedings of SIGIR, pp. 35–42 (2001)
2. Barroso, L.A., Dean, J., Holzle, U.: Web search for a planet: The Google cluster architecture. IEEE Micro 23(2), 22–28 (2003)
3. Broder, A.Z., Carmel, D., Herscovici, M., Soffer, A., Zien, J.: Efficient query evaluation using a two-level retrieval process. In: Proceedings of CIKM, pp. 426–434 (2003)
4. Moffat, A., Zobel, J.: Self-indexing inverted files for fast text retrieval. ACM Trans. Inf. Syst. 14(4), 349–379 (1996)
5. Tonellotto, N., Macdonald, C., Ounis, I.: Efficient and Effective Retrieval using Selective Pruning. In: Proceedings of WSDM (2013)
6. Macdonald, C., Tonellotto, N., Ounis, I.: Learning to Predict Response Times for Online Query Scheduling. In: Proceedings of SIGIR, pp. 621–630 (2012)
7. Carterette, B., Pavlu, V., Fang, H., Kanoulas, E.: Million Query Track 2009 Overview. In: Proceedings of TREC (2009)

Indexes for Jumbled Pattern Matching in Strings, Trees and Graphs

Ferdinando Cicalese[1], Travis Gagie[2,3],
Emanuele Giaquinta[2,*], Eduardo Sany Laber[4],
Zsuzsanna Lipták[5], Romeo Rizzi[5],
and Alexandru I. Tomescu[2,3,**]

[1] Department of Computer Science, University of Salerno, Italy
[2] Department of Computer Science, University of Helsinki, Finland
[3] Helsinki Institute for Information Technology, Finland
[4] Department of Computer Science, PUC Rio de Janeiro, Brazil
[5] Department of Computer Science, University of Verona, Italy

Abstract. We consider how to index strings, trees and graphs for jumbled pattern matching when we are asked to return a match if one exists. For example, we show how, given a tree containing two colours, we can build a quadratic-space index with which we can find a match in time proportional to the size of the match. We also show how we need only linear space if we are content with approximate matches.

1 Introduction

Suppose we are given a connected graph G on n coloured nodes and a multiset M of colours and asked to find a connected subgraph of G whose nodes' colours are exactly those in M, if such a subgraph exists. This problem is commonly referred to as *jumbled pattern matching*, and has recently aroused much interest in the case of strings [7,5,8,6,17,13,14,2]: There we are looking for substrings of the text which have the same multiplicity of each character as the query, also referred to as its *Parikh vector*. (The boolean version is the Parikh fingerprint or character set [1,10].) Parikh vectors appear frequently in applications in computational biology [3,5,11,4], as do jumbled patterns in graphs [16].

Even when G is a tree, there can be exponentially many such matching subgraphs. When G is a path, however, there are $\mathcal{O}(n)$ matches and we can find them all in $\mathcal{O}(n)$ time [7]. When G is a path containing a constant number of colours—i.e. the nodes are coloured with only two colours—then in $\mathcal{O}(n^2)$ time we can build a $o(n^2)$-space index with which we can determine in $o(n)$ time whether there is a match [15]. When G is a path containing only two colours, in $\mathcal{O}(n^2 / \log^2 n)$ time we can build an $\mathcal{O}(n)$-bit index with which we can determine in $\mathcal{O}(1)$ time whether there is a match [8,6,17,13]. Moreover, in $\mathcal{O}(n^2 / \log^2 n)$ time we can build an index of size $\mathcal{O}(n \log n)$-bits with which we can find all the

---
* Supported by Academy of Finland grant 118653 (ALGODAN).
** Supported by Academy of Finland grant 250345 (CoECGR).

O. Kurland, M. Lewenstein, and E. Porat (Eds.): SPIRE 2013, LNCS 8214, pp. 56–63, 2013.

matches using $\mathcal{O}(|M|)$ worst-case time per match [13]. We can build an approximation of this index in $\mathcal{O}(n^{1+\epsilon})$ time with the quality of the approximation depending on ϵ [9]. Throughout this paper our model is the word-RAM with $\Omega(\log n)$-bit words and we measure space in words unless stated otherwise.

Determining whether there is a match is NP-complete even when G is a tree [16]. It is also NP-complete when G contains only two colours, but takes polynomial time when G both has bounded treewidth and contains only a constant number of colours [12]. When G contains only two colours there exists an $\mathcal{O}(n)$-bit index with which we can determine in $\mathcal{O}(1)$ time whether there is a match [13]. Building this index is NP-hard in general but, since finding a match is self-reducible, takes polynomial time when G has bounded treewidth and $\mathcal{O}(n^2/\log^2 n)$ time when G is a tree. At the cost of increasing the space to $\mathcal{O}(n)$ words, this index can be generalized to return a subset of the nodes in the matches that is also a hitting set for all the matches, using $\mathcal{O}(\log n)$ time worst-case time per match. In the worst case, however, this subset of nodes is of little use in finding even a single complete match.

We start by presenting some basic tradeoffs in Section 2, to establish what can be done naïvely on graphs. In Sections 3 to 6 we assume G contains only two colours. In Section 3 we consider the case when G is a path — i.e., a binary string — and describe an $\mathcal{O}(n)$-space index with which we can find a match in $\mathcal{O}(\log n)$ time. In Section 4 we consider the case when G is a tree and, based on our index for binary strings, describe an $\mathcal{O}(n^2)$-space index with which we can find a match in $\mathcal{O}(|M|)$ time. If we are concerned only with multisets of size at most $n^{1/2}$, then we can reduce the space bound to $\mathcal{O}(n)$. In Section 5 we show that we can achieve the same space bound if we are content with approximate matches. In Section 6 we partially extend our results from trees to graphs.

2   Basic Tradeoffs

Suppose G is a graph containing a constant number c of colours and we will be given M as the vector of length c whose components are the frequencies of the colours, the Parikh vector for M; note that the sum of its entries equals $|M|$. Since for any $1 \leq m \leq n$, there are $\binom{m+c-1}{c-1} = \mathcal{O}(m^{c-1})$ possible multisets of size m and it takes $\mathcal{O}(m)$ space to store pointers to a match for such a multiset, there exists an $\mathcal{O}(n^{c+1})$-space index with which we can find a match in $\mathcal{O}(|M|)$ time. When G has bounded treewidth we can build this index in polynomial time, and we can reduce the space bound to $\mathcal{O}(n)$ at the cost of increasing the query time to $|M|^{\mathcal{O}(1)}$. To do the latter, we store G itself and pre-compute and store pointers to matches only for multisets of size at most $n^{1/(c+1)}$. Given a multiset M with $|M| > n^{1/(c+1)}$, we search G in $n^{\mathcal{O}(1)} = |M|^{\mathcal{O}(1)}$ time.

For any positive constant ϵ, we can build an $\mathcal{O}(n\log^c n)$-space approximate index with which, if M has an exact match, then in $\mathcal{O}(1)$ time we can find a substring whose Parikh vector differs from M's by at most a factor of $1 + \epsilon$ in each component. (This index does not tell us whether M has an exact match, however, since we may find such a substring even when it does not.) Without loss of generality, assume we are concerned only with multisets in which each character appears at least once; we can reduce the general case to $2^c = \mathcal{O}(1)$ instances of this one. We store a c-dimensional grid with each side having length $\lfloor \log_{1+\epsilon} n \rfloor + 1$. For each point $(x_0, \ldots, x_{c-1})$ in this grid, we store pointers to the nodes in a connected subgraph whose Parikh vector is component-wise between $\left((1+\epsilon)^{x_0}, \ldots, (1+\epsilon)^{x_{c-1}}\right)$ and $\left((1+\epsilon)^{x_0+1}, \ldots, (1+\epsilon)^{x_{c-1}+1}\right)$. This takes a total of $\mathcal{O}(n \log^c n)$ space. Given the Parikh vector $(v_0, \ldots, v_{c-1})$ of M, we return the subgraph stored for the point $\left(\lfloor \log_{1+\epsilon} v_0 \rfloor, \ldots, \lfloor \log_{1+\epsilon} v_{c-1} \rfloor\right)$ in the grid, if that subgraph exists. We summarize these basic tradeoffs in the following proposition:

Proposition 1. *When G is a graph containing a constant number c of colours there exists an $\mathcal{O}(n^{c+1})$-space index with which we can find a match in $\mathcal{O}(|M|)$ time. For any positive constant ϵ there exists an $\mathcal{O}(n \log^c n)$-space index with which in $\mathcal{O}(|M|)$ time we can find an approximate match in which each colour's frequency is within a factor of $1 + \epsilon$ of its frequency in M. When G has bounded treewidth we can build these indexes in polynomial time and, moreover, we can reduce the space of the exact index to $\mathcal{O}(n)$ at the cost of increasing the query time to $|M|^{\mathcal{O}(1)}$.*

When G is a path — which we can think of as a string over an alphabet of c characters — we can improve these bounds. Since G contains $\mathcal{O}(n^2)$ substrings and we can specify any substring by its two endpoints, we can build an $\mathcal{O}(n^2)$-space index with which we can find a match in $\mathcal{O}(1)$ time. Calculation shows we can reduce the space bound to $\mathcal{O}(n)$ at the cost of increasing the query time to $\mathcal{O}(|M|^c)$, and we can store an approximate index in $\mathcal{O}(\log^c n)$ space.

Suppose G is a string over a constant-size alphabet and $0 < \epsilon \leq 1$. Then in $\mathcal{O}(n^{1+\epsilon})$ expected time we can build an index with which, given a multiset M of characters, we can find all occ matches of M in $\mathcal{O}(|M|^{1/\epsilon} + occ)$ worst-case time. To do this, we store G itself and, for $1 \leq m \leq n^\epsilon$, we make a pass over G and store, for each multiset of size m that has a match in G, a list of all the locations of that multiset's matches. Notice the lists for multisets of size m are disjoint and have total length $n - m + 1$; therefore, with dynamic perfect hashing we use a total of $\mathcal{O}(n^{1+\epsilon})$ expected time and $\mathcal{O}(n^{1+\epsilon})$ space. Given a multiset M with $|M| \leq n^\epsilon$, we return our pre-computed list of the locations of M matches in $\mathcal{O}(|M| + occ)$ time, or $\mathcal{O}(occ)$ time if we are given M as a Parikh vector. Given a multiset M with $|M| > n^\epsilon$, we search G in $\mathcal{O}(n) = \mathcal{O}(|M|^{1/\epsilon})$ time.

As an aside, we note that we can extend our approximate indexes to support approximate scaled-then-permuted pattern matching (see [7]). To do this, for each point $(x_0, \ldots, x_{c-1})$ in the grid for which there is no subgraph whose Parikh vector is between $\left((1+\epsilon)^{x_0}, \ldots, (1+\epsilon)^{x_{c-1}}\right)$ and $\left((1+\epsilon)^{x_0+1}, \ldots, (1+\epsilon)^{x_{c-1}+1}\right)$, we store pointers to the nodes in a connected subgraph (if there is one) whose Parikh vector is a multiple of a one between $\left((1+\epsilon)^{x_0}, \ldots, (1+\epsilon)^{x_{c-1}}\right)$ and $\left((1+\epsilon)^{x_0+1}, \ldots, (1+\epsilon)^{x_{c-1}+1}\right)$. The query time is still proportional to the size of the match returned but that may now be larger than $|M|$.

3   An Index for Binary Strings

Suppose G is a binary string, i.e., $G[1..n] \in \{0,1\}^*$. It holds for any m: If there are p copies of 1 in $G[i..i + m - 1]$ and r copies of 1 in $G[k..k + m - 1]$, then for every value q between p and r there is a position j between i and k such that $G[j..j + m - 1]$ contains q copies of 1. This observation was the basis for the index in [8] and is the basis for ours as well.

We store an $\mathcal{O}(1)$-time rank data structure for G and, for $1 \le m \le n$, we store the endpoints of two substrings of length m in G with the most and with the fewest copies of 1. This takes a total of $\mathcal{O}(n)$ space. Given a Parikh vector (v_0, v_1), we look up the left endpoints i and j of the substrings of length $v_0 + v_1$ in G with the most and with the fewest copies of 1. We set i and j as the initial endpoints for a binary search: at each step, we use two rank queries to find the number q of 1s in $G\left[\left\lfloor \frac{i+j}{2} \right\rfloor .. \left\lfloor \frac{i+j}{2} \right\rfloor + v_0 + v_1 - 1\right]$; if $q = v_1$ then we stop and report this substring by its endpoints; if $q < v_1$ then we set $i = \lfloor (i + j)/2 \rfloor$ and continue; if $q > v_1$ then we set $j = \lfloor (i + j)/2 \rfloor$ and continue. This search takes a total of $\mathcal{O}(\log n)$ time.

Theorem 1. *When G is a path containing only two colours, we can build an $\mathcal{O}(n)$-space index with which we can find a match in $\mathcal{O}(\log n)$ time.*

4   Exact Indexes for Trees with Two Colours

Suppose G is a tree containing only two colours, black and white. Gagie, Hermelin, Landau and Weimann [13] noted that the observation in Section 3 can be extended to connected graphs: For any m, if there are connected subgraphs H_p and H_r in G with m nodes each and p and r white nodes, respectively, then for every value q between p and r, there is a connected subgraph H_q with m nodes and q white nodes.

To see why, notice that we can construct a sequence of connected subgraphs with m nodes such that the sequence starts with H_p and ends with H_r and any consecutive pair of subgraphs in the sequence differ on two nodes. To build this sequence, we find a path between H_p and H_r. We root H_p and H_r, which are trees themselves, at the first and last nodes in the path (or at a shared node, if they are not disjoint). One by one, we remove nodes bottom-up in H_p and add nodes along the path; remove nodes nearest to H_p in the path and add nodes further along the path; then remove nodes from the path and add nodes top-down in H_r.

Suppose p and r are the minimum and maximum numbers of white nodes in any connected subgraphs of size m, and we store a path consisting of the nodes in H_p in bottom-up order, followed by the nodes in the path, followed by the nodes in H_r in top-down order. If we apply Theorem 1 to this path, then we obtain an $\mathcal{O}(n)$-space index with which, given the Parikh vector for a multiset M with $|M| = m$, we can find a match in the graph G in $\mathcal{O}(\log n + |M|)$ time. Notice that, if $|M| < \log n$, then we can simply store an $\mathcal{O}(\log^2 n)$-space lookup table with which we can find a match in $\mathcal{O}(|M|)$ time. Therefore, applying this construction for $1 \leq m \leq n$, we obtain the following theorem:

Theorem 2. *When G is a tree containing only two colours, we can build an $\mathcal{O}(n^2)$-space index with which we can find a match in $\mathcal{O}(|M|)$ time.*

When $m \approx n$, we need $\mathcal{O}(n)$ space to store subgraphs with the minimum and maximum numbers of white nodes and the path between them. When $m \ll n$, however, those subgraphs are small and most of the space is taken up by the path. We now show how we can store G such that we can support fast rank queries on paths.

Lemma 1. *We can store G in $\mathcal{O}(n)$ space such that q rank queries on the path between any two nodes take a total of $\mathcal{O}(\log n + q)$ time.*

Proof. We compute the heavy-path decomposition [18] of G and store $\mathcal{O}(1)$-time rank data structures for each of the heavy paths, which takes $\mathcal{O}(n)$ space. The path between any two nodes u and v is a sequence of $\mathcal{O}(\log n)$ intervals of heavy paths. Given u and v, for each of these intervals we compute the number of white nodes in that interval and to either side of it in the heavy path; this takes a total of $\mathcal{O}(\log n)$ time and rank queries on heavy paths. With this information we can perform any rank query on the path from u to v using a single rank query on a heavy path. □

If we store G with Lemma 1 and store subgraphs with the minimum and maximum numbers of white nodes only for $1 \leq m \leq n^{1/2}$, then our index takes only $\mathcal{O}(n)$ space but supports queries only for $|M| \leq n^{1/2}$. When $|M| > n^{1/2}$ we can use an algorithm by Gagie et al. to find a match in $\mathcal{O}(|M|n) = \mathcal{O}(|M|^3)$ time.

Corollary 1. *When G is a tree containing only two colours, we can build an $\mathcal{O}(n)$-space index with which we can find a match in $\mathcal{O}(|M|)$ time when $|M| \leq n^{1/2}$ and in $\mathcal{O}(|M|^3)$ time otherwise.*

5 An Approximate Index for Trees with Two Colours

In this section we present our most technical result, which is how to store in $\mathcal{O}(n)$ space an approximate index for a tree containing only two colours. Again, an approximate match is one whose Parikh vector differs from M's by a factor of at most $1 + \epsilon$ in each component. (In contrast, with Proposition 1 we would use $\mathcal{O}(n \log^2 n)$ space.) Without loss of generality, assume we are only concerned with multisets in which there are at least as many black nodes as white nodes; we can build a symmetric index for the other case. Notice that in this case, if we can find a connected subgraph H with the same size as the given multiset M and in which the number of white nodes is within a factor of $1 + \epsilon$ of the number in M, then the number of black nodes in H is also within a factor of $1 + \epsilon$ of the number in M.

Our main idea is to store an $\mathcal{O}(n)$-space data structure with which, given a size m, we can find two connected subgraphs with size m that have approximately the minimum and maximum numbers of white nodes. Suppose we store a subgraph with the minimum number of white nodes for each size that is a power of two and for each size such that the minimum number of white nodes is a factor of $1 + \epsilon$ greater than the number in the preceding stored subgraph. That is, we store a sequence of $\lg n$ subgraphs with total size $\mathcal{O}(n)$ and a sequence of $\log_{1+\epsilon} n$ subgraphs with total size $\mathcal{O}(n \log n)$. The latter sequence of subgraphs has total size $\mathcal{O}(n \log n)$ in the worst case because the minimum number of white nodes may stay low until we reach size nearly n and then increase rapidly, causing us to store about $\log_{1+\epsilon} n$ subgraphs each of size nearly n. However, we can store this sequence of subgraphs in a total of $\mathcal{O}(n)$ space using the following lemma. Similarly, we also store a subgraph with the maximum number of white nodes for each size that is a power of two and for each size such that the maximum number of white nodes is a factor of $1 + \epsilon$ greater than the number in the preceding stored subgraph; this also takes $\mathcal{O}(n)$ total space if we store the subgraphs with the following lemma.

Lemma 2. *We can store G in $\mathcal{O}(n)$ space such that, if G contains a connected subgraph of size m with w white nodes, then we can represent some such subgraph in $\mathcal{O}(w)$ space such that recovering this subgraph takes $\mathcal{O}(m)$ time.*

Proof. We store the adjacency lists for G's nodes, with each list ordered such that black neighbours precede white neighbours. With this representation, we can expand a subgraph by adding only black nodes as long as this is possible, using $\mathcal{O}(1)$ time per added node.

Let H be a connected subgraph of size m with w white nodes. We store pointers to the white nodes in H, which takes $\mathcal{O}(w)$ space. Since G is a tree, we can find the unique paths between these nodes in a total of m time; notice these paths are contained in H and consist of black nodes. If the subgraph consisting of the white nodes and these paths has fewer than m nodes, then we add black nodes until it has m nodes, which takes a total of $\mathcal{O}(m)$ time. It is possible to add enough black nodes without adding any white nodes because, e.g., we could add the remaining black nodes in H. $\square$

If we are given a multiset M such that we have subgraphs of size $|M|$ sampled, then we can proceed as in the proof of Theorem 2 and find an exact match if there is one. If we do not have subgraphs of size $|M|$ sampled, then we use our sampled subgraphs to build subgraphs $H_{\min}$ and $H_{\max}$ of size $|M|$ with approximately minimum and maximum numbers of white nodes, then proceed almost as in the proof of Theorem 2: if the number of white nodes $H_{\min}$ is larger but within a factor of $1 + \epsilon$ of the number in M, then we return $H_{\min}$; if the number in $H_{\min}$ is more than a factor of $1 + \epsilon$ larger than the number in M, then there is no exact match and we return nothing; if the number of white nodes $H_{\max}$ is smaller but within a factor of $1 + \epsilon$ of the number in M, then we return $H_{\max}$; if the number in $H_{\max}$ is less than a factor of $1 + \epsilon$ smaller than the number in M, then there is no exact match and we return nothing; in all other cases, we proceed as in Theorem 2.

To build $H_{\min}$ we take the next larger subgraph with a minimum number of white nodes and discard nodes until it has size $|M|$ while leaving it connected. This next larger subgraph has size less than $2|M|$, because we sampled for every size that is a power of two; has at most $1 + \epsilon$ times more white nodes than the subgraph of size $|M|$ with the minimum number of white nodes, because we sampled whenever the minimum number of white nodes increased by a factor of $1 + \epsilon$; and is a tree, because it is a connected subgraph of a tree. It follows that discarding nodes takes $\mathcal{O}(|M|)$ time and, since discarding nodes cannot increase the number of white nodes, $H_{\min}$ contains at most $1 + \epsilon$ times the minimum number of white nodes. To build $H_{\max}$ we take the next smaller subgraph with a maximum number of white nodes and add nodes until it has size $|M|$. By symmetric arguments, this takes $\mathcal{O}(|M|)$ time and, since adding nodes cannot decrease the number of white nodes, the maximum number of white nodes in a subgraph of size $|M|$ is at most $1 + \epsilon$ times the number in $H_{\max}$. Finding the path from $H_{\min}$ to $H_{\max}$ takes $\mathcal{O}(|M|)$ time using the representation from Lemma 1.

Theorem 3. *When G is a tree containing only two colours, for any positive constant ϵ we can build an $\mathcal{O}(n)$-space index with which in $\mathcal{O}(|M|)$ time we can find an approximate match in which each colour's frequency is within a factor of $1 + \epsilon$ of its frequency in M.*

6   Indexes for Graphs with Two Colours

Suppose G is a graph containing only two colours, black and white. Theorem 2 applies in this case as well, if we consider spanning trees of H_p and H_r instead of the connected subgraphs themselves, but we can build the index in polynomial time only in special cases, such as when G has bounded treewidth.

Theorem 4. *When G is a graph containing only two colours, there exists an $\mathcal{O}(n^2)$-space index with which we can find a match in $\mathcal{O}(|M|)$ time.*

When G has bounded treewidth we can find a match in $n^{\mathcal{O}(1)}$ time, so we can prove a weaker version of Corollary 1 for graphs. To do this, we build a spanning tree for G and apply Lemma 1 to that spanning tree. We can build the resulting index in polynomial time.

Corollary 2. *When G is a graph with bounded treewidth containing only two colours, we can build an $\mathcal{O}(n)$-space index with which we can find a match in $\mathcal{O}(|M|)$ time when $|M| \leq n^{1/2}$ and in $|M|^{\mathcal{O}(1)}$ time otherwise.*

We cannot quite extend Theorem 3 to graphs because Lemma 2 does not apply (since we may not be able to recover the correct paths between the white nodes in the subgraph H). However, if we store connected subgraphs explicitly instead of with Lemma 2, then calculation shows the index takes $\mathcal{O}(n \log n)$ space. Again, this index can be built in polynomial time when G has bounded treewidth.

Theorem 5. *When G is a graph containing only two colours, for any positive constant ϵ we can build an $\mathcal{O}(n \log n)$-space index with which in $\mathcal{O}(|M|)$ time we can find an approximate match in which each colour's frequency is within a factor of $1 + \epsilon$ of its frequency in M.*

References

1. Amir, A., Apostolico, A., Landau, G.M., Satta, G.: Efficient text fingerprinting via Parikh mapping. J. Discrete Algorithms 1(5-6), 409–421 (2003)
2. Badkobeh, G., Fici, G., Kroon, S., Lipták, Z.: Binary jumbled string matching for highly run-length compressible texts. Inf. Process. Lett. 113(17), 604–608 (2013)
3. Benson, G.: Composition alignment. In: Benson, G., Page, R.D.M. (eds.) WABI 2003. LNCS (LNBI), vol. 2812, pp. 447–461. Springer, Heidelberg (2003)
4. Böcker, S.: Sequencing from compomers: Using mass spectrometry for DNA de novo sequencing of 200+ nt. J. of Computational Biology 11(6), 1110–1134 (2004)
5. Böcker, S.: Simulating multiplexed SNP discovery rates using base-specific cleavage and mass spectrometry. Bioinformatics 23(2), 5–12 (2007)
6. Burcsi, P., Cicalese, F., Fici, G., Lipták, Z.: Algorithms for jumbled pattern matching in strings. Int. J. Found. Comput. Sci. 23(2), 357–374 (2012)
7. Butman, A., Eres, R., Landau, G.M.: Scaled and permuted string matching. Inf. Process. Lett. 92(6), 293–297 (2004)
8. Cicalese, F., Fici, G., Lipták, Z.: Searching for jumbled patterns in strings. In: Proc. Prague Stringology Conference (PSC 2009), pp. 105–117 (2009)
9. Cicalese, F., Laber, E., Weimann, O., Yuster, R.: Near linear time construction of an approximate index for all maximum consecutive sub-sums of a sequence. In: Kärkkäinen, J., Stoye, J. (eds.) CPM 2012. LNCS, vol. 7354, pp. 149–158. Springer, Heidelberg (2012)
10. Didier, G., Schmidt, T., Stoye, J., Tsur, D.: Character sets of strings. J. Discrete Algorithms 5(2), 330–340 (2007)
11. Eres, R., Landau, G.M., Parida, L.: Permutation pattern discovery in biosequences. Journal of Computational Biology 11(6), 1050–1060 (2004)
12. Fellows, M.R., Fertin, G., Hermelin, D., Vialette, S.: Upper and lower bounds for finding connected motifs in vertex-colored graphs. J. Comput. Syst. Sci. 77(4), 799–811 (2011)
13. Gagie, T., Hermelin, D., Landau, G.M., Weimann, O.: Binary jumbled pattern matching on trees and tree-like structures. In: Bodlaender, H.L., Italiano, G.F. (eds.) ESA 2013. LNCS, vol. 8125, pp. 517–528. Springer, Heidelberg (2013)
14. Giaquinta, E., Grabowski, S.: New algorithms for binary jumbled pattern matching. Inf. Process. Lett. 113(14-16), 538–542 (2013)
15. Kociumaka, T., Radoszewski, J., Rytter, W.: Efficient indexes for jumbled pattern matching with constant-sized alphabet. In: Bodlaender, H.L., Italiano, G.F. (eds.) ESA 2013. LNCS, vol. 8125, pp. 625–636. Springer, Heidelberg (2013)
16. Lacroix, V., Fernandes, C.G., Sagot, M.-F.: Motif search in graphs: Application to metabolic networks. IEEE/ACM Trans. Comput. Biology Bioinform. 3(4), 360–368 (2006)
17. Moosa, T.M., Rahman, M.S.: Sub-quadratic time and linear space data structures for permutation matching in binary strings. J. Discr. Alg. 10, 5–9 (2012)
18. Sleator, D.D., Tarjan, R.E.: A data structure for dynamic trees. J. Comput. Syst. Sci. 26(3), 362–391 (1983)

Adaptive Data Structures for Permutations and Binary Relations*

Francisco Claude[1,2,3] and J. Ian Munro[3]

[1] Akori S.A.
Santiago, Chile
[2] Escuela de Informática y Telecomunicaciones
Universidad Diego Portales, Chile
[3] David R. Cheriton School of Computer Science
University of Waterloo, Canada

Abstract. We present new data structures for representing binary relations in an adaptive way, that is, for certain classes of inputs we achieve space below the general information theoretic lower bound, while achieving reasonable space complexities in the worst case. Our approach is derived from a geometric data structure [Arroyuelo et al., TCS 2011]. When used for representing permutations, it converges to a previously known adaptive representation [Barbay and Navarro, STACS 2009]. However, this new way of approaching the problem shows that we can support range searching in the adaptive representation. We extend this approach to representing binary relations, where no other adaptive representations using this chain decomposition have been proposed.

1 Introduction

Binary relations and permutations arise in many applications in computer science. Examples include text indexing [12] and graph representations [8], among others. These fundamental objects have been heavily studied [11,4,5,6], and very efficient data structures supporting a wide range of operations have emerged. However, most of them remain bounded by the information theoretic lower bound in their space consumption, even in the cases where the objects have exploitable properties; for example Web Graphs [7]. Some exceptions are all the developments for compressed suffix arrays [12] and the work by Barbay and Navarro [6] on more general permutations.

In this paper, we present space-efficient data structures that have adaptive space and time complexities. Our approach comes from a geometric perspective, and for permutations, converges to the representation by Barbay and Navarro [6]. However, our new approach brings a new perspective, showing how to support range searching operations. We show that the work of [6] serves to represent binary relations with Theorem 1, and also prove an alternative tradeoff based on our own formulation of their structure.

* First author funded in part by Google U.S./Canada PhD Fellowship. Second author funded in part by NSERC and the Canada Research Chairs Programme.

O. Kurland, M. Lewenstein, and E. Porat (Eds.): SPIRE 2013, LNCS 8214, pp. 64–71, 2013.

The paper is organized as follows. In Section 2 we present related work on representing permutations and binary relations, we also include some background on data structure for range searching proposed by Arroyuelo et al. [1]. In Section 3 we present our adaptive representation for permutations, and show how this representation converges to the one by Barbay and Navarro. Next, in Section 4, we extend the representation for permutations to cover binary relations in general, presenting two different approaches. Then, in Section 4.2, we present one simple application of our structure. Finally, in Section 5, we present our conclusions.

2 Related Work

We present the related work in the next two subsections. The first one presents previous results on representing permutations and binary relations; the second covers recent results on adaptive range searching.

2.1 Permutation and Binary Relations

The most common queries for a permutation Π over $[n]$[1] are: (1) $\pi(i)$: obtain the value of $\Pi[i]$; (2) $\pi^{-1}(j)$: find i such that $j = \Pi[i]$; (3) $\pi^k(i)$: apply π k times, similarly we define $\pi^{-k}(j)$; and (4) $\mathcal{R}_\Pi(i_1, i_2, j_1, j_2)$: find elements i such that $i_1 \leq i \leq i_2$ and $j_1 \leq \pi(i) \leq j_2$.

One efficient representation for arbitrary permutations is that of Munro et al. [11]. This representation achieves $(1 + \epsilon)n \lg n(1 + o(1))$ bits. It supports π in $O(1)$ time and π^{-1} in $O(\frac{1}{\epsilon})$ time. They also showed that $\pi^{\pm k}$ can be supported in the same time as the time required to perform both π and π^{-1} by using only $O(n)$ extra bits. This extension applies to any representation, and thus, to our results too.

The $\mathcal{R}$ operation is less commonly required, but also of interest. For instance, consider a position-restricted search using a suffix array. The suffix array is a permutation and searching for a pattern P between positions p_1 and p_2 is just the result of doing a range query over the range of suffixes starting with P (i.e., $[i_1, i_2]$) and those pointing to positions in $[p_1, p_2]$. Mäkinen and Navarro showed how to use wavelet trees to solve this operation and all others in $O(\lg n)$ time within $n \lg n(1 + o(1))$ bits of space [10].

Prior to this paper the only adaptive representation for permutations was that proposed by Barbay and Navarro [6]. They show many possible decompositions into monotonic sequences and subsequences, and give their space/time complexities in term of the entropy of such sequences. As we will see later, we converge to the same structure at the end of Section 3.

A natural representation for a permutation is a binary matrix of $n \times n$ where we mark the coordinates (i, j) with a 1 iff $\pi(i) = j$. We will use this conceptual representation in our construction. For a binary relation $\mathcal{B}$ over two sets $[n_1]$

---

[1] We use $[n]$ to represent $\{1, 2, \ldots, n\}$.

Table 1. Operations and implementations supported by the representation of Barbay et al. [4,5]. The space requirement is $t(\lg n_2 + o(\lg n_2))$ bits.

Operation	Implementation 1	Implementation 2
$\text{rowrank}_B(i,j)$: number of 1s in row i up to position j (included).	$O(\lg \lg n_2 \lg \lg \lg n_2)$	$O(\lg \lg n_2)$
$\text{rowselect}_B(i,p)$: p-th 1 in row i, or ∞ if $\text{rowrank}_B(i,n_2) < p$.	$O(\lg \lg n_2)$	$O(1)$
$\text{rowcount}_B(i)$: number of 1s in row i.	$O(1)$	$O(1)$
$\text{colrank}_B(i,j)$: number of 1s in column j up to position i (included).	$O(\lg \lg n_2)$	$O(\lg \lg n_2 \lg \lg \lg n_2)$
$\text{colselect}_B(p,j)$: p-th 1 in column j, or ∞ if $\text{colrank}_B(n_1,j) < p$.	$O(1)$	$O(\lg \lg n_2)$
$\text{colcount}_B(j)$: number of 1s in column j.	$O(1)$	$O(1)$
$\text{relaccess}_B(i,j)$: true iff $(i,j) \in B$.	$O(\lg \lg n_2)$	$O(\lg \lg n_2)$

and $[n_2], n_2 \leq n_1$, with $t = |B|$ elements in the relation, we can also use the same conceptual representation. B is represented as a matrix of n_1 rows by n_2 columns with t ones. A one in position (i,j) indicates that i relates to j in B.

Barbay et al. [4,5] presented a structure for representing binary relations that requires $t(\lg n_2 + o(\lg n_2))$ bits of space and supports the operations, offering two tradeoffs, as shown in Table 1.

In a follow-up work, Barbay et al. [7] proved a set of reductions for many operations on binary relations, and presented two structures supporting a core of operations allowing to answer efficiently this extended set. An important operation that allows us to support many of the operations in the extended set is relrange_B. The operation relrange_B takes a range $[i_1, i_2] \times [j_1, j_2]$ and returns all the coordinates in that range containing a one. From the structures proposed in Barbay et al.'s work [7], the first structure achieves $t \lg n_2 + o(t \lg n_2)$ bits, slightly different from the previous proposal in the lower order term. The second structure achieves $\lg(1 + \sqrt{2})tH(B) + o(tH(B))$ bits, where H corresponds to the general information theoretic lower bound, supporting most operations of interest in $O(\lg n_2)$ time per element retrieved. In this case $H(B) = t \lg \frac{n_1 n_2}{t}$ corresponds to the information theoretic lower bound for representing a binary relation with the characteristics of B.

2.2 Monotonic Decomposition of Sequences

Arroyuelo et al. [1] presented an adaptive data structure for range searching that decomposes the set of points into non-crossing ascending and descending chains. Let k be the number of chains generated by the decomposition, the search time for a range query is $O(\lg k \lg n + k' + output)$ time, where k' corresponds to the number of chains intersecting the query rectangle and output is the number of points in the answer. The main idea behind the search strategy is to first search for a chain that crosses the query rectangle (or discard all of them). Since the chains do not cross, we can binary search the chains, at $O(\lg n)$ cost each probe. Once a chain is found, we have to traverse neighbouring chains until leaving the rectangle in order to retrieve all points.

The decomposition into non-crossing chains can be computed in polynomial time if we are given an optimal decomposition into monotonic subsequences [1]. The optimal decomposition into monotonic subsequences is NP-Hard [15], yet it is interesting that the optimal decomposition for a permutation of length n is bounded by $c\sqrt{n}$, where $c \leq 2$, and that we can get a constant factor approximation in polynomial time [16]. In this work we consider the optimal decomposition and show how this allows for a representation that is adaptive in the number of monotonic subsequences into which a permutation or binary relation can be decomposed. The results as stated apply also for the case when we compute a constant factor approximation, thus making the data structure feasible in practice.

3   Representing Permutations

Our representation works by decomposing the permutation into ascending and descending subsequences. A simple way to visualize this is to consider the representation of the permutation in a grid. Every row represents the index i, the columns represent the value of $\Pi[i]$. It is easy to see that the inverse permutation corresponds just to the transposed matrix. In order to simplify the presentation of this work, we will only consider ascending subsequences, the results extend easily to the general case.

First, we show how to represent a chain using bitmaps that support rank, select and access operations.

Definition 1. *A chain* $[(x_1, y_1), (x_2, y_2) \ldots, (x_n, y_n)]$ *is ascending iff* $x_i \leq x_{i+1}$, $1 \leq i < n$ *and* $y_i \leq y_{i+1}$, $1 \leq i < n$.

From this definition it is easy to prove our first result, stated in the following lemma. In order to present this result in a general way we use $S(n, m)$ as the space requirement (in bits) for representing a bitmap of length n with m ones that supports rank in t_r, select in t_s, and access in t_a time. We use t_b as $max(t_r, t_s, t_a)$.

Lemma 1. *Given an ascending chain* $\mathcal{C} = [(i_1, j_1), (i_2, j_2), \ldots, (i_m, j_m)]$, *of length m, where the values do not exceed n, we can represent the chain in* $2S(n, m)$ *bits and support the following queries:*

- getj$_{\mathcal{C}}(i)$: *gets j such that* $(i, j) \in \mathcal{C}$ *or* $\perp$ *if such pair does not exist. We also define* geti$_{\mathcal{C}}(j)$ *in an analogous way. Both queries are supported in time* $O(t_b)$.
- range$_{\mathcal{C}}(i_1, i_2, j_1, j_2)$: *find the* $(i, j) \in \mathcal{C}$ *such that* $(i, j) \in [i_1, i_2] \times [j_1, j_2]$ *or* $\perp$ *if such a point does not exist. This runs in time* $O(t_b)$.

We can represent each chain using Lemma 1, this leads to the following theorem:

Lemma 2. *Let m_i be the number of elements in chain i. The total space of the structure for a permutation that can be decomposed into χ chains is $2\sum_{i=1}^{\chi} S(n, m_i)$, and supports range queries in $O(t_b \lg \chi + t_b \chi' + output)$, where χ' is the number of chains that intersect the range. The next table summarizes some of the tradeoffs we can achieve.*

Bitmap Representation	Total Space	t_b
Pătraşcu [14]	$2n \lg \chi + O\left(\chi \lg n + \frac{\chi n}{\lg^c n}\right)$	$O(c)$
Okanohara and Sadakane [13]	$2n \lg \chi + O(\chi \lg n + n)$	$O\left(\lg \frac{n}{m_i} + \frac{\lg^4 m_i}{\lg n}\right)$

The complexity for range queries follows from the work by Arroyuelo et al. [1]. The computation of the final space is similar to the one used in proof of Theorem 2.

This representation of course is only useful for very small values of χ, otherwise the structure can be asymptotically bigger than the information theoretic minimum. This can be improved by the following observation.

Observation 1. *Given a set of χ bitmaps of length n, where the total number of ones in the set is n and no two bitmaps contain a 1 in the same position, we can represent them as a sequence of length n over an alphabet of size χ. Furthermore, any sequence representation supporting rank, select, and access in times t_r, t_s, and t_a, allows us to support the same operations in each individual bitmap within the same time.*

This not only allows us to lower the space, but it also simplifies the π and π^{-1} queries. To know which chain contains the value associated with $\pi(i)$ we just need to know which bitmap is referred to in the sequence representation by accessing its position.

It is interesting that we can represent our structure using two sequences (x and y coordinates), and they correspond exactly to S_π and $S_{\pi^{-1}}$. Furthermore, they also correspond to the representation proposed by Barbay and Navarro [6], which was originally proposed using wavelet trees, but can be modified to work with any representation, offering a wider set of tradeoffs [2]. Currently, the most interesting tradeoff is that of of Barbay at el. [3,2].

Another point to highlight, is that this shows that the original structure of Barbay and Navarro also supports adaptive range searching. This particular searching algorithm has proven to be efficient in practice [9]. This allows to state the following corollary.

Corollary 1. *Given (i) a permutation Π, that can be decomposed into χ monotonically ascending and descending chains, and (ii) a sequence representation that requires $S(n, \sigma)$ for representing a sequence of length n over an alphabet of size σ supporting rank, select and access queries in $O(t_b)$ time, there exists a structure requiring $2S(n, \chi)$ bits that supports computing π and π^{-1} in $O(t_b)$ and range search queries in $O(t_b \lg \chi + t_b \chi' + output)$, where χ' is the number of chains that touch the query rectangle, and output the size of the output.*

4   Representing Binary Relations

We use the same approach on the grid representing the binary relation. Given a binary relation $\mathcal{B}$, the pair (i,j) is marked iff i relates to j in $\mathcal{B}$. We follow the notation of the previous sections. Recall that σ is the number of rows, n the number of columns, and t the number of pairs in $\mathcal{B}$.

We assume all columns and rows have at least one element, as we can trivially map the problem when we accept empty row/columns adding a bitmap of length $n + \sigma$ supporting rank, select, and access.

We focus mostly on three operations: (1) Iterating over $\mathrm{rowselect}_{\mathcal{B}}(i, p = 1 \ldots \mathrm{rowcount}_{\mathcal{B}}(i)))$; (2) Iterating over $\mathrm{colselect}_{\mathcal{B}}(i, p = 1 \ldots \mathrm{colcount}_{\mathcal{B}}(i)))$; and (3) Obtaining all pairs in $[i_1, i_2] \times [j_1, j_2]$ (i.e., $\mathrm{relrange}_{\mathcal{B}}$).

The technique presented in Section 3 does not apply directly to this case. The main problem is that a chain could contain many elements that are in the same row or column, and would result in multiple chains in the same position in a bitmap. We first give a representation that matches the result for the permutations in time and space and then we show how to potentially improve the space by using a more elegant technique. This new approach has a worse query time.

4.1   Using Permutations

We show how to transform a binary relation into a permutation by just considering a simple row/column addition algorithm that moves points around and allows one to answer the queries of interest.

The main idea is to create multiples copies of rows and columns having more than one point and then distribute the points across them so that each of them has only one point, leading to a permutation on t elements. This is inefficient in terms of space, but it allows us to match the performance of our structure for permutations. In order to be able to extract the original information we need to add $2t + o(t)$ bits, stored in B_1 and B_2. These bitmaps tell the length, in unary, of each expanded row/column. Using these two bitmaps, we can answer $\mathrm{rowcount} = X_a.\mathtt{select}(1, x+1) - X_a.\mathtt{select}(1, x) - 1$ and $\mathrm{colcount} = X_b.\mathtt{select}(1, y+1) - X_b.\mathtt{select}(1, y) - 1$ in constant time. We omit the algorithm pseudo-code for lack of space. We also omit the proof that our procedure generates a permutation.

This yields a theorem similar to that of Barbay et al. [4], but supporting a different subset of operations.

Theorem 1. *A binary relation $\mathcal{B} \subseteq \{(i,j) | i \in [n_1], j \in [n_2]\}$, where $t = |\mathcal{B}|$, that can be decomposed into k monotonic chains, can be represented as a permutation of length t with $(n_1 + n_2)(1 + o(1))$ extra bits. Furthermore, the resulting permutation can be decomposed into k monotonic chains, and the operations* rowselect, colselect *and* relrange *can be mapped to π, π^{-1} and $\mathcal{R}$ operations on the permutation, respectively. The counting operations can be solved using bitmaps B_1 and B_2.*

4.2   Using Chains Directly

An alternative method can be obtained by decomposing the binary relation directly. A chain could now contain more than one occurrence of a given row or column, and because of that, the transformation that converges to the structure by Barbay and Navarro does not work. At this point, the departure from the original proposal by Barbay and Navarro pays off, allowing the representation of a class wider than that of permutations.

We first take a look at the space consumption of our structure when decomposing $\mathcal{B}$ into chains. For that, we present an alternative representation for the chains. For simplicity, we will use the bitmaps representation by Pătraşcu [14], yet the results translate in a similar way as for Lemma 2, and thus, we can offer a wide set of bounds.

Lemma 3. *An ascending chain of length m with at most $\bar{n} = n + \sigma$ points in $[n] \times [\sigma]$ can be represented in $2m \lg \frac{\bar{n}}{m} + O\left(m + \frac{\bar{n}}{\lg^c n}\right)$ bits of space, but now* geti, getj *and* range *take $O(c \lg m)$ time.*

If we represent the structure using these chains, we obtain the following theorem:

Theorem 2. *A binary relation $\mathcal{B}$ over $[n_1] \times [n_2]$, where $t = |\mathcal{B}|$, can be represented in $2t \lg \frac{nk}{t} + 2t \lg k + O(\frac{kn}{\lg^c n} + k \lg t)$ bits, where $n = \max(n_1, n_2)$. Within this space, we can list elements in $O(r)$ time per datum retrieved, and answer range queries in $O(r(\lg k + k') + \text{output})$ time, where k is the number of chains, k' the number of chains hitting the query rectangle, output the size of the output, and $r = \max(c \lg k, c \lg \lg n)$.*

We could try merging the sequences marking with a bitmap where each position starts, and this would lead to the result obtained in Theorem 1.

Another observation regarding the representation presented in Theorem 2 is that we can answer range minimum queries (RMQs) over the binary relation in the same time as for relrange.

Lemma 4. *By adding $O(t)$ extra bits to the representation from Lemma 2, or Theorems 1 and 2, and adding weights to each pair in the permutation/relation, we can support range minimum queries in the same complexity as the one required for answering* relaccess.

5   Conclusions and Future Work

We presented an alternative formulation for the representation by Barbay and Navarro. This new approach allows to show how to support range searching and provides some different tradeoffs. We then extended the results to the case of binary relations. We proposed two alternatives, both achieving interesting tradeoffs supporting navigation and range searching. It is worth noting that for *easy instances* we obtain smaller and faster representations, which is clearly an interesting behaviour.

References

1. Arroyuelo, D., Claude, F., Dorrigiv, R., Durocher, S., He, M., López-Ortiz, A., Munro, J.I., Nicholson, P.K., Salinger, A., Skala, M.: Untangled monotonic chains and adaptive range search. TCS 412(32), 4200–4211 (2011)
2. Barbay, J., Claude, F., Gagie, T., Navarro, G., Nekrich, Y.: Efficient fully-compressed sequence representations. Algorithmica (to appear, 2013)
3. Barbay, J., Gagie, T., Navarro, G., Nekrich, Y.: Alphabet partitioning for compressed rank/select and applications. In: Cheong, O., Chwa, K.-Y., Park, K. (eds.) ISAAC 2010, Part II. LNCS, vol. 6507, pp. 315–326. Springer, Heidelberg (2010)
4. Barbay, J., Golynski, A., Munro, J.I., Rao, S.S.: Adaptive searching in succinctly encoded binary relations and tree-structured documents. TCS 387(3), 284–297 (2007)
5. Barbay, J., He, M., Munro, J.I., Rao, S.S.: Succinct indexes for strings, binary relations and multi-labeled trees. In: SODA, pp. 680–689 (2007)
6. Barbay, J., Navarro, G.: Compressed representations of permutations, and applications. In: STACS, pp. 111–122 (2009)
7. Barbay, J., Claude, F., Navarro, G.: Compact rich-functional binary relation representations. In: López-Ortiz, A. (ed.) LATIN 2010. LNCS, vol. 6034, pp. 170–183. Springer, Heidelberg (2010)
8. Claude, F., Navarro, G.: Fast and compact Web graph representations. TWEB 4(4), article 16 (2010)
9. Claude, F., Munro, J.I., Nicholson, P.K.: Range queries over untangled chains. In: Chavez, E., Lonardi, S. (eds.) SPIRE 2010. LNCS, vol. 6393, pp. 82–93. Springer, Heidelberg (2010)
10. Mäkinen, V., Navarro, G.: Rank and select revisited and extended. TCS 387, 332–347 (2007)
11. Munro, J.I., Raman, R., Raman, V., Rao, S.S.: Succinct representations of permutations. In: Baeten, J.C.M., Lenstra, J.K., Parrow, J., Woeginger, G.J. (eds.) ICALP 2003. LNCS, vol. 2719, pp. 345–356. Springer, Heidelberg (2003)
12. Navarro, G., Mäkinen, V.: Compressed full-text indexes. ACM Computing Surveys 39(1), article 2 (2007)
13. Okanohara, D., Sadakane, K.: Practical entropy-compressed rank/select dictionary. In: ALENEX (2007)
14. Pătraşcu, M.: Succincter. In: FOCS, pp. 305–313 (2008)
15. Wagner, K.: Monotonic coverings of finite sets. Elektron. Informationsverarb. Kybernet. 20, 633–639 (1984)
16. Yang, B., Chen, J., Lu, E., Zheng, S.Q.: A comparative study of efficient algorithms for partitioning a sequence into monotone subsequences. In: Cai, J.-Y., Cooper, S.B., Zhu, H. (eds.) TAMC 2007. LNCS, vol. 4484, pp. 46–57. Springer, Heidelberg (2007)

Document Listing on Versioned Documents[*]

Francisco Claude[1,2,3] and J. Ian Munro[3]

[1] Akori S.A.
Santiago, Chile
[2] Escuela de Informática y Telecomunicaciones
Universidad Diego Portales, Chile
[3] David R. Cheriton School of Computer Science
University of Waterloo, Canada

Abstract. Representing versioned documents, such as Wikipedia history, web archives, genome databases, backups, is challenging when we want to support searching for an exact substring and retrieve the documents that contain the substring. This problem is called *document listing*.

We present an index for the document listing problem on versioned documents. Our index is the first one based on grammar-compression. This allows for good results on repetitive collections, whereas standard techniques cannot achieve competitive space for solving the same problem.

Our index can also be addapted to work in a more standard way, allowing users to search for word-based phrase queries and conjunctive queries at the same time.

Finally, we discuss extensions that may be possible in the future, for example, supporting ranking capabilities within the index itself.

1 Introduction

Highly repetitive collections are becoming more and more common. We have a lot of versioned information on the Web; good examples of this are software repositories and Wikipedia. It is also expected that in the future we will have to provide storage for genome sequences of many individuals of the same species, perhaps millions of people. This last scenario is interesting because within the same species, the sequences share close to 99.99%, making the collection highly repetitive [15].

Being capable of storing archive data with historic information on how documents evolve is a challenging task by itself, but we also need to provide searching capabilities to make this information easily available for people when needed. In this work we focus on the *document listing* problem for such collections.

Formally speaking, the document listing problem is defined as follows: Given a collection of documents $\mathcal{D} = \{T_1, T_2, \ldots, T_d\}$, and a query string P, we want to retrieve the documents that contain P as a substring. We could add a ranking

---

[*] First author funded in part by Google U.S./Canada PhD Fellowship. Second author funded in part by NSERC and the Canada Research Chairs Programme.

O. Kurland, M. Lewenstein, and E. Porat (Eds.): SPIRE 2013, LNCS 8214, pp. 72–83, 2013.
© Springer International Publishing Switzerland 2013

function f, such that we retrieve the documents ordered by $f(T_i, P)$, and even limit the size of the resulting set by a given parameter k, these are called top-k queries. Some examples of ranking functions include TF-IDF and `closeness` [1].

One important point to clarify is the difference between standard text-indexing and the document listing problem. Usually text indexes allow you to search for a pattern in a text and report the position where the pattern occurs. If we concatenate all the documents and use a classical index, we can retrieve the documents that contain the pattern. The drawback is that we are forced to iterate over all occurrences of the pattern, which means we can pay a huge overhead for just one document if it contains the pattern multiple times. This is the main difference that renders classical text indexes unsuitable for certain instances of the problem. For natural language, the problem has been usually simplified by using an inverted index. For every word w in the language, we have a list $L[w] = \{T_{i_1}, T_{i_2}, \ldots, T_{i_\ell}\}$ listing all documents that contain w, plus extra information to compute the ranking function. Answering a query $\mathcal{Q} = \{P_1, P_2, \ldots, P_q\}$ corresponds to obtaining a subset of the elements in $\cap_{i=1}^{q} L[P_i]$.

This solution has been shown to be effective in space, retrieval time, and quality, but it lacks the freedom we would expect in other domains. For example, if we consider a collection of DNA sequences, the concept of a word is not well defined. For this reason, solutions in one domain may be completely useless in others. We also see this phenomenon in languages where the separation among words is not clearly defined, or hard to determine automatically.

The main idea behind our proposal is as follows: Given a collection of documents, we compress the whole set of texts using a grammar-compressor [13,21,4]. The resulting file is indexed using the result of [7]. Then we augment the structure with a set of inverted lists for non-terminal symbols. This inverted lists store the documents that contain each non-terminal. The queries are answered by first asking the text index to produce the minimum set of non-terminals that match the pattern for which we have to look into their inverted lists.

Once we have all the inverted lists, we compute the union of those, generating the final result, an inverted list for the pattern that was given as a query.

The main contributions of this paper are:

- We show how to extend a grammar-compressed index to support document listing in a simple and clean way. The index also supports access to any document of the collection, verbatim, so it completely replaces the original input. Building our index on top of any grammar-compressor allows us to achieve good space for repetitive sequences, which is the case of versioned documents. In addition to achieving good space [8,5], a straight-forward grammar representation allows for fast decompression, and therefore, access to the content being indexed [6,8,5].
- The resulting structure supports retrieving the inverted list for an arbitrary pattern. This is particularly interesting, since all the algorithms developed for plain posting lists can be applied to the output of our searches. This allows to easily extend our result to support conjunctive queries.

- We can apply the same result for words in natural language, allowing a new index. This index does not support full-text document listing, but solves the problem of searching for phrases, a problem that is also hard to handle with traditional inverted indexes. Due to lack of space, we ommit the experimental results for this particular application of our result.
- Our final index does not only allow document listing. We discuss how to extend it to compute other pieces of information commonly used by ranking functions: Term frequencies for each document and positional information on where patterns occur inside each document.

2   Related Work

Most of the items in our index are built using grammar compression and indexes. A grammar-compressed representation of a sequence corresponds to a context-free grammar that generates one single text, the one being compressed. For purposes of this work, the following definition suffices.

Definition 1 (Grammar-compressed seq.). *Given a grammar* $\mathcal{G} = (\mathcal{X} = \{X_1, X_2, \ldots, X_n\}, \sigma, \Gamma : \mathcal{X} \to \mathcal{X}^+ \cup \sigma, s)$, *where:*

- $\mathcal{X}$ *represents the set of non-terminal symbols.*
- σ *corresponds to the set of terminal symbols.*
- Γ *is the set of rules that transform a non-terminal into a sequence of non-terminals or just one terminal symbol. We do not allow cycles in the rules, and that is enough to make sure the grammar generates only one sequence.*
- *s corresponds to the identifier of the start symbol* X_s.

We define $\mathcal{F}(X_i)$ *as the result of recursively replacing all non-terminals until obtaining a sequence of terminal symbols. We also refer to* $\mathcal{F}(X_i)^R$ *as* $\mathcal{F}(X_i)$ *read from right to left (i.e., reversed).*

We say that $\mathcal{G}$ *compresses* $T = t_1 t_2 \ldots t_u$, *iff* $\mathcal{F}(X_s) = T$.

We call N *the sum of the sizes of all the right sides in the grammar, that is*

$$N = \sum_{i=1}^{n} |\Gamma(X_i)|$$

We also refer to the height of the grammar as the longest path from the starting symbol to a terminal symbol in the parse tree.

We rely on the grammar-based index proposed by Claude and Navarro [7] to support one of the steps in our searching procedure. We explain in more detail the pieces needed in Section 2.1.

2.1   Grammar Indexes

We first explain the basics of the index proposed by Claude and Navarro [7]. The index takes as input a free-context grammar that generates a single sequence. We call $\mathcal{G}$ the grammar, composed of a set of non-terminals $\mathcal{X} = \{X_1, X_2, \ldots, X_n\}$, an initial symbol X_s and a set of rules Γ, that map non-terminals to a sequence of non-terminals or just one single terminal symbol.

The grammar is first preprocessed to remove duplicate rules, and embed rules that are mentioned only once inside the rule that mentions them. This does not increase the size of the grammar, but allows to bound some of the running times further.

The main result of [7] is summarized in Theorem 1. We next explain the structures we need in this paper, omitting some of the details for the sake of readability.

Theorem 1. *[7] Let a sequence $T[1..u]$ be represented by a context free grammar with n symbols, size N and height h. Then, for any $0 < \epsilon \leq 1$, there exists a data structure using at most $2N \lg n + N \lg u + \epsilon n \lg n + o(N \lg n)$ bits that finds the occ occurrences of any pattern $P[1..m]$ in T in time $O((m^2/\epsilon) \lg \left(\frac{\lg u}{\lg n}\right) + (m + occ) \lg n)$. It can extract any substring of length ℓ from T in time $O(\ell + h \lg(N/h))$. The structure can be built in $O(u + N \lg N)$ time and $O(u \lg u)$ bits of working space.*

For the construction of the index, we first preprocess the grammar and re-assign the identifiers of each non-terminal so that they are sorted lexicographically by the reverse of the string they generate, i.e., $\mathcal{F}(X_i)^R$. We number the non-terminals in sorted order, that is, $\mathcal{F}(X_i)^R \leq \mathcal{F}(X_j)^R$ iff $i \leq j$. We then create a bitmap Y where we assign a 1 to position i iff X_i generates just a single terminal symbol. We augment this bitmap to support the following operations:

- $access_Y(i)$: retrieves the bit at position i in Y.
- $rank_Y(b, i)$: counts the number of times bit b appears up to position i in Y.
- $select_Y(b, j)$: retrieves the position of the j-th occurrence of bit b in Y.

We can represent the bitmap Y and support all three operations in constant time using the method of Raman, Raman, and Rao [17]. This representation requires $nH_0(Y) + o(n)$, where $H_0(Y)$ represents the zero order entropy of the bitmap[1].

By using Y we can know whether a rule generates more non-terminals or just one single terminal symbol. Given X_i, if $access_Y(i) = 1$, then we know it generates a terminal symbol. Furthermore, if we assume terminal symbols are contiguous, we know that X_i generates $rank_Y(1, i)$. It is also possible to obtain the non-terminal X_j that generates symbol a by computing $j = select_Y(1, a)$.

In addition, for each proper suffix of each rule, we assign an id, and then reassign them according to the lexicographical order of the strings generated by those proper suffixes. We will call this SuffPerm. In other words, SuffPerm stores at position i the i-th proper suffix of a rule in lexicographical order.

---

[1] The zeroth order entropy of a bitmap of length n with m ones is defined as $\frac{m}{n} \lg \frac{n}{m} + \frac{n-m}{n} \lg \frac{n}{n-m}$. This is bounded above by 1.

Finally, we create a labeled binary relation $\mathcal{R}$ that maps $\texttt{SuffPerm}[i]$ with j through a label k if rule j appears before the suffix represented by $\texttt{SuffPerm}[i]$ in rule k.

We want to support range searching in $\mathcal{R}$. Wavelet trees [12] are a good alternative, access takes $O(\lg n)$ time and range searching takes $O(\lg n)$ per element reported. Wavelet trees, in this context, require $n \lg n(1+o(1))$ bits of space. The time can be further improved to $O(\lg n / \lg \lg n)$ (access and element reported by the range search) within the same space bounds as the standard wavelet trees [2].

In the original paper [7], the grammar is represented as a tree, where we have $N - n$ leaves. In order to have efficient navigation and access to the rules, the tree is represented using the method of Benoit et al. [3], adding a simple trick to allow fast access to the definition of any non-terminal symbol [7]. In our case a simple plain representation of the grammar is enough, we do not need to navigate the parse tree upwards, and the theoretic solution for fast access works slower than traversing a plain representation in practice.

Given a pattern $P = p_1 p_2 \ldots p_m$, we can find two different types of occurrences inside the grammar. The first kind, called *primary occurrences*, are those non-terminals that contain the pattern because two or more rules generated by it, after being concatenated, generate the pattern. The second kind, called *secondary occurrences* are those non-terminals that contain P because they generate a single rule that contain P. Note that actually one non-terminal may be both at the same time, primary and secondary, but for that, the non-terminal must have at least two different occurrences of P.

To find the primary occurrences of a pattern $P = p_1 p_2 \ldots p_m$, we try the m possible partitions: $p_1 \cdot p_2 \ldots p_m$, $p_1 p_2 \cdot p_3 \ldots p_m$, up to $p_1 \ldots p_{m-1} \cdot p_m$. For each partition $P = P_1 \cdot P_2$, we perform a binary search on the rules to determine which ones finish with P_1. Then we perform a binary search over the suffixes of rules, $\texttt{SuffPerm}$, to find suffixes of rules that begin with P_2. Finally, using the binary relation $\mathcal{R}$, we can perform a range search to retrieve the non-terminals that contain elements that start with P_2 preceded by elements that end with P_1.

Secondary occurrences are obtained by following up the primary occurrences in the parse tree. As we will explain later, we only care about primary occurrences in this work, that is why we do not deal with an efficient representation for the parse tree to track secondary occurrences.

Claude and Navarro show how to represent $\texttt{SuffPerm}$ in little space on top of the binary relation, and also how to extract prefixes of suffixes of rules in linear time. We do not need the technical details of these results, it suffices to know the running time of each step. The binary search for P_1 requires $O(m \lg n)$ time. The binary search for P_2 requires $O(m \lg N)$ time. Finally, retrieving the primary occurrences requires $O(\lg n / \lg \lg n)$ time per element retrieved.

Retrieving all occ_p primary occurrences requires $O(m^2 \lg N + \text{occ}_p \lg n / \lg \lg n)$ time.

2.2   Re-pair

Due to its simplicity, we chose Re-Pair as the grammar compression [13] for evaluating our index. It is important to point out that other grammar compressors may achieve better results, yet their implementation for large scale is still an issue. It is also possible to trade compression speed and space for compression ratio using an approximate version [6].

We post-process the result of Re-Pair to make the final grammar smaller. For each rule X_i that generates a set of non-terminals, if it is mentioned only once in the grammar by rule X_j, we expand X_i where X_j mentions it, and remove X_i. We repeat this process until each rule is mentioned at least twice in the grammar.

This is required by the index, but it also has the nice property that matches the dictionary compression algorithm proposed by González and Navarro [11], that has shown to improve the final result considerably (see [11,6]).

3   The Index

In this section we describe how we build the index, augment it to support document listing, and finally how queries are answered.

3.1   Construction for Primary Occurrences

We take the whole collection $\mathcal{D} = \{T_1, T_2, \ldots, T_d\}$, and generate a single sequence

$$T = \$_0 T_1 \$_1 T_2 \$_2 \ldots \$_{d-2} T_{d-1} \$_{d-1} T_d,$$

where $\$_i$ are symbols that do not appear anywhere else in the collection.

When we compress this sequence with Re-Pair, we are sure that no rule spans from one document to the other, since the $\$_i$ symbols cannot form pairs that appear twice. We then remove the $\$_i$ elements, and generate one rule per document, containing all the elements left between the $\$$s in X_s. After that, we replace X_s by a new rule that generates the new rules we just created, in order. This allows us to have direct access to a rule that generates the whole content for any document. Our grammar, after this preprocessing, has the following form:

- X_s generates d non-terminals, $X_{t_1}, X_{t_2}, \ldots, X_{t_d}$, where $\mathcal{F}(X_{t_i}) = T_i$.
- X_{t_i} generates the symbols between $\$_{i-1}$ and $\$_i$ in the original X_s generated by Re-Pair.

When building the index, we leave X_s outside the permutation SuffPerm. This does not only save space, but makes sure that whenever we find a primary occurrence, it is contained inside a single document, and not formed by the concatenation of two.

To access the i-th document in the collection, we just expand the i-th non-terminal generated by s. This allows us to retrieve documents in time proportional to their length (amortized if we don't use the result from [7]).

Note that we can adapt other grammar-based compressors to this scheme. An interesting option is to just simply compress each document separately with a compressor that generates an SLP (rules restricted to generate two non-terminals or just one terminal), and then apply the merge algorithm of Wan [20]. This will generate a grammar that satisfies the conditions above, and by applying the same preprocessing before constructing the index, we can optimize the output even further.

3.2   Adding Inverted Lists

For each non-terminal, we store an inverted list of the documents containing that non-terminal. Note that this requires at most $n \times d$ bits, and we expect n to be small. Yet this is still not satisfactory. If two versions share much of their content, they will appear in a very similar set of lists, since they will be formed by the same non-terminals.

To exploit this, we again use grammar-compression on the sequence of lists. We could use any space-efficient representation of lists, but for repetitive ones, this particular solution has proven to work well in practice [8,5].

We refer to $L[X_i]$ to the list of documents containing non-terminal X_i and will call L the set of inverted lists. We represent the inverted lists in the same way as we represent the documents, this allows to access an entire list in time proportional to its length.

It is interesting to relate the size of this inverted lists to the size of the original sequence. It turns out that under reasonable assumptions, these lists can be represented space efficiently. We see the inverted lists as a grid, where coordinate (i, j) is a 1 iff non-terminal i is contained in document j. Let t be the number of points in this grid. We need $t \lg \frac{nd}{t} + O(t)$ bits to represent the grid[2].

We know that $n \leq t$, therefore, the space is bounded by $t \lg d$, which is the same as the solution by Välimaki and Mäkinen requires for the document array [19]. We can further bound the space by considering the worst possible space for the grid. The space is maximized when $t = \frac{nd}{e}$. In this case, the total space required by the grid is $O(t)$ bits.

On the other hand, we can also bound the length of the text in terms of t. We know that each point on the grid represents at least one occurrence of a rule in the collection, therefore, $u \geq t$. This means that the total extra space for the grid is bounded by the length of the collection in bits, in other words $\frac{D}{\lg \sigma}$ bits.

3.3   Full-Text Document Listing

Having built the grammar-index, and the inverted lists, the searching becomes quite straight-forward. We search for the nonterminals that contain primary occurrences of the pattern, and compute the union of the inverted lists associated to those nonterminals.

At this stage we need to compute the union of sets, in contrast with the usual operation we encounter between inverted lists, which is the intersection. Furthermore, our case is a bit more complicated. We have a grammar-compressed

---

[2] This is a simple information theoretic lower bound, there exist representations that achieve this [9], and some that do better on repetitive cases [5], as in our case.

version of the lists, and thus we want to make use of this fact, both to keep the space low, and to improve the query time.

Given a set of non-terminals representing the primary occurrences of the pattern, we will create a dynamic dictionary containing those elements, called *seen*, and a queue containing the same elements, we call this queue *remaining*. The merge procedure generates a dictionary containing all the elements, and is shown in Algorithm 1.

Data: Set $V = \{v_1, v_2, \ldots, v_n\}$, Lists $\mathcal{G} = (\mathcal{X}, \Gamma, \sigma, s)$
Result: $R = (d_{i_1}, d_{i_2}, \ldots, d_{i_k})$
1 **remaining** $\leftarrow \emptyset$
2 **seen** $\leftarrow \emptyset$
3 $R \leftarrow \emptyset$
4 **for** $v \in V$ **do**
5 $\quad$ **remaining** $\leftarrow$ **remaining** $\cup \{\mathcal{X}_v\}$
6 $\quad$ **seen** $\leftarrow$ **seen** $\cup \{\mathcal{X}_v\}$
7 **while remaining** $\neq \emptyset$ **do**
8 $\quad$ $x \leftarrow$ **GetMax(remaining)**
9 $\quad$ **remaining** $\leftarrow$ **remaining** $- \{x\}$
10 $\quad$ **if** x *is terminal* **then**
11 $\quad\quad$ $R \leftarrow R \cup \{x\}$
12 $\quad$ **for** x^j *in* $\Gamma(x)$ **do**
13 $\quad\quad$ **if** $x^j \notin$ **seen** **then**
14 $\quad\quad\quad$ **seen** $\leftarrow$ **seen** $\cup \{x^j\}$
15 $\quad\quad\quad$ **remaining** $\leftarrow$ **remaining** $\cup \{x^j\}$
16 **return** L

Algorithm 1. Computing the union of the lists for a set of non-terminals

The worst case running time of this algorithm is $O(\text{occ}_p \times output)$. Section 4 shows that occ_p is in general small, and also that our heuristic of keeping track of previously seen non-terminals allows us to save processing time; it exploits the regularities seen between the lists. If two lists contain basically the same elements, we will only explore one of them, since we will encounter a non-terminal we have already seen.

It is quite straight-forward to see why we only find primary occurrences. Secondary occurrences contain documents we already reported as primary occurrences, so processing only primary occurrences maintains the correctness of the result while cutting down the time.

3.4 Adding Ranking Information

The index can be augmented with extra information in a similar way as inverted lists, with a couple of restrictions. We can augment the inverted lists that associate each non-terminal symbol with the documents that contain it with score values. In particular, frequencies offer a property that is easy to exploit here.

When we augment the lists L with frequencies, we can just add up all the values associated with primary occurrences of a certain document and we will obtain precisely the number of occurrences of the pattern in the whole document. We include the details of this algorithm in the Appendix.

We may not need to store the frequencies for each possible occurrence of a document in the inverted lists. We could store an approximation of the frequency to approximate the term frequency and save space, by storing values from a smaller universe.

We can also use the result of Claude and Navarro [7] to support locating the occurrences of the pattern in the collection. This allows to obtain positional information for the query when required. Another option here is to approximate the locations of multiple patterns depending on the primary occurrences. This line of work is out of the scope of this article.

4   Experimental Results

4.1   Practical Considerations

For the practical implementation, we did not implement the real-time access to prefixes/suffixes of rules as described in [7]. We just store the grammar as a set of arrays describing each rule. We also do not need the tree in practice, since we are not tracking occurrences upwards.

The binary relation is represented using a wavelet tree, as implemented in LIBCDS[3]. We also make use of the arrays implemented in the library. We use Navarro's implementation of Re-Pair [4], which runs in linear time. As containers we use the standard C++ STL containers. For sets we use `set`, and for unsorted sequences, we use `vector`.

4.2   Experimental Setup

To test our index we downloaded the first part of Wikipedia in English[5], and sampled documents from it uniformly at random. For each document selected, we extracted all its versions. This was done using anonymous' library.

We also generated synthetic collections composed of symbols A, C, G and T. This is to mimic the compression of genome databases. The process of generation is the following: Generate a random sequence T_1 of length n, and then generate $d - 1$ copies of T_1 and mutate $x\%$ of it.

Table 1 shows the main characteristics of our datasets. The compression ratio may not be very descriptive given that the sequences are highly repetitive. For this reason, we include the compression ratio achieved by anonymous' Re-Pair implementation. This does not include any post-processing, and just represents the original sequences, therefore, it is only a guideline on how much the text could be compressed.

---

[3] Available at http://libcds.recoded.cl

[4] Available at http://www.dcc.uchile.cl/gnavarro/software/

[5] enwiki-20110722-pages-meta-history1.xml

Table 1. Datasets

Dataset	size	# docs	versions/doc (avg)	mutation rate	Re-Pair
Wiki1	69MB	8	582	-	0.36MB
Wiki2	600MB	20	772.85	-	3.45MB
Wiki3	1.5GB	36	831.08	-	5.50MB
DNA1	1000MB	1	1000	0.01%	4.5MB
DNA2	1000MB	1	1000	0.005%	2.09MB
DNA3	1000MB	1	1000	0.0026%	1.17MB

Table 2. Space required for our index for each dataset, separated by components

Collection	T	Lists	SuffPerm	$\mathcal{R}$	Total	Compr.
Wiki1	0.39MB	0.49MB	0.39MB	0.39MB	1.66MB	2.43%
Wiki2	1.75MB	2.14MB	1.69MB	1.71MB	7.29MB	1.22%
Wiki3	3.19MB	4.37MB	3.12MB	3.06MB	13.73MB	0.90%
DNA1	3.21MB	4.76MB	2.94MB	3.03MB	13.95MB	1.40%
DNA2	1.99MB	2.80MB	1.78MB	1.91MB	8.47MB	0.85%
DNA3	1.26MB	1.59MB	1.15MB	1.23MB	5.23MB	0.52%

We generated queries by taking a version uniformly at random, and then choosing a substring uniformly at random from that particular version.

The machine used for generating the indexes and measuring time has 2 Intel(R) Xeon(R) CPU X5660 processors running at 2.80GHz, 11TB of hard drive and 24GB of RAM. The machine is running Ubuntu Linux 11.04 with kernel 2.6.38-13-generic for x86_64. All our code is implemented and C++ and was compiled using gcc version 4.5.2 with flags -O3 -DNDEBUG. Our code is available for download from http://fclaude.recoded.cl/projects.

4.3   Full-Text Document Listing

Table 2 shows the sizes of our index for the different collections. We can see that our indexes, for the Wikipedia samples and the DNA synthetic data, are around 4 to 4.5 times the size of the collection when we compress it using Re-Pair. This

Table 3. Time per element retrieved in microseconds for patterns of length $m = 4, 8, 16, 32$, averaged over $10,000$ queries

Collection	$m = 4$	$m = 8$	$m = 16$	$m = 32$
Wiki1	0.60	1.36	3.37	7.38
Wiki2	0.51	0.72	1.72	4.03
Wiki3	0.54	0.83	2.40	6.23
DNA1	20.03	1.86	3.05	6.05
DNA2	12.42	1.35	2.17	4.06
DNA3	8.05	1.06	1.59	2.90

means, within this space, we are replacing the collection and supporting search operations on top of it. Table 3 shows the time in microseconds per element retrieved. This was averaged over 10,000 queries.

4.4   Comparison to Related Work

The document listing problem was first solved in linear space by Muthukrishnan [14]. Sadakane [18] proposed a different time/space tradeoff, and later Mäkinen and Välimäki [19] and Navarro et al. [16] proposed practical solutions to the problem. All these solutions are not designed for repetitive collections. Only recently, Gagie et al. [10] proposed a solution in this scenario. We measured their results with default parameters for the Wiki collections. They offer a different tradeoff than our solution. We provide superior space, our index is 3.56, 8.22, and 10.86 times smaller for Wiki1, Wiki2, and Wiki3 respectively. On the other hand, their query time is much lower, 11–17 times faster for Wiki1, 18–22 times faster for Wiki2, and 20–31 for Wiki3. We measured patterns of length 4, 8 and 16, since the patterns of length 32 produced inconsistent results in their index, showing less occurrences than documents reported. We also excluded patterns of length 16 from Wiki3 for the same reason. When compared to the solution by Navarro et al. [16], we are 16 to 62 times smaller, considering only Wiki1 and a prefix of Wiki2.

5   Conclusions

We have presented a new index for representing highly repetitive collections. This index can be used in two different scenarios: (1) Indexing a collection to support document listing of exact substrings; (2) Indexing a collection and support phrase searches for words existing in the collection.

The results show that while providing competitive time complexities, we achieve space considerably smaller than previous results. This opens a new line for storing historic information on documents while supporting efficient search operations.

It is easy to relate to our index in terms on the inverted lists. In the symbol-based version, we can build the inverted index for any possible substring using our index. Furthermore, when we tokenize the text, and index the word identifiers, our index is just a grammar-compressed representation of the inverted lists, augmented with extra information to support phrase search operations on top of it, allowing to produce the inverted list of an arbitrary phrase.

Our work also leaves some challenging open problems. First, the union of all non-terminals that represent primary occurrences has no good theoretical bound, yet is reasonable in practice. Is it possible to modify the structure or the grammar in order to provide a reasonable bound, say we do not visit more than k symbols per element in the resulting set? Another interesting problem not considered in this work, is whether we could support approximate searches, allowing to retrieve the phrases or substrings that are most similar to the query. This is important, since typos may have a huge effect in the result.

References

1. Baeza-Yates, R.A., Ribeiro-Neto, B.: Modern Information Retrieval. Addison-Wesley Longman Publishing Co., Inc., Boston (1999)
2. Barbay, J., Claude, F., Navarro, G.: Compact binary relation representations with rich functionality. CoRR abs/1201.3602 (2012)
3. Benoit, D., Demaine, E., Munro, J.I., Raman, R., Raman, V., Rao, S.S.: Representing trees of higher degree. Algorithmica 43(4), 275–292 (2005)
4. Charikar, M., Lehman, E., Liu, D., Panigrahy, R., Prabhakaran, M., Sahai, A., Shelat, A.: The smallest grammar problem. IEEE Trans. Inf. Theo. 51(7), 2554–2576 (2005)
5. Claude, F., Fariña, A., Martínez-Prieto, M., Navarro, G.: Indexes for highly repetitive document collections. In: CIKM, pp. 463–468 (2011)
6. Claude, F., Navarro, G.: A fast and compact Web graph representation. In: Ziviani, N., Baeza-Yates, R. (eds.) SPIRE 2007. LNCS, vol. 4726, pp. 118–129. Springer, Heidelberg (2007)
7. Claude, F., Navarro, G.: Improved grammar-based compressed indexes. In: Calderón-Benavides, L., González-Caro, C., Chávez, E., Ziviani, N. (eds.) SPIRE 2012. LNCS, vol. 7608, pp. 180–192. Springer, Heidelberg (2012)
8. Claude, F., Fariña, A., Martínez-Prieto, M.A., Navarro, G.: Compressed q-gram indexing for highly repetitive biological sequences. In: BIBE, pp. 86–91 (2010)
9. Farzan, A., Gagie, T., Navarro, G.: Entropy-bounded representation of point grids. In: Cheong, O., Chwa, K.-Y., Park, K. (eds.) ISAAC 2010, Part II. LNCS, vol. 6507, pp. 327–338. Springer, Heidelberg (2010)
10. Gagie, T., Karhu, K., Navarro, G., Puglisi, S.J., Sirén, J.: Document listing on repetitive collections. In: Fischer, J., Sanders, P. (eds.) CPM 2013. LNCS, vol. 7922, pp. 107–119. Springer, Heidelberg (2013)
11. González, R., Navarro, G.: Compressed text indexes with fast locate. In: Ma, B., Zhang, K. (eds.) CPM 2007. LNCS, vol. 4580, pp. 216–227. Springer, Heidelberg (2007)
12. Grossi, R., Gupta, A., Vitter, J.S.: High-order entropy-compressed text indexes. In: SODA, pp. 841–850. Society for Industrial and Applied Mathematics, Philadelphia (2003)
13. Larsson, J., Moffat, A.: Off-line dictionary-based compression. Proc. of the IEEE 88(11), 1722–1732 (2000)
14. Muthukrishnan, S.: Efficient algorithms for document retrieval problems. In: FOCS, pp. 657–666 (2002)
15. Navarro, G.: Indexing highly repetitive collections. In: Smyth, B. (ed.) IWOCA 2012. LNCS, vol. 7643, pp. 274–279. Springer, Heidelberg (2012)
16. Navarro, G., Puglisi, S.J., Valenzuela, D.: Practical compressed document retrieval. In: Pardalos, P.M., Rebennack, S. (eds.) SEA 2011. LNCS, vol. 6630, pp. 193–205. Springer, Heidelberg (2011)
17. Raman, R., Raman, V., Rao, S.: Succinct indexable dictionaries with applications to encoding k-ary trees and multisets. In: SODA, pp. 233–242 (2002)
18. Sadakane, K.: Succinct data structures for flexible text retrieval systems. Journal of Discrete Algorithms 5(1), 12–22 (2007)
19. Välimäki, N., Mäkinen, V.: Space-efficient algorithms for document retrieval. In: Ma, B., Zhang, K. (eds.) CPM 2007. LNCS, vol. 4580, pp. 205–215. Springer, Heidelberg (2007)
20. Wan, R.: Browsing and searching compressed documents. Ph.D. thesis, The University of Melbourne (2003)
21. Ziv, J., Lempel, A.: Compression of individual sequences via variable length coding. IEEE Trans. Inf. Theo. 24(5), 530–536 (1978)

Order-Preserving Incomplete Suffix Trees and Order-Preserving Indexes

Maxime Crochemore[4,6], Costas S. Iliopoulos[4,5], Tomasz Kociumaka[1,*],
Marcin Kubica[1], Alessio Langiu[4], Solon P. Pissis[7,8,**],
Jakub Radoszewski[1,***], Wojciech Rytter[1,3,†], and Tomasz Waleń[2,1]

[1] Faculty of Mathematics, Informatics and Mechanics,
University of Warsaw, Warsaw, Poland
{kociumaka,jrad,rytter,walen}@mimuw.edu.pl
[2] Laboratory of Bioinformatics and Protein Engineering,
International Institute of Molecular and Cell Biology in Warsaw, Poland
[3] Faculty of Mathematics and Computer Science,
Copernicus University, Toruń, Poland
[4] Dept. of Informatics, King's College London, London, UK
{maxime.crochemore,c.iliopoulos,alessio.langiu}@kcl.ac.uk
[5] Faculty of Engineering, Computing and Mathematics,
University of Western Australia, Perth, Australia
[6] Université Paris-Est, France
[7] Laboratory of Molecular Systematics and Evolutionary Genetics,
Florida Museum of Natural History, University of Florida, USA
[8] Scientific Computing Group (Exelixis Lab & HPC Infrastructure),
Heidelberg Institute for Theoretical Studies (HITS gGmbH), Germany
solon.pissis@h-its.org

Abstract. Recently Kubica et al. (*Inf. Process. Let.*, 2013) and Kim et al. (*submitted to Theor. Comp. Sci.*) introduced order-preserving pattern matching: for a given text the goal is to find its factors having the same 'shape' as a given pattern. Known results include a linear-time algorithm for this problem (in case of polynomially-bounded alphabet) and a generalization to multiple patterns. We give an $O(n \log \log n)$ time construction of an index that enables order-preserving pattern matching queries in time proportional to pattern length. The main component is a data structure being an incomplete suffix tree in the order-preserving setting. The tree can miss single letters related to branching at internal nodes. Such incompleteness results from the weakness of our so called *weak character oracle*. However, due to its weakness, such oracle can answer queries on-line in $O(\log \log n)$ time using a sliding-window approach. For most of the applications such incomplete suffix-trees provide the same functional power as the complete ones. We also give an $O(\frac{n \log n}{\log \log n})$ time algorithm constructing complete order-preserving suffix trees.

* Supported by Polish budget funds for science in 2013-2017 as a research project under the 'Diamond Grant' program.
** Supported by the NSF–funded iPlant Collaborative (NSF grant #DBI-0735191).
*** The author receives financial support of Foundation for Polish Science.
† Supported by grant no. N206 566740 of the National Science Centre.

O. Kurland, M. Lewenstein, and E. Porat (Eds.): SPIRE 2013, LNCS 8214, pp. 84–95, 2013.
© Springer International Publishing Switzerland 2013

1   Introduction

We introduce order-preserving suffix trees that can be applied for pattern matching and repetition discovery problems in the order-preserving setting. In particular, this setting can be used to model finding trends in time series which appear naturally when considering e.g. the stock market or melody matching of two musical scores, see [11].

Two strings x and y of the same length over an integer alphabet are called *order-isomorphic* (or simply isomorphic), written $x \approx y$, if

$$\forall_{1 \leq i,j \leq |x|} \; x_i \leq x_j \Leftrightarrow y_i \leq y_j.$$

Example 1. $(5, 2, 7, 5, 1, 4, 9, 4, 5) \approx (6, 4, 7, 6, 3, 5, 8, 5, 6)$, see Fig. 1.

The notion of order-isomorphism was introduced in [11] and [14]. Both papers independently study the *order-preserving pattern matching problem* that consists in identifying all consecutive factors of a string x that are order-isomorphic to a given string y. If $|x| = n$ and $|y| = m$, an $O(n + m \log m)$ time algorithm for this problem is presented in both papers. Under a natural assumption that the characters of y can be sorted in linear time, the algorithm can be implemented in $O(n + m)$ time. Moreover, in [11] the authors present extensions of this problem to multiple-pattern matching based on the algorithm of Aho and Corasick.

The problem of order-preserving pattern matching has evolved from the combinatorial study of patterns in permutations. This field of study is concentrated on pattern avoidance, that is, counting the number of permutations not containing a subsequence which is order-isomorphic to a given pattern. Note that in this problem the subsequences need not to be consecutive. The first results on this topic were given by Knuth [12] (avoidance of 312), Lovász [16] (avoidance of 213) and Rotem [17] (avoidance of both 231 and 312). On the algorithmic side, pattern matching in permutations (as a subsequence) was shown to be NP-complete [3] and a number of polynomial-time algorithms for special cases of patterns were developed [1,9,10].

We introduce an index for order-preserving pattern matching. The preprocessing time is $O(n \log \log n)$ and queries are answered in $O(m)$ time for a pattern of length m over polynomially bounded integer alphabet Σ. The index is based on incomplete order-preserving suffix trees (incomplete op-suffix-trees, in short). We also introduce (complete) order-preserving suffix trees (op-suffix-trees) and show how they can be constructed using their incomplete counterpart in $O(n \log n / \log \log n)$ time. We provide randomized (Las Vegas) algorithms for the word-RAM model with $\Omega(\log n)$ word size.

In the literature there are a number of results in the related field of indexing for parameterized pattern matching. This problem is solved using parameterized suffix trees, a notion first introduced by Baker [2] who proposed an $O(n \log n)$ time construction algorithm. The result was then improved by Cole and Hariharan [5] to $O(n)$ construction time. Recently, Lee et al. [15] presented an online algorithm with the same time complexity. What Cole and Hariharan [5] proposed was actually a general scheme for construction of suffix trees for so-called quasi-suffix families with a constant time character oracle. This result can also be applied in the order-preserving setting, however the resulting index has larger construction time, $O(n \log n)$ or $O(n \log n / \log \log n)$ depending on the codes used.

Structure of the Paper. In Sections 2 (preliminary notation) and 3 we give a formal definition of a complete and an incomplete op-suffix-tree and describe their basic properties. Then in Sections 4 and 5 we show an $O(n \log \log n)$ construction of an incomplete op-suffix-tree. The former section contains an algorithmic toolbox that is also used in further parts of the paper. Applications of our data structure for order-preserving pattern matching and longest common factor problems are presented in Section 6. Finally in Section 7 we obtain a construction of complete op-suffix-trees.

2 Order-Preserving Code

Let $w = w_1 \dots w_n$ be a string of length n over an integer alphabet Σ. We assume that Σ is polynomially bounded in terms of n, i.e. $\Sigma = \{1, \dots, n^c\}$ for an integer constant c. We denote the length of a string w by $|w| = n$. By $w[i \, . . \, j]$ we denote the factor $w_i \, . . \, w_j$, and by suf_i – the i-th suffix of w, that is, $w[i \, . . \, n]$.
For any $i \in \{1, \dots, n\}$ define:

$$\alpha_w(i) = i - j \quad \text{where} \quad w_j = \max\{w_k \; : \; k < i, \; w_k \le w_i\},$$

if there is no such j then $\alpha_w(i) = i$, similarly define:

$$\beta_w(i) = i - j \quad \text{where} \quad w_j = \min\{w_k \; : \; k < i, \; w_k \ge w_i\},$$

and $\beta_w(i) = i$ if no such j exists. If several equally good values of j exist, we select the greatest possible value of j that is smaller than i.
We introduce codes of strings in a similar way as in [14]:

$$Code(w) = ((\alpha_w(1), \beta_w(1)), (\alpha_w(2), \beta_w(2)), \dots, (\alpha_w(|w|), \beta_w(|w|))).$$

We also denote $LastCode(w) = (\alpha_w(|w|), \beta_w(|w|))$. The following property is a consequence of Lemma 2 in [14].

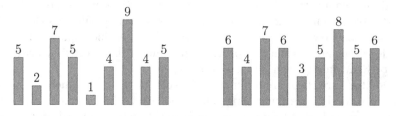

Fig. 1. Example of two order-isomorphic strings. Their codes are equal to $(1, 1)$ $(2, 1)$ $(2, 3)$ $(3, 3)$ $(5, 3)$ $(4, 2)$ $(4, 7)$ $(2, 2)$ $(5, 5)$.

Lemma 2. *Let x and y be two strings of length t and $x' = x[1 .. t - 1]$, $y' = y[1 .. t - 1]$. Then:*

(a) $x \approx y \Leftrightarrow x' \approx y' \wedge (y_i \leq y_t \leq y_j)$, where $i = t - \alpha_x(t)$, $j = t - \beta_x(t)$;
(b) $x \approx y \Leftrightarrow x' \approx y' \wedge LastCode(x) = LastCode(y)$.

Proof. Part (a) is an equivalent formulation of Lemma 2 in [14]. Part (b) is a technical consequence of part (a).   □

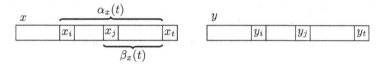

Fig. 2. An illustration of Lemma 2, part (a): $x[1 .. t] \approx y[1 .. t]$ is equivalent to $x[1 .. t - 1] \approx y[1 .. t - 1]$ and $y_i \leq y_t \leq y_j$

Part (b) of Lemma 2 implies that the codes provide an equivalent characterization of order-isomorphism:

Lemma 3. $x \approx y \Leftrightarrow Code(x) = Code(y)$.

The codes of strings can be computed efficiently. Applying Lemma 1 from [14] to strings over polynomially-bounded alphabet we obtain:

Lemma 4. *For a string w of length n, $Code(w)$ can be computed in $O(n)$ time.*

3   Order-Preserving Suffix Trees

Let us define the following family of sequences:

$$SufCodes(w) = \{ Code(suf_1)\#, \ Code(suf_2)\#, \ \ldots, \ Code(suf_n)\# \},$$

see Fig. 3. The *order-preserving suffix tree* of w (*op-suffix-tree* in short), denoted *opSufTree(w)*, is a compacted trie of all the sequences in *SufCodes(w)*.

Example 5. Let $w = (1, 2, 4, 4, 2, 5, 5, 1)$. All *SufCodes(w)* are given in Fig. 3.

The nodes of *opSufTree(w)* with at least two children are called branching nodes, together with the leaves they form explicit nodes of the tree. All the remaining nodes (that 'disappear' due to compactification) are called implicit nodes. For a node v, its explicit descendant (denoted as *FirstDown(v)*) is the top-most explicit node in the subtree of v (possibly *FirstDown(v)* = v). By *Locus*$_{Code(x)}$ we denote the (explicit or implicit) locus of $Code(x)$ in *opSufTree(w)*. Only the explicit nodes of *opSufTree(w)* are stored. The tree contains $O(n)$ leaves, hence its size is $O(n)$.

The leaf corresponding to $Code(suf_i)\#$ is labeled with the number i. Each branching node stores its depth and one of the leaves in its subtree. Each edge suffixes of w: $SufCodes(w)$:

1 2 4 4 2 5 5 1	(1,1) (1,2) (1,3) (1,1) (3,3) (2,6) (1,1) (7,7)	#
2 4 4 2 5 5 1	(1,1) (1,2) (1,1) (3,3) (2,5) (1,1) (7,3)	#
4 4 2 5 5 1	(1,1) (1,1) (3,1) (2,4) (1,1) (6,3)	#
4 2 5 5 1	(1,1) (2,1) (2,3) (1,1) (5,3)	#
2 5 5 1	(1,1) (1,2) (1,1) (4,3)	#
5 5 1	(1,1) (1,1) (3,1)	#
5 1	(1,1) (2,1)	#
1	(1,1)	#

Fig. 3. $SufCodes(w)$ for $w = (1, 2, 4, 4, 2, 5, 5, 1)$

stores the code only of its first character. The codes of all the remaining characters of any edge can be obtained using a *character oracle* that can efficiently provide the code $LastCode(suf_i[1..j])$ for any i, j.

Each explicit node v stores a suffix link, $SufLink(v)$, that may lead to an implicit or an explicit node (see an example in Fig. 4). The suffix link is defined as:

$$SufLink(Locus_{Code(x)}) = Locus_{Code(DelFirst(x))},$$

where $DelFirst(x)$ results in removing the first character of x, see Fig. 5.

Observation 6. $Code(x) = Code(y) \Rightarrow Code(DelFirst(x)) = Code(DelFirst(y))$.

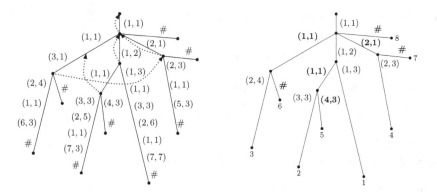

Fig. 4. The uncompacted trie of $SufCodes(w)$ for $w = (1, 2, 4, 4, 2, 5, 5, 1)$ (to the left) and its compacted version, the complete op-suffix-tree of w (to the right). The dotted arrows (left figure) show suffix links for branching nodes, note that one of them leads to an implicit node. Labels in the right figure that are in bold are present also in the incomplete op-suffix-tree.

We also introduce an *incomplete* order-preserving suffix tree of w, denoted $T(w)$, in which the character oracle is not available and each explicit node v can have one outgoing edge that does not store its first character (*incomplete edge*). This edge is located on the longest path leading from v to a leaf.

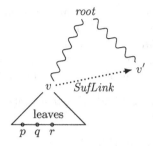

Fig. 5. Let γ be the text spelled out on a path from the root to v in the uncompacted op-suffix trie of w. Similarly, let γ' be the text on a path to $v' = SufLink(v)$. Observe that not necessarily γ' is a suffix of γ, but $\gamma' = Code(DelFirst(x))$, where $x = w[p \mathinner{.\,.} p+k-1]$ or $x = w[q \mathinner{.\,.} q+k-1]$ or $x = w[r \mathinner{.\,.} r+k-1]$, where p, q, r are the labels on the leaves in the subtree rooted in v.

Example 7.
Let $w = (1, 2, 4, 4, 2, 5, 5, 1)$. The op-suffix-tree of w is presented in Fig. 4.

4 Algorithmic Toolbox

We use a predecessor data structure to compute the last symbols of the code of a sequence changing in a queue-like manner.

Lemma 8. [Weak Character Oracle] *An initially empty sequence x over $\{1, \ldots, n\}$ can be maintained in a data structure $\mathcal{D}(x)$ of size $O(|x|)$ so that the following queries are supported in $O(\log \log n)$ expected time:*

compute LastCode(x);   append a single letter to x;   and DelFirst(x).
Only the second operation is valid if x is empty.

Proof. The main tool here is the y-fast tree, a data structure for dynamic predecessor queries. The following fact has been shown in [19].

Claim. Let N be an integer such that $\omega = \Omega(\log n)$, where ω is the machine word-size. There exists a data structure that uses $O(|X|)$ space to maintain a set X of key-value pairs with keys from $\{1, \ldots, N\}$ and supports the following operations in $O(\log \log N)$ expected time:

find(k): find the value associated with k, if any,
predecessor(k): return the pair $(k', v) \in X$ with the largest $k' \leq k$,
successor(x): return the pair $(k', v) \in X$ with the smallest $k' \geq k$,
remove(k): remove the pair with key k,
insert(k, v): insert (k, v) to X removing the pair with key k, if any.

The y-fast trees are now used as follows. The keys are the symbols present in x while the values associated with them are the locations of their last occurrences represented as a time-stamps (that is, the ordinal numbers of the push operations used to append them). Then the *LastCode*() query is answered using one predecessor and one successor query.   □

Our second tool is the dynamic weighted ancestor data structure proposed by Kopelowitz and Lewenstein [13] and originally motivated by problems related to ordinary suffix trees. A *weighted tree* is a rooted tree with integer weight assigned to each node, such that a monotonicity condition is satisfied: the weight of a node is strictly greater than the weight of its parent. The *weighted ancestor query* is:

> given a node v and a weight g find $WeightedAnc(v, g)$ – the highest ancestor of v with weight at least g.

The following lemma is proved in [13].

Lemma 9. *Let N be an integer such that $\omega = \Omega(\log N)$, where ω is the machine word-size. There exists a data structure which maintains a weighted tree T with weights $\{1, \ldots, N\}$ in $O(|T|)$ space and supports the following operations in $O(\log \log N)$ expected time:*

- *answer $WeightedAnc(v, g)$,*
- *insert a leaf with weight g and v as a parent,*
- *insert a node with weight g by subdividing the edge joining v with its parent.*

The weights of inserted nodes must meet the monotonicity condition.

5   Constructing Incomplete Order-Preserving Suffix Tree

We design a version of Ukkonen's algorithm [18] in which suffix links are computed using weighted ancestor queries, see Fig. 6. The weights of explicit nodes represent their depths. In this case for a node u, by $WeightedAnc(u, d)$ we denote its (explicit or implicit) ancestor at depth d.

Our algorithm works online. While reading the string w it maintains:

- the incomplete op-suffix-tree $T(w)$ for w;
- the longest suffix $\mathfrak{F}$ of w such that $Code(\mathfrak{F})$ corresponds to a non-leaf node of $T(w)$, together with the data structure $\mathcal{D}(\mathfrak{F})$; $\mathfrak{F}$ is called the *active suffix*;
- the node (explicit or implicit) $Locus_{Code(\mathfrak{F})}$, called the *active node*.

In the algorithm all implicit nodes are represented in a canonical form: the explicit descendant (*FirstDown*) and the distance to this descendant (depth difference). Each explicit node stores a dynamic hash table (see [5,8]) of its explicit children, indexed by the labels of the respective edges. Note that the explicit child corresponding to the incomplete edge is stored outside of the hash table.

When w is extended by one character, say a, we traverse the *active path* in $T(w)$: we search for the longest suffix $\mathfrak{F}'$ of $\mathfrak{F}$ such that $Locus_{Code(\mathfrak{F}'a)}$ appears in the tree, and for each longer suffix $\mathfrak{F}''$ of $\mathfrak{F}$ we create a branch leading to a new leaf node $Locus_{Code(\mathfrak{F}''a)}$. The active path is found by jumping along suffix links, starting at the active node. The end point of the active path provides the new active node, and $\mathfrak{F}'a$ becomes the active suffix.

To compute the last symbol of $Code(\mathfrak{F}a)$ we use the following observation.

Observation 10. *Due to Lemma 8 we can compute* $LastCode(\mathfrak{F} \cdot a)$ *in* $O(\log \log n)$ *expected time, where* $\mathfrak{F}$ *is the active suffix.*

We also use two auxiliary subroutines.

Function *Transition*$(v, (p, q))$. This function checks if v has an (explicit or implicit) child v' such that the edge from v to v' represents the code (p, q). It returns the node v' or **nil** if such a node does not exist. We check, using hashing, if any of the labeled edges outgoing from v starts with the code (p, q), for (at most one for v) incomplete edge we can check if its starting letter code equals (p, q) by checking two inequalities from part (a) of Lemma 2.

Function *Branch*$(v, (p, q))$. This function creates a new (open) transition from v with the code (p, q). If v was implicit then it is made explicit, at this moment the edge leading to its already existing child remains incomplete.

Algorithm *Construct incomplete opSufTree*(w)
Initialize T as incomplete *opSufTree* for w_1;
$v := root$; $\mathfrak{F} :=$ empty string;
for $i := 2$ **to** n **do**
    $a := w_i$; $\mathfrak{F} := \mathfrak{F} \cdot a$;
    while *Transition*$(v, LastCode(\mathfrak{F})) =$ **nil do**
        Branch$(v, LastCode(\mathfrak{F}))$;
        if $v = root$ **then break**;
        $\mathfrak{F} := DelFirst(\mathfrak{F})$;
        $u := FirstDown(v)$; { u is the first explicit node below v, including v }
        $u' := SufLink(u)$; { u' can be an implicit node }
        $v' := WeightedAnc(u', |v| - 1)$; { weighted ancestor query }
        $SufLink(v) := v'$; $v := v'$;
    $v := Transition(v, LastCode(\mathfrak{F}))$;
return T;

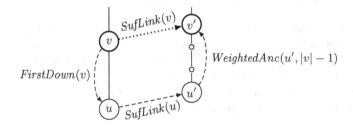

Fig. 6. Computation of $SufLink(v)$. Here u is explicit.

Remark 11. [Why incomplete?] At first glance it is not clear why incomplete edges appear. Consider the situation when we jump to an implicit node $v' = SufLink(v)$ and we later branch in this node. The node v' becomes explicit and the existing edge from this node to some node u' becomes an *incomplete edge*. Despite incompleteness of the edge (v', u') the equality test between the (known) last code letter of the active string and the first (unknown) code letter of the label of this edge can be done quickly due to part (a) of Lemma 2.

In the pseudocode above we perform $O(n)$ operations in total. This follows from the fact that each step of the while-loop creates a new edge in the tree. The operations involving $\mathfrak{F}$ and the operation *WeightedAnc* are performed in $O(\log \log n)$ time and all the remaining operations require constant time only. We obtain the following result.

Theorem 12. *The incomplete op-suffix-tree $T(w)$ for a string w of length n can be computed in $O(n \log \log n)$ expected time.*

6   Incomplete Suffix Tree as Order-Preserving Index

The most common application of suffix trees is pattern matching with time complexity independent of the length of the text.

Theorem 13. *Assume that we have $T(w)$ for a string w of length n. Given a pattern x of length m, one can check if w contains a factor order-isomorphic to x in $O(m)$ time and report all occurrences of such factors in $O(m + Occ)$ time, where Occ is the number of occurrences reported.*

Proof. First we compute the code of the pattern. This takes $O(m)$ time due to Lemma 4. To answer a query, we traverse down $T(w)$ using the successive symbols of the code. At each step we use the function $Transition(v, (p, q))$.

This enables to find the locus of $Code(x)$ in $O(m)$ time. Afterwards all the occurrences of factors that are order-isomorphic to x can be listed in the usual way by inspecting all leaves in the subtree of $Locus_{Code(x)}$. □

The motivating application of the standard suffix trees was finding the longest common factor of two strings. An analog of this problem in the order-preserving setting is especially important, since it provides a way to find common trends in time series. In this problem, given two strings w and x, we need to find the longest factor of x that is order-isomorphic to a factor of w. We show the usefulness of the suffix links in incomplete op-suffix-tree.

Theorem 14. *Let w be a string of length n. Having $T(w)$, one can find the order-preserving longest common factor of w and x, the latter string of length m, in $O(m(\log \log m + \log \log n))$ expected time.*

Proof. The main principle of the algorithm is the same as in the standard setting (see Corollary 6.12 in [6]). However, it needs to be enhanced using our algorithmic tools.

Let $pref(x)$ be the longest prefix of x such that $Code(pref(x))$ corresponds to a node in $T(w)$. Let suf_i^x be the i-th suffix of x. The algorithm computes $pref(suf_1^x)$, $pref(suf_2^x)$ etc. and finds the maximum depth among their loci.

At each point the data structure $\mathcal{D}(pref(suf_i^x))$ for the current suffix is stored. First, the locus of $pref(suf_1^x)$ is found by iterating $Transition(v, (p, q))$, as in the order-preserving pattern matching (Theorem 13). To proceed from $pref(suf_i^x)$ to $pref(suf_{i+1}^x)$, we remove the first letter ($DelFirst$), which also corresponds to a jump along a suffix link, and then keep traversing down the $T(w)$ using $Transition(v, (p, q))$.

By Lemmas 8 and 9, we obtain the required time complexity.   □

7   Constructing Complete Order-Preserving Suffix Tree

In Section 5 we presented an $O(n \log \log n)$ time construction of an incomplete op-suffix-tree. To obtain a complete op-suffix-tree, we need to put labels on incomplete edges and to provide a character oracle. Note that, using a character oracle working in $f(n)$ time, we can fill in the missing labels in $O(nf(n))$ time.

Observation 15. *The op-suffix-tree of a string of length n can be constructed in $O(n \log n)$ time.*

Proof. After $O(n \log n)$ preprocessing one can compute $LastCode(suf_i[1..j])$ for any i, j in $O(\log n)$ time. We use range trees for that, see [7]. Then we can fill in separately each missing label in the incomplete tree in $O(n \log n)$ time.   □

Below we show a slightly faster construction. For this, however, we need a different encoding of strings that also preserves the order. A very similar code was already presented in [11]. For any $i \in \{1, \ldots, n\}$ define:

$$prev_w^{<}(i) = |\{k \,:\, k < i, w_k < w_i\}|, \quad prev_w^{=}(i) = |\{k \,:\, k < i, w_k = w_i\}|.$$

The *counting code* of a string w is defined as:

$$Code'(w) = ((prev_w^{<}(1), prev_w^{=}(1)), \ldots, (prev_w^{<}(|w|), prev_w^{=}(|w|))).$$

We also define $LastCode'(w) = (prev_w^{<}(|w|), prev_w^{=}(|w|))$.

Example 16. The counting code of each of the strings in Fig. 1 is $(0, 0) (0, 0) (2, 0)$ $(1, 1) (0, 0) (2, 0) (6, 0) (2, 1) (4, 2)$.

The following lemma states that $Code'$ is also an order-preserving code. In this version of the paper we omit the proof, since it is basically present in [11].

Lemma 17. $x \approx y \Leftrightarrow Code'(x) = Code'(y)$.

The main advantage of the new order-preserving code is the existence of an $O(\log n/\log\log n)$ time character oracle with $o(n\log n/\log\log n)$ time construction. To design the oracle we use a geometric approach: the computation of $LastCode'$ for w corresponds to counting points in certain orthogonal rectangles in the plane.

Observation 18. *Let us treat the pairs (i, w_i) as points in the plane. Then we have $LastCode'(suf_i[1..j]) = (a,b)$, where a is the number of points that lie within the rectangle $A = [i, i+j-2] \times (-\infty, w_{i+j-1})$ and b is the number of points in the rectangle $B = [i, i+j-2] \times [w_{i+j-1}, w_{i+j-1}]$, see Fig. 7.*

The orthogonal range counting problem is defined as follows. We are given n points in the plane and we are to count the number of points in axis-aligned rectangles given as queries.

An efficient solution to this problem was given by Chan and Pătraşcu, see Theorem 2.3 in [4] which we state below as Lemma 19. We say that a point (p, q) dominates a point (p', q') if $p > p'$ and $q > q'$.

Lemma 19. *We can preprocess n points in the plane in $O(n\sqrt{\log n})$ time, using a data structure with $O(n)$ words of space, so that we can count the number of points dominated by a query point in $O(\log n/\log\log n)$ time.*

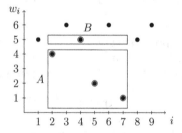

Fig. 7. Geometric illustration of the sequence $w = (5, 4, 6, 5, 2, 6, 1, 5, 6)$. The elements w_i are represented as points (i, w_i). The computation of $LastCode'(suf_2[1..7]) = (3, 1)$ corresponds to counting points in rectangles A, B.

Theorem 20. *The op-suffix-tree of a string of length n using the counting code can be constructed in $O(n\log n/\log\log n)$ expected time.*

Proof. Due to Lemma 3 and the corresponding Lemma 17, the *skeleton* of the op-suffix-tree for each of the order-preserving codes is the same. Hence, to construct the op-suffix-tree for the counting code, we compute the skeleton of the suffix tree using the algorithm for incomplete op-suffix-tree. Afterwards we use the character oracle to insert the first characters on each edge of the skeleton.

Due to Observation 18 and Lemma 19 after $O(n\sqrt{\log n})$ time and $O(n)$ space preprocessing one can compute $LastCode'(suf_i[1..j])$ for any i, j in $O(\log n/\log\log n)$ time.   □

References

1. Albert, M.H., Aldred, R.E.L., Atkinson, M.D., Holton, D.A.: Algorithms for pattern involvement in permutations. In: Eades, P., Takaoka, T. (eds.) ISAAC 2001. LNCS, vol. 2223, pp. 355–366. Springer, Heidelberg (2001)
2. Baker, B.S.: Parameterized pattern matching: Algorithms and applications. J. Comput. Syst. Sci. 52(1), 28–42 (1996)
3. Bose, P., Buss, J.F., Lubiw, A.: Pattern matching for permutations. Inf. Process. Lett. 65(5), 277–283 (1998)
4. Chan, T.M., Patrascu, M.: Counting inversions, offline orthogonal range counting, and related problems. In: Charikar, M. (ed.) SODA, pp. 161–173. SIAM (2010)
5. Cole, R., Hariharan, R.: Faster suffix tree construction with missing suffix links. SIAM J. Comput. 33(1), 26–42 (2003)
6. Crochemore, M., Hancart, C., Lecroq, T.: Algorithms on Strings. Cambridge University Press, USA (2007)
7. de Berg, M., Cheong, O., van Kreveld, M., Overmars, M.: Computational Geometry. Algorithms and Applications, 3rd edn. Springer, Heidelberg (2008)
8. Dietzfelbinger, M., Karlin, A.R., Mehlhorn, K., Meyer auf der Heide, F., Rohnert, H., Tarjan, R.E.: Dynamic perfect hashing: Upper and lower bounds. SIAM J. Comput. 23(4), 738–761 (1994)
9. Guillemot, S., Vialette, S.: Pattern matching for 321-avoiding permutations. In: Dong, Y., Du, D.-Z., Ibarra, O. (eds.) ISAAC 2009. LNCS, vol. 5878, pp. 1064–1073. Springer, Heidelberg (2009)
10. Ibarra, L.: Finding pattern matchings for permutations. Inf. Process. Lett. 61(6), 293–295 (1997)
11. Kim, J., Eades, P., Fleischer, R., Hong, S.-H., Iliopoulos, C.S., Park, K., Puglisi, S.J., Tokuyama, T.: Order preserving matching. CoRR, abs/1302.4064 (2013); Submitted to Theor. Comput. Sci.
12. Knuth, D.E.: The Art of Computer Programming, 2nd edn. Fundamental Algorithms, vol. I. Addison-Wesley (1973)
13. Kopelowitz, T., Lewenstein, M.: Dynamic weighted ancestors. In: Bansal, N., Pruhs, K., Stein, C. (eds.) SODA, pp. 565–574. SIAM (2007)
14. Kubica, M., Kulczynski, T., Radoszewski, J., Rytter, W., Walen, T.: A linear time algorithm for consecutive permutation pattern matching. Inf. Process. Lett. 113(12), 430–433 (2013)
15. Lee, T., Na, J.C., Park, K.: On-line construction of parameterized suffix trees for large alphabets. Inf. Process. Lett. 111(5), 201–207 (2011)
16. Lovász, L.: Combinatorial problems and exercices. North-Holland (1979)
17. Rotem, D.: Stack sortable permutations. Discrete Mathematics 33(2), 185–196 (1981)
18. Ukkonen, E.: On-line construction of suffix trees. Algorithmica 14(3), 249–260 (1995)
19. Willard, D.E.: Log-logarithmic worst-case range queries are possible in space theta(n). Inf. Process. Lett. 17(2), 81–84 (1983)

Compact Querieable Representations
of Raster Data[*]

Guillermo de Bernardo[1], Sandra Álvarez-García[1], Nieves R. Brisaboa[1],
Gonzalo Navarro[2], and Oscar Pedreira[1]

[1] Databases Lab., University of A Coruña, Spain
[2] Department of Computer Science, University of Chile, Chile

Abstract. In Geographic Information Systems (GIS) the attributes of
the space (altitude, temperature, etc.) are usually represented using a
raster model. There are no compact representations of raster data that
provide efficient query capabilities. In this paper we propose compact representations to efficiently store and query raster datasets in main memory. We experimentally compare our proposals with traditional storage
mechanisms for raster data, showing that our structures obtain competitive space performance while efficiently answering range queries involving
the values stored in the raster.

1   Introduction

The raster model is widely used to represent spatial attributes [14]. A raster is
essentially a matrix representing a region of the space, in which the space is split
into cells and a value of the spatial attribute is stored for each of these cells. An
uncompressed raster representation requires much space (e.g., a $50,000 \times 50,000$
grid of integers requires around 10 GB), so it is typically stored in secondary
memory. Compressed raster representations are mainly designed to reduce storage, and are based on well-known compression techniques such as run-length
encoding or LZW [15]. In these representations the full file must be decompressed even to display a small region of the space. Some representations split
the raster into fixed size tiles and compress each tile independently, providing
some level of direct access to regions and taking advantage of the locality of
values to enhance compression (for example, $GeoTIFF^1$ images can be used to
represent raster data and they support this partition into tiles with different
compression techniques including LZW).

Geographic Information Systems (GIS) [16,14] routinely make use of raster
data to represent various kinds of information. They usually need not only direct
access to regions (e.g. to display a local map), but also need to *find* the cells whose

[*] GdB, NB, SAG and OP were funded by MICINN (PGE and FEDER) grants
TIN2009-14560-C03-02, TIN2010-21246-C02-01 and CDTI CEN-20091048, and by
Xunta de Galicia (co-funded with FEDER) ref. 2010/17. GN was founded by Mil-
lennium Nucleus Information and Coordination in Networks ICM/FIC P10-024F.

[1] http://trac.osgeo.org/geotiff/

O. Kurland, M. Lewenstein, and E. Porat (Eds.): SPIRE 2013, LNCS 8214, pp. 96–108, 2013.

value is within some range. A classic example is the visualization of pressure or temperature bands, which require retrieving the coordinates that have values within a given range. Another example is retrieving the regions of a raster with an altitude above a given threshold to find zones with snow alert, or below a value to find regions with risk of floods. However, usual raster representations lack *indexing* capabilities on the values stored in the raster. These representations need to traverse the complete raster in order to return the cells that contain a given value, even when the results may be restricted to a small subregion of the space.

One solution is to consider the raster as a 3-dimensional matrix and use computational geometry solutions to answer all these queries as range reporting queries [5]. However, these solutions require superlinear space and therefore they are not suitable to the large datasets involved. Reading the raster row-wise and storing the sequence of values we could use a compressed sequence representation [10,9,1] to return the cells with a given value (or a range of values [10,13]) efficiently, but further restricting the search to a spatial range is not efficiently handled. Furthermore, these sequence representations achieve at best the zero-order entropy space of the sequence, and this is not a significant space reduction in many cases.

In this paper we present several proposals that aim at providing at the same time a compact representation of raster data and efficient support of queries involving spatial windows and intervals of values. We design our structures to solve queries such as retrieving all the values of a given area, retrieving all the coordinates with a given value, or retrieving all the entries of the raster within a spatial window and with values in a given range. Our structures are enhancements of an existing data structure called k^2-tree [4], originally designed to represent sparse binary matrices. Our first contribution is a variant of the k^2-tree that can compress not only large regions of zeros but also regions of ones. This enhancement allows our structure to compress efficiently not only sparse matrices but also binary images that contain large homogeneous regions. We experimentally compare our structure with a Linear Quadtree [8] representation showing its superiority in space and even time. Our second contribution is a generalization of the k^2-tree to represent multi-dimensional data. We call this structure a k^n-tree. We use these new structures to provide different representations of raster data, each with different strengths. We test our proposals experimentally to demonstrate their low space requirements and their ability to efficiently solve queries. Finally, we describe other application domains where our proposals could be of interest.

2   Previous Work: The k^2-tree

The k^2-tree [4] is a data structure for the compact representation of sparse binary matrices. In this paper we use its simplest variant, k=2, so the k^2-tree is similar to a compact Quadtree [7]. It corresponds to a recursive partition of the binary matrix. At each partitioning step, the matrix is divided into k^2 submatrices of equal size. Each submatrix is represented using a single bit: 1 if the submatrix contains at least one 1, or 0 otherwise. The method proceeds recursively for each 1-child until the current submatrix is full of 0s or we reach the cells of the original matrix. This conceptual tree is traversed levelwise and stored in two bit arrays: T stores all the levels except the last one, and L stores the last level. Figure 1 shows an example of k²-tree. In order to navigate the tree we need to build a *rank structure* over T. This structure stores a set of counters that allow us to compute the number of ones in the bitmap up to any position ($rank_1$ operation) in constant time using sublinear space [12]. Given a value 1 at position *pos* in T, its k^2 children will start at position $pos' = rank_1(T, pos) \times k^2$ of $T : L$. This property provides simple navigation over the conceptual tree using only the bitmaps and the additional rank structure over T. A k²-tree can solve single cell queries, row/column queries or general range reporting queries (i.e., report all the 1s in a range) using only rank operations, by visiting all the necessary subtrees.

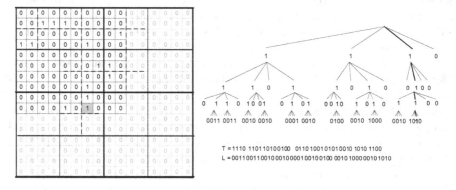

$T = 1110\ 1101\ 10100100\ \ 0110\ 1001\ 01010010\ 1010\ 1100$
$L = 0011\ 0011\ 0010\ 0010\ 0001\ 0010\ 0100\ \ 0010\ 1000\ 00101010$

Fig. 1. Binary matrix and its k²-tree representation, for k=2. The matrix is virtually expanded to the next power of k.

The original k²-tree was designed as a static data structure. A dynamic variant of the k²-tree, called dk²-tree, also exists [3]. The dk²-tree essentially splits the bitmaps T and L in chunks and builds tree structures to store these chunks. The internal nodes of the trees store counters that replace those of static rank structures and allow access to specific positions in T or L. The dk²-tree provides the same query capabilities of the original k²-tree but allows at the same time update operations, including changes in the values of the cells and insertion of new rows/columns at the end of the matrix. See the original paper [3] for further details.

3   Compression of Ones

In this section we propose variants of the k²-tree that are able to compress efficiently these large regions of ones and zeros. We will show 2 variants: the first performs better when the number of ones and zeros in the matrix is not too different; the second variant is designed to be used when the proportion of zeros and ones is very different (without loss of generality, we will consider that the less frequent value is 1).

The idea behind our proposals is to stop the decomposition of the binary matrix when a uniform region is found, be it of zeros or ones. This means that in our k^2-tree we must discriminate among the 3 possible "colors" of a node: white nodes are regions of zeros, black nodes are regions of ones and gray nodes are internal nodes that correspond to regions with zeros and ones. With this information, large regions of ones can be represented using a single node, just like regions of zeros in the original k^2-tree. The variants with compression of ones can be traversed like the original k^2-trees, and provide in addition a way to find the color of a node. Using the additional navigation rules, all the operations supported by original k^2-trees can be implemented directly in a k^2-tree with compression of ones. The algorithms must only be adapted to expand automatically all the results that fall in the submatrix covered by a black node. We now propose two specific representations of the color of nodes.

3.1   2-Bits Variant

In this variant, we mark with a 1 all the regions that contain both zeros and ones (i.e., the internal or gray nodes), whereas uniform regions are assigned a 0. In order to tell apart white from black nodes, we create a second bitmap T' in addition to T, that stores the value of each uniform region (that is, T' contains a bit for each 0 in T that will store the color of that region). The navigational properties of the original k^2-tree still hold: the children of the (gray) node at position p will start at position $p' = rank_1(T, p) \times k^2$, because each bit set to 1 in T represents a gray node and only gray nodes have children. If a position p in T is set to 0, we can check $T'[rank_0(T, p)]$ to see if it corresponds to a region of zeros or of ones. The bitmap L behaves as in original k^2-trees.

Figure 2 shows a k^2-tree with compression of ones and the bitmaps generated for this variant (left). We highlight a black node and the positions where its bits are assigned in the bitmaps T and T'.

3.2   Unbalanced (1-5)-Bits Variant

In this variant, we achieve compression of ones using the same bitmaps of the original k^2-tree. White nodes will be represented with a 0 and gray nodes will be assigned a 1, exactly like in original k^2-trees. Black nodes will be encoded as a gray node with $k^2 = 4$ white children (i.e., they are encoded using 5 bits). This combination, that can not appear in the original k^2-tree (gray nodes, by definition, represent regions with at least a 1), is used to represent regions of ones without the need of additional structures. Figure 2 (right) shows the bit distribution for this variant. To take into account regions of ones, at each step of the k^2-tree traversal, we must check the $k^2 - 1$ siblings of the current node to detect if the current node is white (the current bit is 0, but one of its siblings

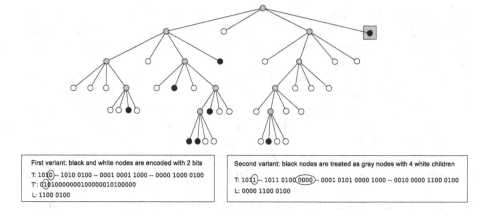

Fig. 2. Example of k²-trees with compression of ones

is 1)² or we are in a region of ones (the current bit is 0 and all its siblings are 0, meaning that the parent node was actually black).

This approach will obtain worse compression than the one based on 2 bits in most cases. However, it may obtain better compression when there are blocks of ones but the zeros are much more frequent. Also, this variant will never use more space than the original k²-tree.

Both variants of k²-trees with compression of ones can also be applied to dynamic k²-trees. Each bitmap used by the static representations is replaced by a tree structure as in the original dk²-tree, to support updates as well as access and rank queries.

3.3   Comparison of k²-tree with Linear QuadTree

The QuadTree (QT) [7] is a well-known spatial index structure for representing binary images. The partitioning principle of QT and k²-tree is the same. At the root of the tree, the matrix is partitioned into four quadrants, which correspond to the four children of the root. In the QT, the quadrants that are not entirely black or white are recursively partitioned following the same principle until we reach a fully black or white region or the unitary cells of the matrix. To access a particular position of the raster, the tree is traversed from the root to the leaves following the appropriate path. Depending on the distribution of black and white cells in the matrix, the QT can represent the binary matrix while saving significant space and providing efficient access.

The Linear QuadTree (LQT) [8] was proposed as a representation of QTs without pointers that can be easily managed in secondary memory and requires less space than the original QTs. At each node, the branches corresponding to

---

² The $k²$ bits are contiguous, so we perform this check in constant time.

the NW, NE, SW, and SE quadrants are labelled with 0, 1, 2, and 3 respectively. A leaf is formed when the submatrix is full of points. LQTs assign a *quadcode* to each leaf of the QT, which describes the (4-ary) path from the root to that leaf. To represent the quadcode of each leaf, a digit is added for each branch that is traversed. An additional symbol is used to represent that a region is not further partitioned (i.e., the subregion covered by that branch is full of ones, i.e., it is a leaf). All these quadcodes are then stored in a B-Tree in secondary memory. To access a position of the raster, we search for its corresponding quadcode or a quadcode that contains it in the B-Tree.

Other representations of QTs that achieve less space requirements have been proposed, such as the FBLQ [6] and the CBLQ [11]. However, they are designed mainly to represent binary images and support union, intersection and difference of two images, which require a full traversal of the raster.

The original k^2-trees cannot be used to represent a wide class of binary images because they do not efficiently compress regions of ones. Our variants of k^2-tree with compression of ones overcome this limitation. As a proof of concept of the capabilities of our proposals we compare our k^2-trees with LQTs.

Experimental Comparison. We compare the space requirements and search performance of k^2-trees and LQTs. Since k^2-trees work in main memory, we implemented in-memory versions of the LQT. To provide fair comparisons, we build two different variants of LQT. The first one (LQT-BTree) stores the quadcodes in a B-Tree maintained in main memory. This representation can handle modifications, so we compare it with a dk^2-tree. The second variant (LQT-Array) stores the sorted sequence of quadcodes directly in an array, and a binary search is used to find them. We compare the LQT-Array with a static k^2-tree with compression of ones. We run all our experiments on an AMD-Phenom-II X4 955@3.2 GHz, with 8GB DDR2 RAM. The operating system is Ubuntu 9.10. All our implementations are written in C and compiled with gcc version 4.4.1 with -O9 optimizations.

We use five collections in our comparison. The first three are binary images obtained from elevation rasters of the Digital Terrain Model MDT05 of the Spanish Geographic Information Center [3]. A threshold is applied to each raster, generating a binary image with 25% of ones, corresponding to the higher values. The last two collections are adjacency matrices of Web graphs [2], and therefore difficult to compress using LQTs because they are very sparse. Table 1 shows the space required by all the representations. Table 2 shows the average time needed to retrieve a cell of the matrix. To obtain this time, we run a million random queries in each dataset and compute the average access time.

Our results show that the space required by k^2-trees is an order of magnitude smaller than the required by our two variants of LQTs, while the access time is also better in most cases, particularly in static k^2-trees.

---

[3] Original rasters are available for download at `http://cnig.es/`

Table 1. Space utilization of k^2-trees and LQTs (in bits per one)

Dataset	rows × cols	#ones	k^2-tree		LQT-Array	LQT-BTree
			Static	Dynamic		
mdt-600	3961×5881	11,647,287	0.02	0.04	0.25	0.31
mdt-700	3841×5841	13,732,734	0.02	0.04	0.17	0.17
mdt-800	3921×6001	21,580,638	0.01	0.02	0.11	0.11
cnr	325,557×325,557	3,216,152	3.14	4.95	41.32	41.46
eu	862,664×862,664	19,235,140	3.81	5.86	49.92	50.07

Table 2. Time to retrieve the value of a cell of the binary matrix, in μs/query

Dataset	k^2-tree		LQT-Array	LQT-BTree
	Static	Dynamic		
mdt-600	0.46	0.86	1.64	1.27
mdt-700	0.50	1.00	1.64	1.27
mdt-800	0.36	0.56	1.64	1.20
cnr	0.86	2.88	3.16	3.06
eu	0.90	3.85	3.76	3.85

4   Multi-dimensional k^2-trees: The k^n-tree

The k^2-tree has been applied in several contexts to the representation of binary relations. Intuitively, the k^2-tree can be extended to solve problems of higher dimensionality extending its space partitioning while maintaining the representation techniques used. We call the extension of the k^2-tree a k^n-tree.

A k^n-tree represents a binary n-dimensional matrix $M_{m_1 \times \ldots \times m_n}$ by recursively partitioning it into k^n n-dimensional submatrices of equal size. This partitioning strategy generates a conceptual tree similar to a k^2-tree in which each node has k^n children. The conceptual tree can then be represented and queried using the same techniques of k^2-trees.

Notice that a 1 in a cell of a binary matrix that has another 1 falling in the same submatrix does not consume any additional space in the k^2-tree, but isolated ones induce a complete branch in the k^2-tree, consuming much space. In other words, the k^2-tree takes advantage of the proximity of the ones in the matrix, because paths in the conceptual tree to each of these ones can be shared for most of the levels. This feature becomes more important as the value of k increases, and also if we build a k^n-tree for a high n, because each node of a k^n-tree will have k^n children that must be represented if the region represented contains ones and zeros. Therefore, a k^n-tree may not be useful for unclustered data without any regularities. We will use the k^n-tree as an efficient method to represent multi-dimensional data in a way that ensures that the data is clustered across the different dimensions. Particularly, we will show its application to the representation of raster matrices.

5   Representation of Raster Data Using k²-tree Variants

In this section we design several structures for the representation of general raster data using the k²-tree variants we have proposed in previous sections.

We assume in our proposals that the raster values have a "realistic" precision[4], so that the number of different values in the third dimension is not too high. We consider that the raster has m different values and denote by v_i the i-th different value of the raster in ascending order.

Our first representation of a raster dataset consists of a collection of k²-trees (k^2-$base$), one for each different value of the raster (i.e., we build a k²-tree K_i that stores all the cells with value v_i). A variant with compression of ones will be used in order to exploit the regularities of spatial attributes. This representation can efficiently answer queries asking for cells with a given value, because these cells are indexed in a single k²-tree. However, queries that ask for cells with values in a range $[v_\ell, v_r]$ require traversing many k²-trees.

We also propose an alternative representation using "accumulated" k²-trees (k^2-acc). In this representation, each k²-tree K_i will store not the cells whose value is v_i but all the cells whose value is smaller or equal to v_i. While this increases the number of cells to represent in each k²-tree, it will also increase the clustering of ones and therefore the ability of each k²-tree to compress these regions. To ask for the value of a given cell in this variant, we can binary search the first k²-tree that contains the desired cell of the raster. This variant also has an advantage when asking for cells with values within a given range, as we only need to query at most two k²-trees: the cells with values in the range $[v_\ell, v_r]$ are the cells of K_r that do not appear in $K_{\ell-1}$. As a counterpart, two k²-trees must be queried instead of one in order to obtain the cells with a given value.

Finally, we propose a k³-tree as a better representation of the raster data, as it is an indexed representation of the spatial coordinates and the values altogether. Our k³-tree stores the tuples $\langle x, y, z \rangle$ such that the coordinate (x, y) of the raster has value z. Notice that for each pair (x, y) only one z value will be set, so we will not find (cubic) regions of ones in this case. Because of this, our k³-tree is based on the original k²-tree codes, without compression of ones. To query for the value of a cell in the k³-tree, we traverse all the paths found in the k³-tree for fixed x and y values. To return all the cells with a given value, the z coordinate is fixed and all the pairs (x, y) of the corresponding k³-tree slice are returned. To ask for cells with a range of values the query is similar, traversing the k³-tree only within the bounds given by the interval $[z_\ell, z_r]$.

5.1   Experimental Framework

To test the efficiency of our proposals we use several real raster matrices obtained from the MDT05 collection. Table 3 gives details about the different fragments

---

[4] When representing spatial attributes, in many cases the values stored can not be considered as an accurate estimation of actual values (e.g. a temperature measurement may be given in a precision of one thousandth of a degree, but this measurement may only be accurate in terms of a degree or a tenth of a degree).

Table 3. Raster datasets used

Dataset	Raster size	#values	Description
mdt-500	4001 × 5841	578	Raster 500
mdt-700	3841 × 5841	472	Raster 700
mdt-medium	7721 × 11081	978	Rasters 47,48, 72 and 73 combined
mdt-large	48266 × 47050	2142	Raster covering the region of Galicia

taken. The number of different values in each raster is also shown in the table, after rounding the elevation values to a precision of 1 meter.

To measure query results, we first determine a sufficiently large number of queries to obtain accurate results for each type of query. Then we build a different set of random queries of each type for each dataset. All the time results shown correspond to CPU time.

5.2 Experimental Results

We show the space required to represent each dataset with all our proposals. To provide an element of comparison we convert the rasters to GeoTIFF[5] format using two different sets of options. The first, *tiff-plain*, is a plain representation without compression, that stores all the values in row order (we use 16-bit integers as the datatype for the representation). The second representation, *tiff-comp*, is optimized for space: the image is divided in tiles of size 256 × 256 and each tile is compressed using a linear predictor and LZW encoding.

Table 4 shows the space utilization of our proposals and the reference Geo-TIFF images, in bits per cell of the raster. As an additional reference, columns 2 and 3 show the base-2 logarithm of the number of different values in each raster and the zero-order entropy H_0 of these values. These columns represent the minimum space that would be required by a representation of the raster as an uncompressed or entropy-compressed sequence, respectively. The k^3-tree clearly obtains the best space utilization amongst our approaches in all the datasets, being very close to the compressed GeoTIFF representation and using much less space than the zero-order entropy. Notice that the GeoTIFF format only offers, at best, random access to the raster data, whereas our proposals are indexed representations that efficiently solve various types of queries.

Table 4. Space utilization of all approaches (in bits/cell)

Dataset	log(#values)	H_0(values)	k^2-base	k^2-acc	k^3-tree	tiff-plain	tiff-comp
mdt-500	9.17	5.43	2.83	2.30	1.83	16.01	**1.52**
mdt-700	8.88	4.39	2.13	2.40	1.38	16.01	**1.12**
mdt-medium	9.93	5.86	3.06	2.72	1.77	16.01	**1.52**
mdt-large	11.06	5.32	3.16	4.37	1.62	16.00	**1.35**

---

[5] http://trac.osgeo.org/geotiff/

Table 5. Retrieving the value of a single cell. Times in μs/query.

Dataset	k^2-base	k^2-acc	k^3-tree	tiff-plain	tiff-comp
mdt-500	66.7	4.6	**2.2**	2.6	491.7
mdt-700	39.6	3.0	**1.8**	2.7	461.9
mdt-medium	76.4	5.3	**2.6**	5.2	499.0
mdt-large	415.3	11.1	**2.8**	87.9	494.8

Next we compare the query times of all our proposals for some queries of interest. We implement the same queries over the GeoTiff images used to compress the rasters. We build simple algorithms on top of the *libtiff* library[6] to retrieve fragments from the GeoTIFF images and run each query type. This comparison is given as a simple sanity check, since libtiff is not designed to process these queries[7]. We only aim to show that these queries are not easy to solve with the traditional formats.

First we measure the time required to retrieve the value of a single cell of the raster. This operation shows the ability of the representations to provide random access to the raster. Table 5 shows the results obtained. The k^3-tree representation obtains the best results among our proposals, showing the efficiency of the multi-dimensional index in this context. The approach based on independent k^2-trees, as expected, behaves much worse than our other proposals, since it has to scan the k^2-trees one by one. The accumulated k^2-tree approach obtains also good results because it can binary search the first k^2-tree that contains the cell. Note that the *tiff-plain* representation should obtain much faster times, but libtiff spends a lot of time copying chunks of the image that are useless for this query. On the other hand, the *tiff-comp* representation presents poor query times in comparison with the uncompressed version, as not only the appropriate tile of the image has to be recovered but also decompressed in order to recover a single cell.

Next we show the efficiency of the representations to select the cells of the raster that contain a specific value. The results are shown in Table 6. Not surprisingly, our representations obtain much better results than the *tiff* representations, because the latter must always traverse the complete raster. In this case, the k^2-base obtains better results, as expected, because only one k^2-tree is accessed and the regions of the ones in the k^2-tree can be decoded efficiently. The k^2-acc has to access two k^2-trees, essentially doubling the query time of the independent k^2-trees. The k^3-tree only needs to traverse the appropriate slice, but it may need to explore many more nodes that correspond to regions with close values.

Finally, we measure the efficiency of window-range queries, that ask for cells of the raster within a spatial window and a range of values. These queries are widely used when processing raster data corresponding to spatial attributes (for

---

[6] http://www.libtiff.org

[7] The library *libtiff* reads the images from disk, but we are considering only CPU time.

Table 6. Retrieving all the cells with a given value. Times in ms/query.

Dataset	k^2-base	k^2-acc	k^3-tree	tiff-plain	tiff-comp
mdt-500	**3.9**	5.8	9.4	39.5	221.4
mdt-700	**3.0**	6.0	7.3	37.5	199.5
mdt-medium	**8.2**	13.6	18.9	142.6	799.0
mdt-large	**110.2**	255.1	196.6	3,838.9	19,913.4

instance, regions with risk of floods or snow alert may be computed from elevation rasters selecting the cells with values above a threshold or in a given interval). In this case, the k^3-tree and the k^2-acc take advantage of their structure to obtain the best times. The k^2-acc only needs to perform a window query in two k^2-trees, and the k^3-tree can restrict the navigation of the tree in all the dimensions to the given bounds.

Table 7. Retrieving cells inside a window and within a range of values. Times in μs/query.

Dataset	Window size	Range length	k^2-base	k^2-acc	k^3-tree	tiff-plain	tiff-comp
mdt-500	10	10	5.9	**1.6**	1.9	24.5	525.7
		50	27.4	**1.9**	2.6	24.4	525.0
	50	10	10.3	**3.6**	5.1	124.0	697.1
		50	51.0	**5.4**	16.2	124.0	699.5
mdt-700	10	10	5.9	**1.6**	**1.6**	24.3	496.0
		50	27.6	**1.8**	2.3	24.5	493.4
	50	10	10.1	**3.6**	4.5	123.7	653.2
		50	49.2	**5.0**	13.6	123.8	649.9
mdt-medium	10	10	6.4	2.4	**2.0**	45.7	531.4
		50	28.4	**2.5**	**2.5**	45.9	533.6
	50	10	9.9	**3.7**	4.2	229.2	705.4
		50	46.5	**4.7**	10.9	228.5	705.4
mdt-large	10	10	10.5	3.9	**2.2**	285.5	519.1
		50	44.2	3.9	**2.5**	287.4	545.8
	50	10	13.1	4.6	**3.2**	1,021.6	693.6
		50	54.5	**5.2**	5.8	1,009.6	691.9

6   Conclusions

We have presented several compact data structures that can represent raster data in reduced space, supporting not only access to random areas in the raster but also advanced queries involving the values stored in the raster. We compare our representations, based on k^2-trees, with existing formats used to store and process raster data. Our experiments show that the k^3-tree can obtain very good space results, being close to the compressed GeoTIFF representation. The k^3-tree also shows competitive times in all the queries tested, being the fastest to retrieve the value of a cell and in some window queries. The variant with independent k^2-trees obtains the best time results to retrieve all the cells with a given value, but it is much slower in queries involving a range of values. The variant with accumulated k^2-trees obtains the best results in most of the queries involving ranges of values. In all the queries tested the results of our proposals are clearly better than the representations based on GeoTIFF images.

We believe that the proposed variants of the k^2-tree could be used in a wider range of application domains. We have shown, as a proof of concept, the applicability of the k^2-tree with compression of ones to the representation of binary images. Variants of k^n-tree could also be used to represent, for instance, spatio-temporal raster datasets and moving region databases. Spatio-temporal raster datasets can be seen as a collection of rasters stored for different time instants, so we can consider the time as a fourth dimension in the matrix that represents the raster and use a k^4-tree to represent space, time and values stored. An example of spatio-temporal raster is a collection of temperature rasters in different days. In this example it is expected that the values stored for cells close in space or for the same cell along time are similar. Moving region databases represent regions of space that change with time. These regions can be encoded with a 3-dimensional matrix that stores spatial coordinates covered by a region along time, so that cells of the matrix determine if the region covered a given position at a given time. In many cases, these regions are continuous and change only slightly between time instants (e.g., the evolution of oil spills along time will yield a 3-dimensional matrix with large uniform regions that will change slowly with time). Therefore, a k^3-tree with compression of ones could exploit these regularities to obtain good compression results, providing also spatio-temporal query support.

References

1. Barbay, J., Gagie, T., Navarro, G., Nekrich, Y.: Alphabet partitioning for compressed rank/select and applications. In: Cheong, O., Chwa, K.-Y., Park, K. (eds.) ISAAC 2010, Part II. LNCS, vol. 6507, pp. 315–326. Springer, Heidelberg (2010)
2. Boldi, P., Vigna, S.: The Webgraph framework I: compression techniques. In: Proc. 13th WWW, pp. 595–602 (2004)
3. Brisaboa, N.R., de Bernardo, G., Navarro, G.: Compressed dynamic binary relations. In: Proc. 22nd DCC, pp. 52–61 (2012)
4. Brisaboa, N.R., Ladra, S., Navarro, G.: k^2-Trees for compact web graph representation. In: Karlgren, J., Tarhio, J., Hyyrö, H. (eds.) SPIRE 2009. LNCS, vol. 5721, pp. 18–30. Springer, Heidelberg (2009)
5. Chan, T.M., Larsen, K.G., Pătrașcu, M.: Orthogonal range searching on the RAM, revisited. In: Proc. 27th SoCG, pp. 1–10 (2011)
6. Chang, H.K., Chang, J.W.: Fixed binary linear quadtree coding scheme for spatial data. In: Proc. 9th VCIP, vol. 2308, pp. 1214–1220 (1994)
7. Finkel, R.A., Bentley, J.L.: Quad trees: A data structure for retrieval on composite keys. Acta Informatica 4, 1–9 (1974)
8. Gargantini, I.: An effective way to represent quadtrees. Communications of the ACM 25(12), 905–910 (1982)

9. Golynski, A., Munro, J.I., Rao, S.S.: Rank/select operations on large alphabets: a tool for text indexing. In: Proc. 17th SODA, pp. 368–373 (2006)
10. Grossi, R., Gupta, A., Vitter, J.S.: High-order entropy-compressed text indexes. In: Proc. 14th SODA, pp. 841–850 (2003)
11. Lin, T.W.: Set operations on constant bit-length linear quadtrees. Pattern Recognition 30(7), 1239–1249 (1997)
12. Munro, J.I.: Tables. In: Chandru, V., Vinay, V. (eds.) FSTTCS 1996. LNCS, vol. 1180, pp. 37–42. Springer, Heidelberg (1996)
13. Navarro, G.: Wavelet trees for all. In: Kärkkäinen, J., Stoye, J. (eds.) CPM 2012. LNCS, vol. 7354, pp. 2–26. Springer, Heidelberg (2012)
14. Rigaux, P., Scholl, M., Voisard, A.: Spatial databases - with applications to GIS. Elsevier (2002)
15. Welch, T.A.: A technique for high-performance data compression. Computer 17(6), 8–19 (1984)
16. Worboys, M., Duckham, M.: GIS: A Computing Perspective, 2nd edn. CRC Press, Inc. (2004)

Top-k Color Queries on Tree Paths*

Stephane Durocher[1], Rahul Shah[2],
Matthew Skala[1], and Sharma V. Thankachan[2]

[1] University of Manitoba, Winnipeg, Canada
{durocher,mskala}@cs.umanitoba.ca
[2] Louisiana State University, Baton Rouge, USA
{rahul,thanks}@csc.lsu.edu

Abstract. We present a data structure for the following problem: Given a tree $\mathcal{T}$, with each of its nodes assigned a color in a totally ordered set, preprocess $\mathcal{T}$ to efficiently answer queries for the top k distinct colors on the path between two nodes, reporting the colors sorted in descending order. Our data structure requires linear space of $O(n)$ words and answers queries in $O(k)$ time.

1 Introduction and Related Work

Given an array $A[1..n]$ of color values in $\{1, 2, \ldots, \sigma\}$ and a function $p(c)$ that defines priorities for the colors, the *array range top-k color query problem* is to report, given indices a and b and a count k, the k distinct color values of highest priority to occur in the array range $A[a..b]$, in descending order of priority. A data structure for this problem must preprocess the array to answer queries efficiently. In this work we generalize the array range top-k color query problem to paths in trees.

Karpinski and Nekrich [15] give a data structure for the array problem with $O(k)$ query time using $O(n \log \sigma)$ bits, which is asymptotically optimal. Other related problems include reporting all distinct colors, counting the number of distinct colors in the query range, finding the most frequent, least frequent, majority or minority color in the query range, and so on. Efficient data structures offering different space-time trade-offs for these kinds of problems are known [1–6, 8–11, 16]. Such problems arise frequently in information retrieval and computational geometry.

Any two nodes in a tree define a unique path between them, just as two elements of an array define a unique range. When the entire tree is a path, tree paths reduce to array ranges. Most array range query problems can thus be generalized to a tree setting, and some of these problems are well studied. Krizanc et al. [16] discuss finding the median and mode on tree paths. He et al. [12] give efficient linear-space solutions for path selection (including median), counting,

---

* Work supported in part by the Natural Sciences and Engineering Research Council of Canada (NSERC) and National Science Foundation (NSF) Grants CCF–1017623 (R. Shah and J. S. Vitter) and CCF–1218904 (R. Shah).

O. Kurland, M. Lewenstein, and E. Porat (Eds.): SPIRE 2013, LNCS 8214, pp. 109–115, 2013.

and reporting, on individual node values (weights) as opposed to classes of distinct node values (colors) [12]. Recent results include succinct data structures for these weighted path query problems [13,19].

Suppose each node in a tree $\mathcal{T}$ of n nodes is assigned a color $c \in \{1, \ldots, \sigma\}$ and each color has a priority $p(c)$. Let $\mathcal{T}[a..b]$ represent the unique path connecting the nodes with preorder indices a and b. Then the *tree path top-k color query problem* is to report, given indices a and b and a count k, the k distinct color values of highest priority to occur in $\mathcal{T}[a..b]$, in descending order according to priority. We will prove the following result in Word-RAM model.

Theorem 1. *There exists a linear-space data structure of $O(n)$ words for answering tree path top-k color queries in $O(k)$ time.*

2   Our Solution

We begin by defining some useful concepts. Let $\mathcal{T}$ be a tree with n nodes. We define the *size* of an internal node v to be the number of leaves in the subtree rooted at v. Then the *heavy path* of the tree $\mathcal{T}$ is the path starting from the root, which each node v on the path is the largest-size child of its parent. Let the root of a heavy path be its highest node, that is, closest to the root of $\mathcal{T}$. The *heavy path decomposition* of the tree $\mathcal{T}$ is the operation where we decompose each off-path subtree of the heavy path recursively; therefore, edges in $\mathcal{T}$ will be partitioned into disjoint heavy paths. Each leaf node belongs to a unique path in the heavy decomposition, and each heavy path contains exactly one leaf. Therefore the number of heavy paths is exactly equal to the number of leaves. This decomposition will allow us to break query paths into a small number of pieces by the following lemma.

Lemma 1 (Sleator and Tarjan [20]). *Any path from the root to a leaf in $\mathcal{T}$ intersects at most $\log n$ paths of the heavy path decomposition.*

We will also use a data structure of Navarro and Nekrich for three-sided two-dimensional top-k queries [18]. Given a set of n points on an $n \times n$ grid, each having a weight, this data structure can report the k points of greatest weight in a query region of the form $[a, b] \times [0, h]$, in $O(h + k)$ time. Furthermore, it reports the points in order of decreasing weight, and it reports them online, that is, in constant time per point after the $O(h)$ time to start the query; k need not be specified in advance.

Any tree path $\mathcal{T}[a..b]$ can be divided into two overlapping paths $\mathcal{T}[a..z]$ and $\mathcal{T}[z..b]$, where z is the lowest common ancestor (LCA) of a and b. If we can answer tree path top-k color queries on these two paths in $O(k)$ time, then we can merge the answers to answer such queries on arbitrary paths in the tree in the same time. Therefore, it suffices to solve the following problem.

Problem 1. Preprocess $\mathcal{T}$ to efficiently answer tree path top-k color queries on paths of the form $\mathcal{T}[a..z]$ where z is an ancestor of a.

A tree path top-k query involves three constraints. Each element returned must be (1) on the query path; (2) among the top k; and (3) of a distinct color. In other words, each color must be reported, and counted towards the top k, only once, even if many elements of that color appear on the query path. We eliminate the duplicates using an adaptation of Muthukrishnan's chaining approach to reporting distinct colors in array ranges [17].

Let $depth(i)$ be the number of nodes on the path from the root to a node i. Let $chain(i)$ be the depth of the lowest ancestor of i that has the same color as i, with $chain(i) = 0$ if there is no such ancestor. If there exists at least one node with color c in $\mathcal{T}[a..z]$, z being an ancestor of a, then there exists a unique node i in $\mathcal{T}[a..z]$ with color c and $chain(i) < depth(z)$. Therefore, Problem 1 can be rephrased as follows:

Problem 2. Preprocess $\mathcal{T}$ to efficiently find, given a node a, one of its ancestors z, and a count k, the top k colors in decreasing order of priority among the nodes in the set $\{i \in \mathcal{T}[a..z] | chain(i) < depth(z)\}$.

For any node given by its preorder rank i, let $\phi(i)$ be the root of the path containing i in the heavy path decomposition of $\mathcal{T}$. Let ℓ_j be the preorder rank of the jth leftmost leaf in $\mathcal{T}$. We can transform $\mathcal{T}$ to another tree $\mathcal{T}'$, which is actually a path, by concatenating the paths $\mathcal{T}[\ell_i, \phi(\ell_i)]$ for each i up to the number of leaves. Then we can define an array $A[1..n]$ containing the priorities of the colors of nodes in $\mathcal{T}'$, in order along the path starting from the root. property is ensured. Any subpath of a path in the heavy path decomposition must correspond to a contiguous range of A.

We build the optimal array range top-k color query data structure of Karpinski and Nekrich on A, using $O(n \log \sigma)$ bits [15]. From each node i in $\mathcal{T}$, we store the index of the corresponding entry in A. The total space consumption is $O(n)$ words assuming $\sigma \leq n$. Because heavy paths are contiguous in A, the special case where both a and z are on the same heavy path can be handled optimally by first finding the corresponding range in A, and then performing a top-k color query on the array range top-k color query data structure.

For the case of a and z not on the same path of the decomposition, we map each node i in $\mathcal{T}$ to a weighted two-dimensional point (x_i, y_i) with weight w_i, letting w_i be the priority of the color of node i, x_i be the index in A corresponding to that node, and y_i be the number of paths of the heavy path decomposition intersected by the path from the root to $chain(i)$. We build the data structure [1] of Navarro and Nekrich for three-sided two-dimensional top-k queries on these weighted points [18, Theorem 2.1]. Because $y_i \leq \log n$, the query time to return k points from this data structure is $O(\log n + k)$.

To answer a general tree path top-k color query, we first find the lowest common ancestors of its endpoints and split the path into two queries of the form

---

[1] Note that an alternate approach is to build the data structure described in [15] over the list of colors corresponding to each heavy path, and later perform top-k queries on all heavy paths intersect with $\mathcal{T}[a..z]$. However, the drawback of this approach is that the same color may be reported from several heavy paths.

$\mathcal{T}[a..z]$, with z an ancestor of a. Then for each of the two, we partition the query path $\mathcal{T}[a..z]$ into at most $\log n$ disjoint subpaths $\mathcal{T}[a_1..\phi(a_1)]$, $\mathcal{T}[a_2..\phi(a_2)]$, ..., $\mathcal{T}[a_r..z]$, where $a_1 = a$, a_i is the parent of $\phi(a_{i-1})$ for $i = 2, 3, \ldots, r$, and $r \leq \log n$ is such that z is on the subpath $\mathcal{T}[a_r..\phi(a_r)]$. This step takes only $O(r) = O(\log n)$ time, by consulting a stored copy of the heavy path decomposition.

The query $\mathcal{T}[a_r..z]$ corresponds to a contiguous range in A, so we can find its top k colors in $O(k)$ time and merge them in $O(k)$ time with the top k colors for $\mathcal{T}[a_1..\phi(a_{r-1})]$. It only remains to query $\mathcal{T}[a_1..\phi(a_{r-1})]$ efficiently.

From the definition in Problem 2, we have that a node i can only be part of the result for $\mathcal{T}[a_1..\phi(a_{r-1})]$ if $chain(i) < depth(\phi(a_{r-1}))$. In fact, it suffices to check that the number of centroidal paths intersected by the path from the root of $\mathcal{T}$ to $chain(i)$ is less than the number intersected from the root to $\phi(a_{r-1})$. Since $\phi(a_{r-1})$ and its parent cannot be on the same path of the decomposition, any ancestor of it must be in a path nearer the root.

Let h be the number of paths of the heavy path decomposition that are intersected by the path from $\phi(a_{r-1})$ to the root, and let $A[s_i..e_i]$ be the range of A associated with $\mathcal{T}[a_i..\phi(a_i)]$. Then for each $i \in \{1, 2, \ldots, r-1\}$, the weighted points $(x_j, y_j) \in [s_i, e_i] \times [0, h]$ correspond to nodes with distinct colors on the path $\mathcal{T}[a_1..\phi(a_{r-1})]$. The union of those lists would include the top k colors in $\mathcal{T}[a_1..\phi(a_{r-1})]$, but we still must merge the lists.

We issue $r - 1$ simultaneous queries to the top-k geometric data structure, corresponding to the query regions $[s_i, e_i] \times [0, h]$ for $i = 1, 2, \ldots, r-1$. The answers can be merged using a max-heap H with its size limited to at most $r - 1 = O(\log n)$ points. We insert the first point returned from each R_i into H. Then we repeat the following steps until we have reported k colors:

1. Extract the highest weighted point in H and report it.
2. If the reported point was from the query box R_i, then fetch the next highest weighted point from R_i and insert it into H.

Since the size of H is always $\log^{O(1)} n$, we can use an atomic heap, which can perform all heap operations in constant time in the Word RAM model [7]. Therefore, the number of heap operations, and the required time, can be bounded by $O(k + \log n)$. Each three-sided two-dimensional top-k query takes $O(\log n)$ time in addition to the number of points it returns. Therefore the total time for query is $O(k + \log^2 n)$.

Lemma 2. *There exists a linear-space data structure for answering tree path top-k color queries in* $O(k + \log^2 n)$ *time.*

For $k = \Omega(\log^2 n)$, that time is optimal. To handle the case of small k, we use other techniques. First, we will choose a subset of the nodes in $\mathcal{T}$, called the *marked* nodes, as follows. Let g be an integer to be chosen later called the *grouping factor*, and mark every node i in $\mathcal{T}$ such that $i \equiv 0 \pmod{g}$. Also mark the lowest common ancestor of any pair of marked nodes. This is a simplified version of the scheme introduced by Hon et al. [14] for identifying marked nodes in a suffix tree. It has the properties given in the following lemma (the proof is deferred to the full version).

Lemma 3. *If we mark all nodes i such that $i \equiv 0 \pmod{g}$, and all lowest common ancestors of pairs of such nodes, then:*

1. *the number of marked nodes is $O(n/g)$;*
2. *the lowest marked ancestor of any node is $O(g)$ nodes above it;*
3. *any* unmarked *node has at most one unique highest marked descendant; and*
4. *for any unmarked node i, the subtree of $\mathcal{T}$ rooted at i and excluding the subtree rooted at its highest marked descendant j (if any) contains $O(g)$ nodes.*

We mark nodes in $\mathcal{T}$ using $g = \log^3 n$. For every marked node i and every j such that j is an ancestor of i and is the root of a path in the heavy path decomposition, we store explicitly a precomputed sorted list of the top $O(\log^2 n)$ colors on the path $\mathcal{T}[i..j]$. The space is bounded by $O((n/\log^3 n) \log^2 n \log n) = O(n)$ words. Using these precomputed lists we can find the top k colors in $\mathcal{T}[a..z]$, where a is a marked node and z is one of its ancestors, by splitting the query, as before, into $\mathcal{T}[a_1..\phi(a_{r-1})]$ and $\mathcal{T}[a_r..z]$. The former is precomputed and the latter corresponds to a contiguous range of A. We can find the top k colors in both of them and merge the lists in $O(k)$ time.

Lemma 4. *There exists a linear-space data structure for answering tree path top-k color queries of the form $\mathcal{T}[a..z]$ in $O(k)$ time, where a is a marked node and z is one of its ancestors.*

Next we must handle the case in which a is not a marked node. For any node i such that i is not marked but its parent is marked, define the *mini-tree* $\mathcal{T}^i$ to be the subtree rooted at i but excluding any descendants of the highest marked descendant of i, if any. By Lemma 3, $\mathcal{T}^i$ is of size $O(g)$. We choose a grouping factor $g' = \log^3 g = \Theta(\log^3 \log n)$ and use it to mark nodes within each mini-tree, and build the data structure of Lemma 4 for each mini-tree. The total space is bounded by $O(n)$ because each node in $\mathcal{T}$ belongs to exactly one mini-tree. By querying from a to the root of its mini-tree, and from the parent of that root to z, and merging the results, we can answer top-k color queries of the form $\mathcal{T}[a..z]$ in $O(k)$ time when a is marked in its mini-tree and z is its ancestor, even if a is not marked in T.

Finally, we generalize the optimal-time solution to arbitrary a. For any i not marked in $\mathcal{T}$ nor in the mini-tree that contains i, let j be the lowest marked node above i in the same mini-tree. The node j is at most $g' = O(\log^3 \log n)$ nodes above i. Therefore the top k colors on the path $\mathcal{T}[i..j]$, for all choices of i, can be stored in $O(n(\log^3 \log n)^2 \log \log \log n)$ bits: there are n choices of i, $O(\log^3 \log n)$ lists, each of length at most $O(\log^3 \log n)$, and only $O(\log \log \log n)$ bits are needed (as indices into the mini-block) to store the entries in the lists. That is a total of $o(n)$ words.

Then to answer an arbitrary query $\mathcal{T}[a..b]$, we first split into two queries $\mathcal{T}[a..z]$ and $\mathcal{T}[b..z]$ with z the lowest common ancestor of a and b. We can find the top k colors from a to its lowest marked ancestor within its mini-tree, and from there to the root of the mini-tree, using the precomputed lists. From the lowest marked ancestor of a in $\mathcal{T}$ to the highest marked ancestor of a below z, we use Lemma 4; and from there to z we do an array range query in A. We do the same with the query $\mathcal{T}[b..z]$. All these queries can be performed, and their results merged, in $O(k)$ time. This completes the proof of Theorem 1.

References

1. Belazzougui, D., Gagie, T., Navarro, G.: Better space bounds for parameterized range majority and minority. In: Dehne, F., Solis-Oba, R., Sack, J.-R. (eds.) WADS 2013. LNCS, vol. 8037, pp. 121–132. Springer, Heidelberg (2013)
2. Belazzougui, D., Navarro, G., Valenzuela, D.: Improved compressed indexes for full-text document retrieval. J. Discrete Algorithms 18, 3–13 (2013)
3. Chan, T.M., Durocher, S., Larsen, K.G., Morrison, J., Wilkinson, B.T.: Linear-space data structures for range mode query in arrays. In: Proc. STACS, vol. 14, pp. 291–301 (2012)
4. Chan, T.M., Durocher, S., Skala, M., Wilkinson, B.T.: Linear-space data structures for range minority query in arrays. In: Fomin, F.V., Kaski, P. (eds.) SWAT 2012. LNCS, vol. 7357, pp. 295–306. Springer, Heidelberg (2012)
5. Durocher, S., He, M., Munro, J.I., Nicholson, P.K., Skala, M.: Range majority in constant time and linear space. Inf. & Comp. 222, 169–179 (2013)
6. Durocher, S., Shah, R., Skala, M., Thankachan, S.V.: Linear-space data structures for range frequency queries on arrays and trees. In: Chatterjee, K., Sgall, J. (eds.) MFCS 2013. LNCS, vol. 8087, pp. 325–336. Springer, Heidelberg (2013)
7. Fredman, M.L., Willard, D.E.: Trans-dichotomous algorithms for minimum spanning trees and shortest paths. J. Comput. Syst. Sci. 48(3), 533–551 (1994)
8. Gagie, T., He, M., Munro, J.I., Nicholson, P.K.: Finding frequent elements in compressed 2D arrays and strings. In: Grossi, R., Sebastiani, F., Silvestri, F. (eds.) SPIRE 2011. LNCS, vol. 7024, pp. 295–300. Springer, Heidelberg (2011)
9. Gagie, T., Kärkkäinen, J., Navarro, G., Puglisi, S.J.: Colored range queries and document retrieval. Theor. Comput. Sci. 483, 36–50 (2013)
10. Gagie, T., Puglisi, S.J., Turpin, A.: Range quantile queries: Another virtue of wavelet trees. In: Karlgren, J., Tarhio, J., Hyyrö, H. (eds.) SPIRE 2009. LNCS, vol. 5721, pp. 1–6. Springer, Heidelberg (2009)
11. Gfeller, B., Sanders, P.: Towards optimal range medians. In: Albers, S., Marchetti-Spaccamela, A., Matias, Y., Nikoletseas, S., Thomas, W. (eds.) ICALP 2009, Part I. LNCS, vol. 5555, pp. 475–486. Springer, Heidelberg (2009)
12. He, M., Munro, J.I., Zhou, G.: Path queries in weighted trees. In: Asano, T., Nakano, S.-I., Okamoto, Y., Watanabe, O. (eds.) ISAAC 2011. LNCS, vol. 7074, pp. 140–149. Springer, Heidelberg (2011)
13. He, M., Munro, J.I., Zhou, G.: Succinct data structures for path queries. In: Epstein, L., Ferragina, P. (eds.) ESA 2012. LNCS, vol. 7501, pp. 575–586. Springer, Heidelberg (2012)
14. Hon, W.-K., Shah, R., Vitter, J.S.: Space-efficient framework for top-k string retrieval problems. In: Proc. FOCS, pp. 713–722 (2009)
15. Karpinski, M., Nekrich, Y.: Top-k color queries for document retrieval. In: Proc. SODA, pp. 401–411 (2011)
16. Krizanc, D., Morin, P., Smid, M.: Range mode and range median queries on lists and trees. In: Ibaraki, T., Katoh, N., Ono, H. (eds.) ISAAC 2003. LNCS, vol. 2906, pp. 517–526. Springer, Heidelberg (2003)

17. Muthukrishnan, S.: Efficient algorithms for document retrieval problems. In: Proc. SODA, pp. 657–666 (2002)
18. Navarro, G., Nekrich, Y.: Top-k document retrieval in optimal time and linear space. In: Proc. SODA, pp. 1066–1077 (2012)
19. Patil, M., Shah, R., Thankachan, S.V.: Succinct representations of weighted trees supporting path queries. J. Discrete Algorithms 17, 103–108 (2012)
20. Sleator, D.D., Tarjan, R.E.: A data structure for dynamic trees. J. Comput. Syst. Sci. 26(3), 362–391 (1983)

A Lempel-Ziv Compressed Structure for Document Listing*

Héctor Ferrada and Gonzalo Navarro

Department of Computer Science, University of Chile
{hferrada,gnavarro}@dcc.uchile.cl

Abstract. Document listing is the problem of preprocessing a set of sequences, called documents, so that later, given a short string called the pattern, we retrieve the documents where the pattern appears. While optimal-time and linear-space solutions exist, the current emphasis is in reducing the space requirements. Current document listing solutions build on compressed suffix arrays. This paper is the first attempt to solve the problem using a Lempel-Ziv compressed index of the text collections. We show that the resulting solution is very fast to output most of the resulting documents, taking more time for the final ones. This makes this index particularly useful for interactive scenarios or when listing some documents is sufficient. Yet, it also offers a competitive space/time tradeoff when returning the full answers.

1 Introduction

The classical Information Retrieval (IR) problems aimed at natural language text collections can be naturally generalized to general sequence collections. Such *general document retrieval* problems are of interest in various areas like bioinformatics, multimedia databases, software repositories, and so on [17]. Moreover, IR on Oriental languages like Chinese and Korean also regards the texts as general sequences, since inverted indexes do not handle well those languages.

In this paper we focus on the simplest document retrieval problem, called *document listing*. Given D *documents*, which are strings $d_1 \ldots d_D$ over an alphabet of size σ, each terminated with a special symbol $, we preprocess them to build an *index*. Later, given a pattern $p[1, m]$ over the same alphabet, we must list the ndoc documents where p appears.

Muthukrishnan [15] solved this problem in optimal time $O(m + \text{ndoc})$, using an index of $O(n)$ words of space, where $n = \sum |d_i|$ is the total length of the documents. This space usage, albeit linear, is very large in practice. Much subsequent research focused on reducing the space requirements. One research line [23,8,7,19] achieved about $O(m + \text{ndoc} \lg D)$ time and $|\text{CSA}| + n \lg D + O(n)$ bits of space, where CSA is a compressed suffix array [18] of T. The CSA has a space close to that of the compressed text and can replace it. They achieve in practice fast document listing, but the extra space $n \lg D$ is still considerable. A second

---

* Partially funded by Fondecyt grant 1-110066, Chile.

O. Kurland, M. Lewenstein, and E. Porat (Eds.): SPIRE 2013, LNCS 8214, pp. 116–128, 2013.

research line [22,11] reduced the space to $|\mathsf{CSA}| + o(n)$ bits, but with the higher listing time $O(m + \mathsf{ndoc}\, \lg^{1+\varepsilon} n)$.

In this paper we propose a novel alternative, which obtains low time and low extra space. We build on the idea of the LZ-index [16,1] so as to produce a document listing index that is small thanks to LZ78 compression, whereas it can list the documents fast. While the theoretical upper bounds we can prove, $5|\mathsf{LZ78}| + O(n \lg^2 \sigma / \lg n)$ bits (where $|\mathsf{LZ78}| \approx |\mathsf{CSA}|$ is the size of the LZ78-compressed text) and $O(m^2 \lg n + \mathsf{ndoc}\, m \lg^2 n)$ time, are not too good, they are overly pessimistic. Indeed, a good part of the occurrences are listed in $O(1)$ time each. We show that the index is very fast to list those first occurrences (which usually form most of the output), becoming slower to output the final ones. This makes it ideal for interactive scenarios, where one wishes to show some results to the user as fast as possible, and there is much more time to produce further results while the user browses the first ones. Another scenario is when only a partial or approximate answer is sufficient, that is, when one simply wants to find several documents where the pattern appears. However, the index also offers a very competitive space/time combination when returning the full set of answers.

2   Related Work

Muthukrishnan [15] solved the document listing problem in optimal time $O(m + \mathsf{ndoc})$, using an index of $O(n)$ words of space. Let $T[1, n]$ be the concatenation of the D documents. Let $A[1, n]$ be the suffix array [13] of T. Muthukrishnan defined the so-called *document array* $E[1, n]$, where $E[i]$ is the identifier of the document containing the suffix $A[i]$. A new array $C[1, n]$ is defined over E as $C[i] = \max\{1 \le k < i, E[k] = E[i]\} \cup \{0\}$, that is, the position of the previous occurrence of $E[i]$ in E, or 0 if there is no previous occurrence. Array C is then preprocessed for range minimum queries (RMQs), which are of the form $\mathrm{RMQ}_C(i, j) = \mathrm{argmin}_{i \le k \le j} C[k]$, that is, it gives the position of the minimum value in $C[i, j]$. RMQs can be solved in constant time after a linear-time preprocessing (see, e.g., [6]). Once the interval $A[sp, ep]$ of the suffixes starting with p is determined, the problem becomes that of listing the distinct values in the interval $E[sp, ep]$. The interval is found in time $O(m)$ using a suffix tree [24]. The ndoc distinct values are listed in time $O(\mathsf{ndoc})$ using the observation that the first occurrence $E[k]$ of each distinct value in $E[sp, ep]$ satisfies $C[k] < sp$. Then the process recursively finds the smallest values of $C[sp, ep]$: It first computes $k = \mathrm{RMQ}_C(sp, ep)$ and reports $E[k]$, then it continues recursively with $C[sp, k-1]$ and $C[k + 1, ep]$. The recursion stops at any branch where $C[k] \ge sp$.

While this solution is time-optimal, it requires much space, $O(n \lg n)$ bits. Subsequent work has focused on reducing the space, giving away the optimality.

Välimäki and Mäkinen [23] proposed a low-space implementation of Muthukrishnan's structure. They used a $2n + o(n)$ bit, constant time RMQ succinct index [6] that still required access to C. They showed that access to C can be implemented by *rank* and *select* queries on E, where $\mathsf{rank}_c(E, i)$ is the number of occurrences of symbol c in $E[1, i]$ and $\mathsf{select}_c(E, j)$ is the position in E of the

jth occurrence of c. Then it holds $C[i] = \mathsf{select}_{E[i]}(E, \mathsf{rank}_{E[i]}(E, i-1))$ if we assume that $\mathsf{select}_c(E, 0) = 0$. By representing E with a multiary wavelet tree [5,9], the space is $n \lg D + o(n)$ bits and the operations are carried out in time $O(1 + \lg D / \lg \lg n)$. Finally, the suffix tree is replaced by a compressed suffix array (CSA), of which there are many choices [18]. A recent one [3] requires $|\mathsf{CSA}| = nH_k(T) + o(nH_k(T)) + O(n)$ bits of space and finds the interval $[sp, ep]$ in time $t_{\mathsf{tsearch}}(m) = O(m)$. A slightly smaller one [2] reaches $|\mathsf{CSA}| = nH_k(T) + o(nH_k(T)) + o(n)$ bits and $t_{\mathsf{tsearch}} = O(m \lg \lg \sigma)$. Here $H_k(T)$ is the empirical kth order entropy of T [12]. Overall, their solution requires $|\mathsf{CSA}| + n \lg D + O(n)$ bits and solves the problem in time $O(t_{\mathsf{tsearch}}(m) + \mathsf{ndoc}(1 + \lg D / \lg \lg n))$.

Gagie et al. [8,7] showed that a wavelet tree [10] can be used for document listing without any need of RMQs, but just a DFS traversal. Their index can use $|\mathsf{CSA}| + n \lg D + o(n)$ bits and their document listing time is $O(t_{\mathsf{tsearch}}(m) + \mathsf{ndoc} \lg(D/\mathsf{ndoc}))$. Navarro et al. [19] achieved nearly 50% compression of the wavelet tree in practice, at the price of nearly doubling the time required (these wavelet-tree based indices also solve more complex queries).

Sadakane [22] initiated another line based on the idea of Muthukrishnan, but avoiding the large $n \lg D$-bit term in the space. He replaced the RMQ solution by a constant-time one that does not need access to C. His structure needed $4n + o(n)$ bits, but more recent ones [6] require $2n + o(n)$ bits. The other use for C is to determine where to stop the recursion. Sadakane used instead a bitmap $V[1, D]$ where the already reported documents are marked. Once a branch of the recursion attempts to report a marked document, it is pruned. Finally, array E is only needed to list the document identifiers. This is done with a bit vector $B[1, n]$ that marks the positions in T where the documents start; then it holds $E[k] = \mathsf{rank}_1(B, A[k])$. Value $A[k]$ is computed by the CSA, for example in time $O(\lg^{1+\varepsilon} n)$ for any constant $\varepsilon > 0$ [2,3]. Bitmap B can be represented in $D \lg(n/D) + O(D) + o(n)$ bits with rank queries supported in constant time [20]. Overall, the data structure requires only $|\mathsf{CSA}| + 2n + D \lg(n/D) + O(D) + o(n) = |\mathsf{CSA}| + O(n)$ bits and $O(t_{\mathsf{tsearch}}(m) + \mathsf{ndoc} \lg^{1+\varepsilon} n)$ time. Hon et al. [11] achieved a further reduction to $|\mathsf{CSA}| + D \lg(n/D) + O(D) + o(n)$ bits, within the same asymptotic time, by running the RMQs over blocks of $\lg^\varepsilon n$ cells (see Navarro [17] for comments on the correctness of this solution).

As it can be seen, all the approaches build on the suffix array. Our new approach uses instead the LZ-index, a compressed text index not based on suffix arrays but on the LZ78 compression [25] of the text.

3   The LZ-Index

The algorithm LZ78 builds a dictionary of phrases (text substrings), with the aim of replacing strings by pointers to their previous occurrences in the text. The dictionary grows as the text is processed, and the result is a sequence of n' distinct phrases ($n' \leq n / \lg_\sigma n$). The phrases are formed by scanning the text left to right. In each step, the method finds the longest prefix of the remaining text that is a phrase of the dictionary. It then creates a new phrase formed by the phrase found plus the symbol following it in the remaining text. This is represented by a pointer to the dictionary and the extra character. The number of bits output by the compressor is $|\mathsf{LZ78}| = n'(\lg n + \lg \sigma) \leq n\,H_k(T) + o(n \lg \sigma)$ for $k = o(\lg_\sigma n)$ [12]. The LZ-index [16] is a compressed text index built on the LZ78 parsing of the text, and it supports locating the occurrences of a pattern $p[1, m]$ in T. The index is formed by the following components.

1. **LZTrie**: a trie composed of all the phrases produced by the LZ78 parsing. Note that the set of phrases is prefix-closed (the prefix of a phrase is also a phrase), so LZTrie has n' nodes. It stores the phrase identifiers of each node.
2. **RevTrie**: a trie storing the reversed phrases. It is not prefix-closed, so there are *empty* nodes not associated to phrases. We collapse unary paths of empty nodes. The trie has $n_{rev} = n' + n_e \leq 2n'$ nodes, where n_e empty nodes remain after collapsing. The phrase numbers of the n' nonempty nodes are stored.
3. **Node**: an array mapping from phrase numbers to their preorder in LZTrie.
4. **Range**: an $n' \times n'$ grid where the rows represent the phrases and the columns the reverse phrases, both in lexicographic order. If the $(k+1)$th text phrase is at row i and the kth at column j, then there is a point at (i, j) in the grid.

Thus the LZ-index uses $4|\mathsf{LZ78}|(1 + o(1))$ bits of space. To search for the occurrences of pattern $p[1, m]$ we divide them into three classes: (1) those completely inside a phrase, (2) those spanning two phrases, (3) those spanning 3 phrases or more. Those are found separately.

- **Type 1**. Search for p^r (the reversed pattern) in RevTrie, arriving at node v^r. Each node u^r descending from v^r (including v^r) corresponds to an occurrence of type 1 where p appears at the end of the phrase. The other occurrences of type 1 are the nodes u' that descend from u in LZTrie, where u corresponds to u^r. Thus, for each node u^r that is nonempty, we read the phrase id f_u of u^r, compute $u = Node(f_u)$, and report all the phrase ids in the subtree of u. This takes $O(m + occ_1)$ time, reporting the occ_1 occurrences of type 1. See Fig. 1, ignoring for now Doc_{lz}, Doc_{rev}, and $LDoc_{rev}$.
- **Type 2**. Partition $p = p_{start} \cdot p_{end}$ in the $m-1$ possible ways, searching for p^r_{start} in RevTrie and for p_{end} in LZTrie. The subtrees found define column and row ranges in the grid *Range*, and each point in the range is a type 2 occurrence. The phrase identifiers are obtained from those stored in LZTrie using the rows of the reported points. Using a linear-space geometric data structure, the total time is $O(m^2)$ for the m searches in LZTrie and RevTrie, $O(m \lg n)$ for the m range searches, and $O(occ_2 \lg n)$ for reporting the occ_2 points found.
- **Type 3**. Since phrases are unique, each $p[i, j]$ equal to a phrase leads to at most one occurrence of type 3. We search LZTrie incrementally for the $O(m^2)$ pattern substrings $p[i, j]$ and find their phrase ids, if any. Then we find concatenations of consecutive phrases that together form a maximal substring $p[i, j] = b_k \ldots b_l$. Finally, we check if the phrases $b_k - 1$ and $b_l + 1$ are equal to the strings $p[1, i-1]$ and $p[j+1, m]$, respectively. For the second we check that the subtree of phrase $p[j+1, m]$ in LZTrie contains $Node(l+1)$.

For the first we check if the column range of the node for $p[1, i-1]^r$ in RevTrie has a point at row $Node(k)$, corresponding to LZTrie (the m searches in RevTrie are computed once). Thus these occurrences require $O(m^2 \lg n)$ time [1].

The total search time for the occ occurrences is $O(m^2 \lg n + occ \lg n)$.

Wavelet Trees. The geometric data structure we use in practice is a wavelet tree [10]. It is a pefect binary tree where the points are sorted in row order at the root and in column order in the bottom. The coordinates are not explicitly stored. At the root, a bitmap marks with a 0 or a 1 whether each point belongs to the left or right half of the grid, respectively. Those on the left/right side of the grid are then recursively subdivided at the left/right child of the root node. The wavelet tree uses in total $n' \lg n'$ bits.

To support range searches, the bitmaps are enhanced with rank/select data structures. Both can be computed in constant time and $o(n')$ extra bits [14]. To find the points in a range $[i, i'] \times [j, j']$ (rows $\times$ columns), we start with $B[i, i']$ in the root bitmap, and project the interval to the left/right children, towards the new interval $[rank_{0/1}(B, i-1) + 1, rank_{0/1}(B, i')]$. We continue splitting the interval, stopping when it becomes empty, or the wavelet tree node has no intersection with the columns $[j, j']$, or it is fully included in $[j, j']$. In the last case, all the values in the current bitmap interval are points in the range. They can be counted directly, or reported one by one by tracking them to the leaves, to know their column values, for example. As any range is decomposed into $O(\lg n')$ wavelet tree nodes that have in total $O(\lg n')$ ancestors, counting the points in the range takes $O(\lg n')$ time and reporting each of them requires $O(\lg n')$ time.

4   A Novel LZ-Index Based Document Listing Structure

We now adapt the LZ-index to carry out document listing instead of reporting all the occurrences of a pattern p. The general search strategy will be as follows. For occurrences of type 1, we store the RMQ of the expansion of RevTrie with the subtree of LZTrie that corresponds to each node. This requires $O(n)$ bits and allows us to apply Muthukrishnan's algorithm [22] directly. For type 2, we enhance the bitmaps of the wavelet tree of *Range* with RMQ data structures for their documents. We can then apply Muthukrishnan's algorithm on any of the $O(\lg n')$ nodes into which the range is decomposed. For occurrences of type 3 we find their documents one by one. The total time will be $O(m^2 \lg n + \mathsf{ndoc}\, m \lg^2 n)$.

Structure. We modify the LZ78 parsing so that no phrase crosses a document boundary. Now consider the LZTrie and RevTrie structures of the original LZ-index resulting from this parsing. We store the following structures.

- ***RevTrie.*** We represent only the topology and the letters of RevTrie and LZTrie, just in order to be able to navigate RevTrie and to search it for patterns in constant time per symbol [1]. The structure requires $3n' \lg \sigma + O(n')$ bits. (In the implementation we do not represent LZTrie, but all the nodes of RevTrie, which in the worst case can be n but in practice are not.)

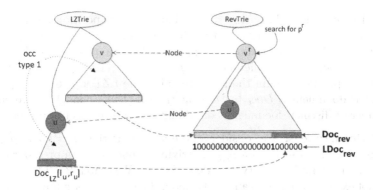

Fig. 1. Our structures for occurrences of type 1

- **Doc.** Let us define Doc_{lz}, the sequence of n' document identifiers of the LZTrie phrases in preorder. We save Doc_{lz} explicitly with $n'\lceil \lg D \rceil$ bits. This is equivalent to the document array [15], but restricted to phrases. Now we define Doc_{rev}, a sequence of n document identifiers built as follows. We traverse RevTrie in preorder, and for each phrase node v^r, let v be the corresponding LZTrie node. Let $Doc_{lz}[l_v, r_v]$ be the range in Doc_{lz} of all the descendants of v (included). We append $Doc_{lz}[l_v, r_v]$ to Doc_{rev}. Doc_{rev} will not be stored, but just its RMQ structure, so as to run Muthukrishnan's algorithm [15] over Doc_{rev}. This RMQ structure answers queries in $O(1)$ time without accessing Doc_{rev} and uses $2n + o(n)$ bits [6].[1] Finally we store a bitmap $LDoc_{rev}[1, n]$, which marks the Doc_{rev} positions where the intervals $Doc_{lz}[l_v, r_v]$ start. In total we store $n'\lceil \lg D \rceil + 3n + o(n)$ bits. Fig. 1 illustrates.
- **Node.** Now this is a mapping from RevTrie to LZTrie. If the node v^r in RevTrie with nonempty preorder i corresponds to the node v in LZTrie with preorder j, then $Node[i] = j$. Array $Node$ uses $n'\lceil \lg n' \rceil$ bits.
- **Range.** An enhanced binary wavelet tree. Each wavelet tree node implicitly represents a sequence of points. Now consider the array of their corresponding documents. We store, in addition to the bitmap, the RMQ structure corresponding to Muthukrishnan's algorithm [15] on its (virtual) array of documents. The total space of $Range$ is then $3n' \lg n' + o(n' \lg n')$ bits.

Overall, our structure requires $4n' \lg n' + n' \lg D + 3n' \lg \sigma + o(n' \lg n') + 3n + o(n) \leq 5nH_k(T) + 3n + o(n \lg \sigma)$ bits (and $\leq 4nH_k(T) + 3n + o(n \lg \sigma)$ if $\lg D = o(\lg n)$). This is close to the original LZ-index size [16]. We describe the document listing procedure now.

Type 1 Occurrences. We search for $p[1, m]^r$ in RevTrie, arriving at node v with preorder j_v. We find the interval $I = Doc_{rev}[s_v, e_v]$ containing all the occurrences of type 1, where $s_v = select_1(LDoc_{rev}, j_v)$ and $e_v = select_1(LDoc_{rev}, j_v + subtree\text{-}size(j_v)) - 1$. Next, we report all the distinct documents in I with

---

[1] The length is n because n is the internal path length (sum of all node depths) in LZTrie. Each LZTrie node is appended to Doc_{rev} once per ancestor it has in LZTrie.

Muthukrishnan's algorithm using RMQs. For each new position $Doc_{rev}[pos]$ reported by an RMQ, we determine the nonempty preorder $j = rank_1(LDoc_{rev}, pos)$ of the RevTrie node holding that position, and the preorder of this node in LZTrie, $i = Node[j]$. The difference $d = pos - select_1(LDoc_{rev}, j)$ provides the offset of this position within the leaf interval of the LZTrie node with preorder i. Thus, the document is $Doc_{lz}[i + d]$. The time is $O(m + ndoc_1)$, where $ndoc_1$ is the number of distinct documents containing at least one occurrence of type 1.

Type 2 Occurrences. We consider all the $m - 1$ partitions $p = p_{start} \cdot p_{end}$. For each one, we search RevTrie for p_{start}^r, arriving at node v^r with preorder interval $[j, j']$. To find the LZTrie interval we do as follows. We search RevTrie for p_{end}^r. If it does not exist, or it leads to an empty node, then p_{end} is not a phrase and there are no phrases starting with p_{end} (as phrases are built incrementally letter by letter). If instead we reach node u^r, with nonempty preorder t, then $i = Node[t]$ is the LZTrie preorder of the corresponding node u, which represents p_{end}. It is also the left end of the preorder interval of the descendants of u. We compute the size of the interval using $LDoc_{rev}$: $\ell = select_1(LDoc_{rev}, t + 1) - select_1(LDoc_{rev}, t)$, then $i' = i + \ell - 1$ and the row interval for the search in $Range$ is $[i, i']$.

Now we identify in $Range$ the $O(\lg n')$ wavelet tree nodes that cover the interval $[j, j']$, and the ranges where interval $[i, i']$ is projected on their bitmaps. Each of these $O(\lg n')$ intervals represent documents with occurrences of type 2, and we list the documents in each by running Muthukrishnan's algorithm over the RMQ structures that enhance the bitmaps. For each document, which is found in $O(1)$ time, we need $O(\lg n')$ time to reach the corresponding leaf and find its identifier in Doc_{lz}. Although unlikely, in the worst case we can output the same document in each of the $O(\lg n')$ intervals for each of the $m - 1$ partitions, which gives $O(m^2)$ time for the RevTrie searches plus a (very pessimistic) worst-case bound of $O(ndoc_2 \, m \lg^2 n')$ time for the $ndoc_2$ occurrences of type 2.

Type 3 Occurrences. We wish to apply the same algorithm of the original LZindex and then output the documents, yet we have fewer data structures now. First, all the searches for all the substrings $p[i, j]$ are carried out in RevTrie, in time $O(m^2)$, and we record the RevTrie and LZTrie preorder values of each (the latter using $Node$ from the RevTrie node). For each i, we store in array A_i the information for the substrings of the form $p[i, j]$, sorted by LZTrie preorder value. Now note that we have not stored phrase numbers, yet we can still use $Range$ to determine the LZTrie preorder t of the phrase following that of $p[i, j]$, which has RevTrie preorder t^r. If we traverse the wavelet tree of $Range$ starting at position t^r in the root bitmap and track it to the leaves, the final position is precisely t. This operation takes $O(\lg n')$ time. Now we can binary search A_{j+1} for LZTrie preorder t, and if we find it corresponding to a phrase $p[j + 1, j']$, we can concatenate $p[i, j]$ to get $p[i, j']$. Therefore we can carry out the same process for finding maximal concatenations [16], in total time $O(m^2 \lg n)$.

Finally, we have to check if $p[1..i - 1]$ precedes the maximal concatenation and if $p[j + 1, m]$ follows it. The first question is equivalent to computing whether the preorder interval for $p[1..i - 1]^r$ in RevTrie is connected with the LZTrie preorder value t of the first phrase in the maximal concatenation. The second question corresponds to computing the LZTrie preorder interval of $p[j+1, m]$ (which can be done using RevTrie, as before) and then asking if the RevTrie preorder value t^r of the last phrase in the maximal concatenation is connected with some point in the LZTrie interval. These tests add up $O(m \lg n)$ time.

This adds up to the promised total time of $O(m^2 \lg n + \mathsf{ndoc}\, m \lg^2 n)$. Note, however, that the occurrences of type 1 are reported very early, in time $O(m + \mathsf{ndoc}_1)$. If the text is generated by an ergodic source, the occurrences of any pattern p appear regularly, every d positions on average (e.g., $d = \sigma^m$ if the symbols are generated uniformly and independently). On the other hand, since $n' \leq n/\lg_\sigma n$, only $O((n/d)m/\lg_\sigma n)$ of those occurrences hit a phrase boundary on average. This means that that a fraction of $1 - O(m/\lg_\sigma n)$ of the occurrences are of type 1, and also $\mathsf{ndoc}_2 = O(\mathsf{ndoc}\, m/\lg_\sigma n) = o(\mathsf{ndoc})$ if $m = o(\lg_\sigma n)$. Thus we report almost all of the occurrences in $O(1)$ time each. If we just lose those $o(\mathsf{ndoc})$ occurrences not of type 1, our time is the optimal $O(m + \mathsf{ndoc})$! We show in the next section that, indeed, our index is particularly competitive to show the first occurrences (those of type 1), which are the most for short patterns.

5   Experimental Results

We consider the following document collections, following previous work [19].

- ClueChin: A 2.3 MB sample of ClueWeb09 (boston.lti.cs.cmu.edu/Data/clueweb09), formed by 23 Web pages in Chinese.
- ClueWiki: A 141 MB sample of ClueWeb09, formed by 3,334 Web pages from the English Wikipedia (same source as the previous).
- KGS: A 75 MB collection of 18,838 sgf-formatted Go game records from year 2009 (www.u-go.net/gamerecords).
- Proteins: A 60 MB collection formed by 143,244 sequences of Human and Mouse proteins (www.ebi.ac.uk/swissprot).

Our machine is an Intel Xeon with 8 processors of 2.4GHz and 12MB cache, with 96GB RAM. It runs Linux 2.6.32-46-server, and we use gcc with full optimization. We choose 40,000 patterns of lengths $m = 3$ and $m = 8$ extracted randomly from the collection.

Table 1 gives the space obtained by our LZ-Index structure on those collections. ClueWiki and KGS are the most compressible ones, reaching 11–12 bpc, whereas ClueChin and Proteins are the least compressible ones. All are, as roughly expected from the space analysis, 4.3–5.3×|LZ78|. We show how |LZ78| relates to n/n', and how it roughly coincides with the output size of *Compress*, a classical LZW Unix compressor.

In the more compressible collections, *RevTrie* uses less than 20% of the space, *Doc* uses slightly more than 30%, *Node* slightly more than 10%, and *Range* uses almost 40%. The distribution varies a bit on the less compressible collections, where the fraction of *Node* and *Range* increases, reaching 50%. Note that component *Range* can be omitted if we only want to list the occurrences of type 1, in which case the index size is reduced by 40%–50%.

Table 1. Space breakdown of the main components of our LZ-Index based structure. The numbers are in bpc. Main components are in bold and their space is the sum of the second-level components (bpc in italics). The percentages are w.r.t. the total LZ-Index size, whose line indicates its ratio over $|LZ78|$. The $|LZ78|$ line, in turn, gives also (n/n'). The last line gives the bpc of a real LZ78-like compression program.

Component	ClueChin		ClueWiki		KGS		Proteins	
RevTrie	2.429	(15%)	1.725	(16%)	2.091	(18%)	2.154	(9%)
topology	*0.396*		*0.182*		*0.247*		*0.461*	
labels	*1.793*		*1.396*		*1.613*		*1.530*	
empty nodes	*0.240*		*0.147*		*0.231*		*0.163*	
Doc	3.594	(22%)	3.529	(33%)	3.864	(33%)	5.777	(25%)
Doc_{lz}	*0.638*		*0.696*		*1.002*		*2.788*	
Doc_{rev} RMQ	*2.331*		*2.336*		*2.360*		*2.348*	
$LDoc_{rev}$	*0.625*		*0.497*		*0.502*		*0.641*	
Node	2.424	(15%)	1.335	(12%)	1.403	(12%)	3.717	(16%)
Range	7.938	(48%)	4.279	(39%)	4.423	(37%)	11.748	(50%)
Total *LZ-Index* ($/|LZ78|$)	16.386	(4.27×)	10.870	(5.21×)	11.831	(5.35×)	23.550	(4.90×)
$|LZ78|$ (avg. phrase length)	3.840	(7.81)	2.088	(17.24)	2.211	(14.93)	4.805	(6.45)
Compress	2.927		2.733		1.851		4.610	

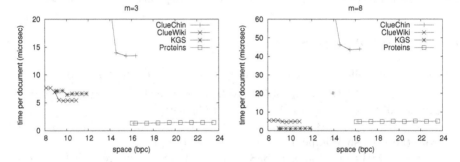

Fig. 2. Space versus listing time per document output. The tradeoff is obtained by not representing RMQ information on the last levels of the wavelet tree of *Range*.

A way to reduce the space without sacrificing functionality is to remove the RMQ structures at the last levels of the wavelet tree of *Range*. In those levels we simply obtain all the documents one by one. The query covers at most two ranges per level, those at the last levels are the smallest, and they are closest to the bottom, so obtaining the document identifiers is faster. Thus, removing those structures should not impact much the time. Fig. 2 confirms that the time is practically unaffected when the first levels are removed, while the space improves noticeably. From now on we will remove the RMQ structures on the last 6 levels of ClueChin, 12 levels of ClueWiki and KGS, and all the levels in Proteins.

Table 2 shows the number of documents listed by the queries. In these relatively small collections we list a good percentage of the documents, with the exception of Proteins, which has many more documents and then a document listing query is selective enough. From the listed documents, many are obtained as type 1 occurrences (75%–100% for $m = 3$ and 50%–95% for $m = 8$). This shows that we could obtain a significant part of the result using just the fastest listing and without representing *Range*.

Table 2. Number of occurrences of each type, for pattern lengths $m = 3$ and $m = 8$. Global percentages are w.r.t. the total number of documents, whereas local percentages (in italics) are w.r.t. the total number of occurrences found.

Occurrences	ClueChin		ClueWiki		KGS		Proteins	
$m = 3$	14.20	(62%)	2,732.41	(82%)	15,799.10	(84%)	12,106.90	(8%)
Type 1	13.60	(96%)	2,727.93	(99%)	15,132.60	(96%)	9,185.01	(76%)
Type 2	0.598	(4%)	25.06	(1%)	667.40	(4%)	2,921.90	(24%)
Type 3	0.002	(0%)	0.001	(0%)	0.022	(0%)	0.015	(0%)
$m = 8$	6.52	(28%)	1,742.52	(52%)	4,285.02	(23%)	89.45	(0%)
Type 1	5.02	(77%)	1,646.97	(95%)	2,943.00	(69%)	46.27	(52%)
Type 2	1.28	(20%)	94.79	(5%)	1,338.74	(31%)	42.49	(48%)
Type 3	0.208	(3%)	0.724	(0%)	3.29	(0%)	0.981	(0%)

Fig. 3 compares our LZ-Index structures in three modes: the full mode where it returns all the occurrences, a mode where it can return all the occurrences but we take the time needed to return only the occurrences of type 1, and use the minimum space for *Range* (called "up to type 1"), and a mode where it can only return the occurrences of type 1 as it does not store *Range* at all (called "only type 1"). We also compare Sadakane's document listing [22] we implemented on top of Sadakane's CSA [21] obtained from *PizzaChili*[2], and showing three points using suffix array sampling steps 32, 64, and 128. Finally, we include the variant using document arrays as plain wavelet trees [23], as RePair-compressed wavelet trees, and an intermediate between both called "alpha", as implemented by their authors [19] and using Sadakane's CSA with no sampling to minimize space (the sampling is not needed here).

It can be seen that Sadakane's technique uses less space than our smallest LZ-Index variant, but it is orders of magnitude slower (except on ClueChin), even on this CSA that is the fastest [4] to compute $A[i]$. The wavelet trees dominate our LZ-Index variants on ClueChin, because it has very few documents and thus the wavelet trees are small and fast. On the other collections, instead, wavelet trees use much more space than our LZ-Index variants. Indeed, in all but the toy collection ClueChin, even the LZ-Index in full mode is a relevant alternative, whereas the approximate ones offer even better space/time performance.

6   Final Remarks

We have introduced the first document listing data structure based on Lempel-Ziv compression. Apart from offering a competitive space/time tradeoff in general, an interesting feature of the index is its ability to retrieve a large number of documents very fast. This makes it an ideal choice in interactive scenarios, where one must show some answers immediately and others can be calculated in the background, and in cases where only some answers are sufficient.

We plan to extend our ideas to top-k document retrieval. Since the bulk of the occurrences are type 1, considering only those for computing top-k would yield very fast an answer that will usually be very accurate.

---

[2] From site pizzachili.di.unipi.it or pizzachili.dcc.uchile.cl

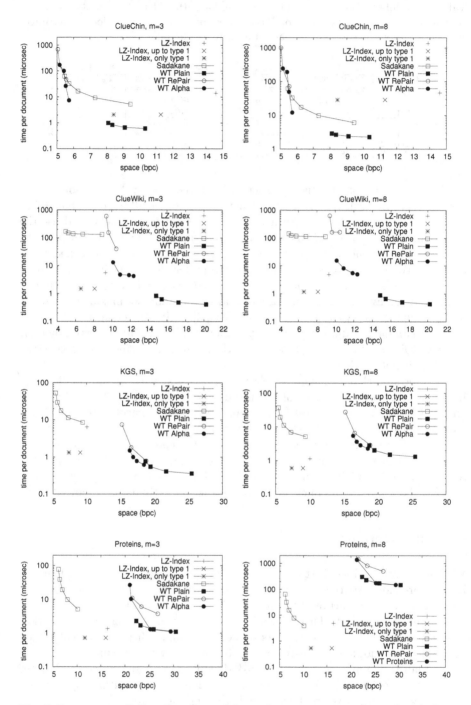

Fig. 3. Space versus listing time (logscale) per document output, for various indexes

References

1. Arroyuelo, D., Navarro, G., Sadakane, K.: Stronger Lempel-Ziv based compressed text indexing. Algorithmica 62(1), 54–101 (2012)
2. Barbay, J., Gagie, T., Navarro, G., Nekrich, Y.: Alphabet partitioning for compressed rank/select and applications. In: Cheong, O., Chwa, K.-Y., Park, K. (eds.) ISAAC 2010, Part II. LNCS, vol. 6507, pp. 315–326. Springer, Heidelberg (2010)
3. Belazzougui, D., Navarro, G.: Alphabet-independent compressed text indexing. In: Demetrescu, C., Halldórsson, M.M. (eds.) ESA 2011. LNCS, vol. 6942, pp. 748–759. Springer, Heidelberg (2011)
4. Ferragina, P., González, R., Navarro, G., Venturini, R.: Compressed text indexes: From theory to practice. ACM J. Exp. Alg. 13, art. 12 (2009)
5. Ferragina, P., Manzini, G., Mäkinen, V., Navarro, G.: Compressed representations of sequences and full-text indexes. ACM Trans. Alg. 3(2), article 20 (2007)
6. Fischer, J., Heun, V.: Space-efficient preprocessing schemes for range minimum queries on static arrays. SIAM J. Comp. 40(2), 465–492 (2011)
7. Gagie, T., Navarro, G., Puglisi, S.J.: New algorithms on wavelet trees and applications to information retrieval. Theor. Comp. Sci. 426-427, 25–41 (2012)
8. Gagie, T., Puglisi, S.J., Turpin, A.: Range quantile queries: Another virtue of wavelet trees. In: Karlgren, J., Tarhio, J., Hyyrö, H. (eds.) SPIRE 2009. LNCS, vol. 5721, pp. 1–6. Springer, Heidelberg (2009)
9. Golynski, A., Grossi, R., Gupta, A., Raman, R., Rao, S.S.: On the size of succinct indices. In: Arge, L., Hoffmann, M., Welzl, E. (eds.) ESA 2007. LNCS, vol. 4698, pp. 371–382. Springer, Heidelberg (2007)
10. Grossi, R., Gupta, A., Vitter, J.S.: High-order entropy-compressed text indexes. In: Proc. 14th SODA, pp. 636–645 (2003)
11. Hon, W.-K., Shah, R., Vitter, J.: Space-efficient framework for top-k string retrieval problems. In: Proc. 50th FOCS, pp. 713–722 (2009)
12. Kosaraju, S., Manzini, G.: Compression of low entropy strings with Lempel-Ziv algorithms. SIAM J. Comp. 29(3), 893–911 (2000)
13. Manber, U., Myers, G.: Suffix arrays: a new method for on-line string searches. SIAM J. Comp. 22(5), 935–948 (1993)
14. Munro, I.: Tables. In: Chandru, V., Vinay, V. (eds.) FSTTCS 1996. LNCS, vol. 1180, pp. 37–42. Springer, Heidelberg (1996)
15. Muthukrishnan, S.: Efficient algorithms for document retrieval problems. In: Proc. 13th SODA, pp. 657–666 (2002)
16. Navarro, G.: Indexing text using the ziv-lempel trie. J. Disc. Alg. 2(1), 87–114 (2004)
17. Navarro, G.: Spaces, trees and colors: The algorithmic landscape of document retrieval on sequences. CoRR, arXiv:1304.6023v1 (2013)
18. Navarro, G., Mäkinen, V.: Compressed full-text indexes. ACM Comp. Surv. 39(1), art. 2 (2007)
19. Navarro, G., Puglisi, S.J., Valenzuela, D.: Practical compressed document retrieval. In: Pardalos, P.M., Rebennack, S. (eds.) SEA 2011. LNCS, vol. 6630, pp. 193–205. Springer, Heidelberg (2011)
20. Raman, R., Raman, V., Rao, S.S.: Succinct indexable dictionaries with applications to encoding k-ary trees, prefix sums and multisets. ACM Trans. Alg. 3(4), art. 43 (2007)

21. Sadakane, K.: New text indexing functionalities of the compressed suffix arrays. J. Alg. 48(2), 294–313 (2003)
22. Sadakane, K.: Succinct data structures for flexible text retrieval systems. J. Disc. Alg. 5(1), 12–22 (2007)
23. Välimäki, N., Mäkinen, V.: Space-efficient algorithms for document retrieval. In: Ma, B., Zhang, K. (eds.) CPM 2007. LNCS, vol. 4580, pp. 205–215. Springer, Heidelberg (2007)
24. Weiner, P.: Linear pattern matching algorithm. In: Proc. 14th IEEE Symposium on Switching and Automata Theory, pp. 1–11 (1973)
25. Ziv, J., Lempel, A.: Compression of individual sequences via variable-rate coding. IEEE Trans. Inf. Theor. 24(5), 530–536 (1978)

Minimal Discriminating Words Problem Revisited

Paweł Gawrychowski[1], Gregory Kucherov[2,3],
Yakov Nekrich[4], and Tatiana Starikovskaya[5]

[1] Max-Planck-Institut für Informatik, Saarbrücken, Germany
gawry@cs.uni.wroc.pl
[2] Laboratoire d'Informatique Gaspard Monge, Université Paris-Est & CNRS,
Marne-la-Vallée, Paris, France
Gregory.Kucherov@univ-mlv.fr
[3] Department of Computer Science, Ben-Gurion University of the Negev,
Be'er Sheva, Israel
[4] Department of Electrical Engineering & Computer Science, University of Kansas,
Lawrence, USA
yakov.nekrich@googlemail.com
[5] School of Applied Mathematics and Information Science,
Higher School of Economics, Moscow, Russia
tat.starikovskaya@gmail.com

Abstract. We revisit two variants of the problem of computing *minimal discriminating* words studied in [5]. Given a pattern P and a threshold d, we want to report (i) all shortest extensions of P which occur in less than d documents, and (ii) all shortest extensions of P which occur only in d *selected* documents. For the first problem, we give an optimal solution with constant time per output word. For the second problem, we propose an algorithm with running time $O(|P| + d \cdot (1 + output))$ improving the solution of [5].

1 Introduction

Given a collection of text documents (character sequences), we are often interested in patterns that characterize a certain subset of these documents, i.e., occur only in the documents of this subset and not in the others. Such patterns (*words*) are called *discriminating* with respect to the corresponding subset. Identifying such patterns can be part of a *machine learning* or *data mining* task over a sample of documents, or can arise in *automated text classification*. In computational biology, patterns that appear in a subset of sequences sharing some biological feature and do not appear in the other sequences of the considered sample can be naturally assumed to be responsible for that feature.

In [5], the authors introduced the problem of *minimal discriminating words* along with the complementary problem of *maximal generic words*. In both of them, it is asked to compute some *extensions* of a given pattern P (which can be an empty word), i.e. strings which have P as a prefix. Consider a collection of strings (documents) $T_1, T_2, \ldots, T_m$ of total length n. Two variants of the minimal discriminating words problem have been considered in [5]. The basic variant is to report,

O. Kurland, M. Lewenstein, and E. Porat (Eds.): SPIRE 2013, LNCS 8214, pp. 129–140, 2013.

given a pattern P and a threshold $d \leq m$, all extensions of P which occur in *at most d* documents and which are minimal, i.e. any proper prefix of a reported extension must occur in more than d documents. A more practically motivated variant, called *minimal discriminating words for specified documents*, is to compute all minimal extensions starting with P which occur only in documents within a given subset $T_{i_1}, T_{i_2}, \ldots, T_{i_d}$. Minimality condition means that any proper prefix of a reported extension must occur in documents other than $T_{i_1}, \ldots, T_{i_d}$.

To exemplify minimal discriminating words, consider $T_1 = baaababb$, $T_2 = babaabab$ and $T_3 = babbaaab$. For $d = 2$, minimal discriminating extensions of $P = aa$ are aaa (discriminates $\{T_1, T_3\}$) and $aaba$ (discriminates $\{T_1, T_2\}$).

The complementary problem of maximal generic words looks for all *maximal* extensions of P occurring in *at least d* documents. In [5] a linear-space solution to the problem of reporting all maximal generic extensions was given. Its running time was optimal time $O(|P| + output)$, where $output$ is the number of reported extensions. The same paper proposed efficient solutions for the two variants of the minimal discriminating words problem, but their time bounds were not optimal: the basic variant of the problem was solved in $O(|P| + \log \log n + output)$ time, and the variant with specified documents was solved in $O(|P| + d \log \log m \cdot (1 + output))$ time. Moreover, in the latter case, the solution of [5] has the following undesirable property: it assumes that each T_i ends with a unique sentinel $\$_i$ that can be a part of a discriminating word even if dropping the sentinel yields a word which is not discriminating. Both solutions use $O(n)$ space.

In this paper, we revisit both variants of the minimal discriminating words problem and improve the bounds of [5]. For the second variant, we also get rid of the unnatural assumption about sentinel symbols occurring in discriminating words. Specifically, we propose $O(n)$ space solutions for the first and for the second problem with $O(|P| + output)$ and $O(|P| + d(1 + output))$ time respectively. Thus, for the first variant, we reach the optimal time bound. For the second variant, our running time does not depend on the size nor the number of documents, but only on the number of selected documents. In particular, when this number is constant, we obtain again an optimal $O(|P| + output)$ time. In both cases our solutions have the desirable property that after first spending $O(|P|)$ or $O(|P| + d)$ time, respectively, to initialize the computation, the worst-case delay between reporting two successive extensions is $O(1)$ or $O(d)$, respectively.

Similar to [5], our solutions are based on the generalized suffix tree of $T_1, T_2, \ldots, T_m$ that can be viewed as a compacted trie for strings $T_1\$_1, T_2\$_2, \ldots, T_m\$_m$. It is well-known that the generalized suffix tree can be computed in $O(n)$ time. For each node v of the generalized suffix tree we store its weight $\mathsf{weight}(v)$ defined as the number of *distinct* documents whose suffixes occur in the subtree rooted at v. All $\mathsf{weight}(v)$ can be computed in $O(n)$ time [3].

The *locus* of a string S in a trie on a set of strings is defined as the highest explicit node labelled by an extension of S. It is important to note that in all our algorithms, each output word is specified by its locus in the generalized suffix tree for $T_1, T_2, \ldots, T_m$ (rather than by "spelling out" the word itself).

2   Minimal Discriminating Words

Suppose that a set of documents $T_1, T_2, \ldots, T_m$ of total length n is given. For a pattern P and a threshold $d \leq m$, we want to find all minimal extensions of P which occur in at most d distinct documents. "Minimal" here means that no proper prefix of a reported extension satisfies this property. We describe a linear-space data structure for this problem.

2.1   General Idea

Consider the generalized suffix tree for $T_1, T_2, \ldots, T_m$. We first delete sentinels $\$_i$ from labels of its edges, and then delete edges with empty labels. Consider the locus of P in the resulting trie, which we call GST. Any descendant u of the locus such that $\mathsf{weight}(u) \leq d$ and $\mathsf{weight}(p(u)) > d$, where $p(u)$ is the parent of u, will be a locus of a desired extension of P. The extension itself will be equal to the label of $p(u)$ extended by the first letter on the edge $(p(u), u)$. (By construction the first letter on any edge of GST is not a sentinel but a letter of the alphabet).

We represent GST with its compacted version GST^c, and a number of arrays. An array A_u corresponding to an edge $(p(u), u)$ of GST^c contains links to nodes which were removed in order to obtain the edge $(p(u), u)$. More precisely, $A_u[\Delta]$ links to the lowest ancestor v of u such that $\mathsf{weight}(v) \geq \mathsf{weight}(u) + \Delta$, for every $\Delta < \mathsf{weight}(p(u)) - \mathsf{weight}(u)$. Note that the total length of the arrays is just $O(n)$ as each entry corresponds to one suffix. Loci of the extensions can be found in the following way: first, find the locus v of P in GST^c and compute all nodes u in its subtree for which $\mathsf{weight}(u) \leq d$ and $\mathsf{weight}(p(u)) > d$. Then for each found node u compute its ancestor u' in GST such that $\mathsf{weight}(u') \leq d$ and $\mathsf{weight}(p(u')) > d$. The last step can be done in constant time using the array A_u associated with $(p(u), u)$: we choose as u' the node $A_u[d - \mathsf{weight}(u)]$, and if $\mathsf{weight}(u') > d$ we replace it by the unique son on the $u - p(u)$ path. Then the node u' will be a locus of a desired extension. We refer to nodes u as above as *extension loci*.

Let v be the locus of P in GST^c. We denote the subtree of GST^c rooted at v by T_v. Leaves of GST^c of weight bigger than d will be called *d-heavy leaves*. First note that if T_v has no d-heavy leaves, every root-to-leaf path contains an extension locus and hence we can find each extension locus in *amortized* constant time by traversing T_v in depth-first order. Below we explain how to overcome the assumption about d-heavy leaves and to achieve *worst-case* constant time per an extension locus. We first prove the following useful lemma.

Lemma 1. *Each extension locus belongs to a maximal subtree of T_v without d-heavy leaves.*

Proof. An extension locus u cannot have a d-heavy leaf in its subtree, otherwise weight of u would be bigger than d. Let u' be the highest ancestor on the path from u to v that does not have a d-heavy leaf in its subtree. Then u belongs to $\mathsf{T}_{u'}$, and $\mathsf{T}_{u'}$ is a maximal subtree of T_v that does not have d-heavy leaves.   □

The algorithm will iterate over the maximal subtrees of T_v without d-heavy leaves and report extension loci for each of them. We give the details below.

2.2   Computing Maximal Subtrees

Let a trie τ_k, $0 \leq k \leq m - 1$, be a compact trie containing labels of all k-heavy leaves. (Note that τ_0 is essentially GST^c.) From the construction it follows that there is one-to-one correspondence between leaves of τ_k and k-heavy leaves. Moreover, for each node u of τ_k there is a node w of GST^c such that the labels of u and w are equal. Such nodes w will be referred to as k-nodes. We say that u and w are of type 1 iff the degree of u is smaller than the degree of w, and that they are of type 2 iff there is at least one node on the path from w to its nearest k-node ancestor. (A node can be of type 1 and of type 2 simultaneously or neither of type 1 nor of type 2.) We store nodes of types 1 and 2 in two lists ordered as in the depth-first traversal of τ_k. Next, let us consider a node of GST. Nodes of type 1 in its subtree form a sublist in the first list. For each node we store pointers to the start and to the end of the corresponding sublist. Pointers associated with nodes of type 2 are defined in a similar way.

Note that the parent p of the root of a maximal subtree T without d-heavy leaves has a d-heavy leaf in its subtree (otherwise, T would not be maximal). That is, p is either a d-node or a node on the path connecting a d-node and its nearest d-node ancestor. At the same time, p has at least one son (the root of T) which does not have a d-leaf, and, consequently, a d-node, in its subtree. Therefore, in the first case p is a d-node of type 1, and in the second case p is on the path connecting a d-node of type 2 and its nearest d-node ancestor.

We now return to the description of the algorithm. We start by computing the locus of P in τ_d in $O(|P|)$ time in a usual way. Then we iterate over nodes of types 1 and 2 in the subtree of the locus using the pointers and the lists and report associated maximal subtrees without d-heavy leaves.

Let w be a d-node of GST^c of type 1, and u be the corresponding node of τ_d. By the definition, the degree of u is smaller than the degree of w, which means that at least one child of w is a root of a maximal subtree without d-heavy leaves. Such children form subranges of the list of all children of w, and we assume that pointers to these subranges are available. (As we show below, the total number of the pointers is linear and they can be precomputed in linear time.) Using the pointers we can output i requested children of w in $O(i)$ time.

If w is a d-node of GST^c of type 2 and w' is its nearest d-node ancestor, then all subtrees hanging off the path from w to w' are maximal subtrees without d-heavy leaves. The subtrees can be found in linear time by iterating over nodes on the path from w to w'.

All in all, retrieving maximal subtrees without d-heavy leaves takes constant time per subtree in the worst case.

Lemma 2. *Tries τ_k and pointers to the subranges of children of k-nodes that do not have k-heavy leaves in their subtrees occupy $O(n)$ space in total and can be constructed in $O(n)$ time.*

Proof. To estimate the space occupied by the tries it is enough to estimate the total number of their leaves. The latter is equal to n, because a string which is a suffix of k documents will correspond to a leaf in τ_1, to a leaf in τ_2, ..., to a leaf in τ_k, that is, k leaves in total. The statement follows.

The tries are built as follows. We first augment GST with a linear-space data structure [8] that allows to answer lowest common ancestor queries in constant time. This step takes $O(n)$ time. We then iterate over leaves of GST^c from the left to the right and for each k compose a lexicographically ordered list L_k of k-heavy leaves' labels. Secondly, we scan L_k and compute the length of the longest common prefix of every two consecutive suffixes in L_k. (The length is equal to the string depth of the lowest common ancestor of the leaves corresponding to the suffixes and hence can be computed in constant time). Once we have L_k and the lengths, we build τ_k in linear time in a usual way. Correspondence between nodes of τ_k and GST^c and hence types of nodes can be established in $O(|\tau_k|)$ time with the help of the lowest common ancestor queries. Finally, the lists of nodes of types 1 and 2 are constructed by depth-first traversal of τ_k.

Let w be a k-node of GST^c and u be the corresponding node of τ_k. The number of subranges formed by children of w without k-heavy leaves in their subtrees does not exceed the degree of u. Therefore, the total number of the subranges does not exceed the total size of the tries, which is $O(n)$. Next, note that a node does not have k-heavy leaves in its subtree if and only if the weight of the heaviest leaf in its subtree $\leq k$. We compute the subranges in two steps. First we traverse GST^c bottom-up and for each node compute the weight of the heaviest leaf in its subtree. Secondly, we scan the list of children of each node and for each k such that the node is a k-node remember the starting and the ending points of maximal subranges with the weights $\leq k$. Construction takes linear time in total.   □

2.3   Computing Extension Loci

Here we show how to report all extension loci in a maximal subtree of T_v without d-heavy leaves. We start with an auxiliary lemma.

Lemma 3. *A compact trie of size n can be partitioned into disjoint node-to-leaf paths of length $O(\log n)$ each.*

Proof. For a node u of the trie we define $h(u)$ to be the length of the shortest downward path to a leaf from u, and $\ell(u)$ to be the number of leaves in the subtree rooted at u. We prove by induction that $\ell(u) \geq 2^{h(u)}$.

If $h(u) = 0$, then u is a leaf and $\ell(u) = 1$. Suppose that the inequality holds for all u such that $h(u) \leq k$. A node u of the compact trie with $h(u) = k+1$ has at least two descendants v_1, v_2 and both $h(v_1)$ and $h(v_2)$ must be at least k, hence $\ell(u) \geq 2^{h(v_1)} + 2^{h(v_2)} \geq 2^{k+1}$, the claim follows.

For each node u of the compact trie we colour the edge from u to its child v with the smallest $h(u)$ red. This colouring induces a partition of all nodes into node-disjoint red paths. From the inequality it follows that the length of any red path is $O(\log n)$.   □

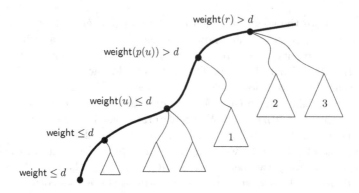

weight(r) $> d$

weight($p(u)$) $> d$

weight(u) $\leq d$

weight $\leq d$

weight $\leq d$

2

3

1

Fig. 1. We push the root of the subtree 1 into S first. When we pop it from S, we push the root of the subtree 2 into S, and so on.

We partition GST^c into disjoint node-to-leaf paths of length $O(\log n)$ using Lemma 3. q-heaps [4] allow to support predecessor queries on logarithmic-size subsets of $[1, n]$ in constant time using linear space and a common precomputed table of size $o(n)$. We use q-heaps to answer predecessor queries on weights of each path of the partition. The total space occupied by q-heaps is $O(n)$.

Lemma 4. *Given a maximal subtree* T *of* T_v *without d-heavy leaves, all extension loci in* τ *can be reported in* $O(1)$ *time per locus.*

Proof. To simplify the description, assume that nodes of GST^c are rearranged so that an edge from a node to its leftmost child is always red.

Let r be the root of T. If weight(r) $\leq d$, then the only extension locus in the subtree is r. (Remember that the parent of r has a d-heavy leaf in its subtree and therefore its weight is bigger than d). If weight(r) $> d$, we start with the node-to-leaf path containing r. Since the weight of any leaf of T is at most d, the path contains an extension locus u. Using one predecessor query, we can find u in $O(1)$ time. We then push the second child of a parent of u into a stack S.

We perform recursive calls for the subtrees rooted at nodes from S. Each time we pop a node w from S we push its right brother into S. If no brothers are left and the parent and the grandparent of w are on the same red path, we push the second child of the grandparent of w into S (see Fig. 1). The algorithm stops when S is empty.

We now show that the algorithm is correct. Note that any subtree hanging off the path below u does not contain extension loci, while each tree hanging off the path above u contains at least one such node. All the latter trees are examined due to the order of recursive calls. Each call takes constant time and returns a requested node. □

To sum up, each extension locus inside a given maximal subtree without d-heavy leaves can be reported in constant time in the worst case. Since, according to Section 2.2, each such subtree of T_v can be identified in worst-case constant time, we obtain the final theorem.

Theorem 1. *For a given pattern P and a threshold d, all minimal discriminating extensions of P can be reported in time $O(|P| + output)$, where output is the number of reported extensions. The underlying indexing data structure occupies $O(n)$ space, where n is the total length of the strings $T_1, T_2, \ldots, T_m$.*

3   Minimal Discriminating Words for Specified Documents

In many applications, we need to compute words that discriminate documents from a *given sample*. Consider a set of documents $T_1, T_2, \ldots, T_m$ of total length n. Given a set of indices $\mathsf{Ind} = \{i_1, i_2, \ldots, i_d\}$ and a pattern P, we want to find all minimal extensions of P occurring *only* in documents $T_i, i \in \mathsf{Ind}$, where "minimal" means that any of their proper prefixes has at least one occurrence in a document which does not belong to this subset.

Here we propose a linear-space data structure which allows to compute such extensions in time $O(|P| + d \cdot (output + 1))$, where *output* is the number of reported extensions.

3.1   General Idea

Consider the generalized suffix tree for $T_1, T_2, \ldots, T_m$. For each suffix of $T_1, T_2, \ldots, T_m$ we create an explicit node labelled by this suffix (if it does not exist already). We denote the resulting tree by GST. Note that the size of GST is $O(n)$. Problems of computing loci of the minimal extensions in the generalized suffix tree and GST are equivalent.

An inner node of GST is called $\$$-*terminating* if all its outgoing edges are labelled by sentinels. If, in addition, the sentinels are $\$_i$, where $i \in \mathsf{Ind}$, then the node is called Ind-*terminating*. From the definition it follows that the locus of any string occurring only in documents T_i, $i \in \mathsf{Ind}$, contains an Ind-terminating node in its subtree, in particular, the locus of any minimal extension contains such node in its subtree. Besides, each Ind-terminating node belongs to a subtree rooted at the locus of some minimal extension, as shown below.

Lemma 5. *Suppose that w is an Ind-terminating node and that its label starts with P. Then the path from w to the root contains a locus of a minimal extension of P occurring only in documents $T_i, i \in \mathsf{Ind}$.*

Proof. The label S of w is an extension of P occurring only in documents T_i, $i \in \mathsf{Ind}$. The locus of the shortest prefix of S occurring only in documents T_i, $i \in \mathsf{Ind}$, will be the locus of a requested extension and will be on the path from w to the root of GST.   ☐

A high-level description of the algorithm is as follows. We start by locating the locus u of P in GST in time $O(|P|)$ and retrieving the interval $[L(u), R(u)]$ of ranks of suffixes ending below u. Rank of a suffix is simply its rank in the lexicographic order, equal suffixes are assigned equal ranks. The algorithm keeps a stack of intervals which it is to process, initialized to contain just $[L(u), R(u)]$. At each step it pops an interval $[a, b]$ from the stack, finds an Ind-terminating node v covering a subrange of $[a, b]$, computes the ancestor w of v labelled by a requested extension of P, and pushes the intervals $[a, L(w) - 1]$ and $[R(w) + 1, b]$ onto the stack. If there is no such Ind-terminating node, the algorithm does nothing. The algorithm terminates when the stack is empty.

To estimate the running time of the algorithm, we note that each of the processed intervals, except for $[L(u), R(u)]$, either corresponds to a reported extension, or is a child of an interval corresponding to a reported extension (and each such interval has two children). Hence the total number of processed intervals will be $O(output + 1)$, where $output$ is the number of reported extensions. Below we show that processing of each interval takes $O(d)$ time. Note that if we want to make sure that the delay between reporting two minimal extensions is $O(d)$, we only need to check if the interval contains an Ind-terminating node before we push it onto the stack.

3.2   Computing an Ind-Terminating Node

Given an interval $[a, b]$, we want to find some Ind-terminating node u such that all leaves in its subtree are of ranks in $[a, b]$, or to show that there is none. Below we show that it can be done in $O(d)$ time.

Consider a trie T on the reverses $T_1^R, T_2^R, \ldots, T_m^R$ of the documents. Each node v of T corresponds to a prefix of some T_j^R, or, equivalently, to a reversed suffix of T_j. We call the node v *active* if the suffix is a label of a \$-terminating node of GST. If the node is also Ind-terminating, we call v Ind-good, otherwise we call it Ind-bad. Note that if a node is Ind-bad, then all its ancestors are Ind-bad. That is, Ind-good nodes are exactly active nodes of maximal subtrees of T without Ind-bad nodes. We compactify T leaving nodes labelled by T_i^R, $1 \le i \le m$, explicit. The resulting trie is denoted by T^c (see Fig. 2).

For an edge e of T^c we define a set $S(e)$ to contain ranks of some suffixes of $T_1, T_2, \ldots, T_m$ in the lexicographic order. The suffixes are exactly the suffixes

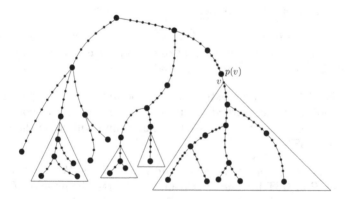

Fig. 2. Maximal subtrees of T without any Ind-bad nodes. Thick nodes exist in T^c.

the reverses of which were the labels of the active nodes removed in order to obtain e.

Lemma 6 (Theorem 4 in [1]). $S(e)$ *can be stored using linear space so that given any interval* $[a, b]$ *we can in* $O(1)$ *time either retrieve some element in* $S(e) \cap [a, b]$ *or detect that there is none.*

Remember that we want to find an Ind-terminating node of GST such that all leaves in its subtree are of ranks in $[a, b]$. The algorithm will search, instead, for an active node of maximal subtrees of T without Ind-bad nodes corresponding to such Ind-terminating node.

Consider a maximal subtree of T without Ind-bad nodes (see Fig. 2). The parent $p(v)$ of its root is labelled by a prefix of T_j^R, for some $j \notin$ Ind, while v is not. Hence, $p(v)$ is either a node of degree bigger than 1, or is labelled by T_j^R. In both cases, $p(v)$ is a node of T^c. It follows that we can decompose a set of active nodes of the subtree into a set of active nodes existing in T^c and sets of active nodes associated with the edges of T^c. Each leaf v of a maximal subtree without Ind-bad nodes corresponds to T_i^R, for some $i \in$ Ind. We will make use of precomputed values lcp(i), where lcp(i) is the length of the longest common prefix of T_i^R and T_j^R, $j \notin$ Ind. Note that all ancestors of v of string depth bigger than lcp(i) belong to the maximal subtree.

We start at a node v labelled by some T_i^R, $i \in$ Ind, and go up until we reach a node of string depth lcp(i) or an already visited node. For each encountered node we check if it is active and if it corresponds to the desired Ind-terminating node. If yes, the algorithm stops. For each edge e we traverse we try to retrieve an element in $S(e) \cap [a, b]$. If there is such an element, the algorithm finds the corresponding Ind-terminating node and stops. We repeat such procedure for each $i \in$ Ind.

The total number of processed nodes and edges is bounded by the total size of the maximal subtrees without Ind-bad nodes. Since each leaf and each inner node of degree one in the subtrees corresponds to T_i^R, $i \in$ Ind, the total size of the subtrees, and hence the time of computing an Ind-terminating node covering a subrange of $[a, b]$, is $O(d)$. It remains to show that the values lcp(i), $i \in$ Ind, can be precomputed efficiently.

Lemma 7. *Given* Ind, *we can compute* lcp(i) *for all* $i \in$ Ind *in* $O(d)$ *total time.*

Proof. To compute the values, we use an array R defining the lexicographic order on $T_1^R, T_2^R, \ldots, T_m^R$, its inverse R^{-1} and an array LCP which contains the length of the longest common prefix of every pair of consecutive (in the lexicographic order) reversed documents. The LCP array is augmented with a range minimum query data structure [2], which allows to compute the minimum value in any interval of LCP in constant time. All these structures are built in the preprocessing phase without knowing Ind.

Consider an index $i \in$ Ind, and let $T_{k_1}^R$ and $T_{k_2}^R$ with $k_1, k_2 \notin$ Ind be the reversed documents closest to T_i^R in the lexicographic order from the left and from the right, respectively. From the properties of the lexicographic order it follows that $\mathsf{lcp}(i)$ is equal to the maximum of the lengths of the longest common prefixes of $T_{k_1}^R$ and T_i^R and of $T_{k_2}^R$ and T_i^R. The lengths can be computed by taking the minimum of the values stored in the array LCP between the entries corresponding to $T_{k_1}^R$ and T_i^R and of $T_{k_2}^R$ and T_i^R respectively. Hence the only question is how to find k_1 and k_2 efficiently.

Consider the occurrences of T_i^R for all $i \in \mathsf{Ind}$ in R, and let $R[a..b]$ be a maximal interval of such occurrences, i.e., both $R[a-1]$ and $R[b+1]$ correspond to reversals outside of Ind. Then $k_1 = a - 1$ and $k_2 = b + 1$ for all $i \in \mathsf{Ind}$ corresponding to occurrences in the interval. Our method identifies such maximal intervals one-by-one and updates the values of $\mathsf{lcp}(i)$ accordingly. To make the identification efficient, we store an additional bit vector B of length m to keep track of the already processed indices from Ind, initially containing all zeros. We loop over Ind, and if a given $i \in \mathsf{Ind}$ is not processed yet, sweep to the left and to the right starting from $R^{-1}[i]$ to identify the maximal interval of R containing i and some other indices from Ind. Then knowing the values of k_1 and k_2 for all indices in the interval we calculate their values of lcp. Finally, we set their corresponding bits in B to one. In the very end we iterate through all $i \in \mathsf{Ind}$ and clear their corresponding bits in B. The algorithm clearly spends just constant time per a single element of Ind. $\qquad\square$

3.3   Computing Ancestor Loci

Here we show how to find the ancestor w of an Ind-terminating node that corresponds to a minimal discriminating word. We assume that for each node v of GST there is a pointer to its highest ancestor with the same weight and that the ranks $L(v)$ and $R(v)$ of the leftmost and the rightmost leaves in the subtree of v can be retrieved in $O(1)$ time. We also store an array D such that $D[i] = k$ if the i-th leaf of GST in the left-to-right order corresponds to a suffix from T_k.

Lemma 8. *Given a node u in* GST *and $M \in [L(u), R(u)]$, for all distinct values j occurring in $D[L(u), R(u)]$ we can find the leftmost occurrence of j after position M and the rightmost occurrence of j before M in $D[L(u), R(u)]$ in $O(\mathsf{weight}(u))$ time, where $\mathsf{weight}(u)$ is the number of distinct documents whose suffixes occur in the subtree rooted at u.*

Proof. We can enumerate all distinct values in an interval of D using the data structure of Muthukrishnan [7]. As follows from the description in [7], the structure reports the leftmost occurrence of each j that occurs in the interval. By reversing the input, we can modify the structure so that the rightmost occurrence of each j is reported, too. We obtain the result by reporting the leftmost occurrence of each distinct j in the interval $D[M, R(u)]$ and the rightmost occurrence of each j in the interval $D[L(u), M]$. $\qquad\square$

In Lemma 5 we showed that any Ind-terminating node w in the subtree rooted at the locus of P has an ancestor v that is a locus of a desired minimal extension. We compute v in two steps.

Using the pointers we can find the highest ancestor w' of w of weight at most d in $O(d)$ time. The interval $D[L(w'), R(w')]$ contains indices of at most d different documents, and we output these indices in time $O(d)$ using Lemma 8. For each index j that occurs in $D[L(w'), R(w')]$ but does not belong to Ind, we find the rightmost occurrence of j before $L(w)$. The maximum (rightmost) position among them is denoted by L'. Similarly, we find the leftmost position of each $j \notin$ Ind after $R(w)$, and denote the leftmost among them by R'. This step takes $O(d)$ time. $[L', R']$ is the maximal segment that contains $[L(w), R(w)]$ and consists only of indices from Ind.

Now consider the node w again. We initialize v to w and jump from v to the highest node v' such that $\mathsf{weight}(v) = \mathsf{weight}(v')$. Let $p(v')$ be the parent of v'. If $L' \leq L(p(v')) \leq R(p(v')) \leq R'$, we set $v = p(v')$ and repeat the same step for the new node v. Otherwise we set $v = v'$ and stop. Observe that each iteration increases the number of different indices occurring in $[L(v), R(v)]$ by at least one and takes just constant time.

Lemma 9. *Given an* Ind-*terminating node w in the subtree of u being the locus of P. The node on the path from w to the root of* GST *that is a locus of a minimal extension of P occurring only in documents T_i, $i \in$* Ind, *can be computed in $O(d)$ time.*

Combining Lemma 9 and the algorithm described in Section 3.2, we obtain the final result.

Theorem 2. *Given a subset of indices $\{i_1, i_2, \ldots, i_d\}$ and a pattern P, all minimal extensions of P which occur only in the documents $T_{i_1}, T_{i_2}, \ldots, T_{i_d}$ can be computed in time $O(|P| + d(\text{output} + 1))$, where output is the number of reported extensions. The underlying indexing data structure occupies $O(n)$ space, where n is the total length of the strings $T_1, T_2, \ldots, T_m$.*

We can also output the loci of minimal extensions in lexicographic order without increasing the query time. We achieve this by keeping intervals $[L(w), R(w)]$ for all found extension loci w in a tree $\mathcal{T}$. We initialize $\mathcal{T}$ to a one-node tree and store the interval $[L(u), R(u)]$ at its root r. If we find a new extension locus w, we replace $[L(u), R(u)]$ with $[L(w), R(w)]$ and append two child nodes to r. Intervals $[L(u), L(w) - 1]$ and $[R(w) + 1, R(u)]$ are stored in the left and the right children of r respectively. Every time when we find an Ind-terminating node v in the interval $[l(\nu), r(\nu)]$ stored in some $\nu \in \mathcal{T}$, we identify the ancestor w of v that is the locus of a minimal extension. Then we replace $[l(\nu), r(\nu)]$ with $[L(w), R(w)]$ and append two child nodes to ν as described above. When all loci are found, we traverse internal nodes of $\mathcal{T}$ in-order to obtain a sorted list L of the intervals $[L(w), R(w)]$ for extension loci w. The traversal of $\mathcal{T}$ takes $O(\text{output})$ time; thus the total asymptotic time necessary to answer a query remains unchanged.

We remark that the data structure of Theorem 2 can be constructed in $O(n \log^\varepsilon n)$ time for any constant $\varepsilon > 0$ with high probability [6]. The preprocessing time is dominated by the cost of constructing data structures $S(e)$.

4 Conclusions

We developed an optimal algorithm for reporting all minimal discriminating words. For the problem of reporting all minimal discriminating words for a specified set of documents, our solution is optimal when $d = O(1)$, but it might still be possible to improve the running time for the case of non-constant value of d.

Another interesting question is whether counting the number of solutions can done faster than reporting them all according to our algorithm. Finally, we also wonder if we can generate k lexicographically smallest solutions in time proportional to k rather than to *output*. Our algorithms can be used to output k distinct solutions with such complexity, but we cannot guarantee that the generated solutions are lexicographically smallest.

References

1. Alstrup, S., Brodal, G.S., Rauhe, T.: Optimal static range reporting in one dimension. In: Proc. of the 33rd Annual ACM Symposium on Theory of Computing, pp. 476–482 (2001)
2. Bender, M.A., Farach-Colton, M.: The LCA problem revisited. In: Gonnet, G.H., Viola, A. (eds.) LATIN 2000. LNCS, vol. 1776, pp. 88–94. Springer, Heidelberg (2000)
3. Hui, L.C.K.: Color set size problem with applications to string matching. In: Apostolico, A., Galil, Z., Manber, U., Crochemore, M. (eds.) CPM 1992. LNCS, vol. 644, pp. 230–243. Springer, Heidelberg (1992)
4. Fredman, M.L., Willard, D.E.: Trans-dichotomous algorithms for minimum spanning trees and shortest paths. J. Comput. Syst. Sci. 48(3), 533–551 (1994)
5. Kucherov, G., Nekrich, Y., Starikovskaya, T.: Computing discriminating and generic words. In: Calderón-Benavides, L., González-Caro, C., Chávez, E., Ziviani, N. (eds.) SPIRE 2012. LNCS, vol. 7608, pp. 307–317. Springer, Heidelberg (2012)
6. Mortensen, C.W., Pagh, R., Patrascu, M.: On dynamic range reporting in one dimension. In: Proc. of the 37th Annual ACM Symposium on Theory of Computing, pp. 104–111 (2005)
7. Muthukrishnan, S.: Efficient algorithms for document retrieval problems. In: Proc. of the 13th Annual ACM-SIAM Symposium on Discrete Algorithms. SIAM (2002)
8. Schieber, B., Vishkin, U.: On finding lowest common ancestors: Simplification and parallelization. SIAM Journal on Computing 17, 111–123 (1988)

Adding Compression and Blended Search to a Compact Two-Level Suffix Array

Simon Gog and Alistair Moffat

Department of Computing and Information Systems,
The University of Melbourne, Australia 3010

Abstract. The suffix array is an efficient in-memory data structure for pattern search; and two-level variants also exist that are suited to external searching and can handle strings larger than the available memory. Assuming the latter situation, we introduce a factor-based mechanism for compressing the text string that integrates seamlessly with the in-memory index search structure, rather than requiring a separate dictionary. We also introduce a mixture of indexed and sequential pattern search in a trade-off that allows further space savings. Experiments on a 4 GB computer with 62.5 GB of English text show that a two-level arrangement is possible in which around 2.5% of the text size is required as an index and for which the disk-resident components, including the text itself, occupy less than twice the space of the original text; and with which *count* queries can be carried out using two disk accesses and less than two milliseconds of CPU time.

Keywords: string search, pattern matching, suffix array, Burrows-Wheeler transform, succinct data structure, disk-based algorithm, experimental evaluation.

1   Introduction

The problem of *string search* is well known: given a text $T[0 \ldots n-1]$ over some alphabet Σ of size σ, and a pattern $P[0 \ldots m-1]$, identify whether P occurs in T. Four different query modes are possible: reporting whether or not the pattern occurs at all (*existence* queries); reporting how many times the pattern occurs (*count* queries); reporting the locations in T at which it occurs (*locate* queries); and reporting, for each such occurrence, the context in T within which P appears (*context* queries).

One way of preprocessing T to facilitate fast search is to construct a suffix array [11]. Array $SA[0 \ldots n]$ is a suffix array for T if and only if $T_{SA[i]} < T_{SA[j]}$ whenever $i < j$, where $T_k = T[k \ldots n]$ is the kth suffix of T, the ordering between strings is lexicographic, and where T is assumed to be augmented by a sentinel $T[n]$ that is smaller than every element in Σ. Using binary search in SA, the range $SA[lb \ldots rb]$ corresponding to P can be found in $O(m \log n)$ time. Each SA entry in that range is the location in T of an occurrence of P ; and so *locate* and *context* queries can be answered in $O(m \log n + k)$ and $O(m \log n + km)$ time respectively, where k is the number of matching locations.

Stored as a sequence of integers, SA requires $n \log n$ bits, in addition to the $n \log \sigma$ bits required by T (note that $\log x$ should be interpreted as $\lceil \log_2 x \rceil$ when appropriate). The need for random-access to the suffix array and text during pattern search means that both structures need to be held in memory. If that is not possible, either compressed indexing

O. Kurland, M. Lewenstein, and E. Porat (Eds.): SPIRE 2013, LNCS 8214, pp. 141–152, 2013.

structures such as the FM-INDEX must be used [5], or a two-level blocked suffix array arrangement must be employed. The drawback of the former is that random-access memory is still required, and so while using 250 MB for an FM-INDEX for 1 GB of ASCII text may be viable, using 2.5 GB to index 10 GB may not be. On the other hand, the drawback of two-level suffix arrays structures has been – until recently – that they require even more disk space than a straight suffix array.

Our Contribution: The *Reduced On-disk Suffix Array* (ROSA) structure of Gog et al. [6] supports pattern search on large strings using an in-memory index of around 2–3% of the text size, and less disk space than a plain suffix array. Here we further reduce the disk space requirement of the ROSA, by:

- using the block prefix strings associated with the set of ROSA suffix blocks as a phrase-book for representing T in compressed form as a sequence of factors in a manner that still allows random access decoding;
- storing the addresses of phrases (rather than characters) in the suffix array, reducing the number of bits required for suffix pointers;
- quantizing the set of addresses used as suffix pointers, further trading disk space for moderately increased query execution costs.

As is demonstrated by the experiments described below, *count* queries on multi-gigabyte English strings can be carried out via an index of around 2.5% of $|T|$; total disk space, including T, of under $2|T|$; two disk accesses; and less than two milliseconds of CPU.

2   Two-Level Suffix Arrays

The drawback of using suffix arrays on high-latency storage devices has been recognized by a range of authors [1,2,4,8,9,13]; Gog et al. [6] summarize that work. Our assumption here is that memory is only available for an index structure that is some small percentage of the text size, and that disk operations to fetch secondary data incur a fixed latency cost plus a transfer cost proportional to the amount of data transferred. That is, searching and retrieval cost are assessed experimentally, rather than via the machine-independent external memory model.

Variable-Sized Blocks: Sinha et al. [13] describe the LOF-SA, a two-level suffix array structure. The distinguishing characteristic of the LOF-SA compared to other two-level structures is that it makes use of variable-sized suffix blocks. An upper bound b determines how the suffix array is partitioned into blocks; the requirement is that no block may contain more than b pointers, and that the suffixes in each block share a common prefix. That is, a block is formed for a block prefix string (BPS) v if and only if $size(node(v)) \leq b$ and $size(parent(node(v))) > b$, where $node(v)$ is the suffix tree node corresponding to v, and $size(z)$ is the number of leaves in the subtree rooted at node z. Sinha et al. suggest the use of a trie as an in-memory search structure; and add an LCP and other auxiliary information to each of the suffix pointers, to allow suffix blocks to be searched without needing multiple accesses to T. A LOF-SA structure for a text of $n = |T|$ symbols ($n < 2^{32}$) requires $12|T|$ bytes of disk storage; in follow-up work, the use of byte-codes reduced the space to approximately $7|T|$ bytes [12].

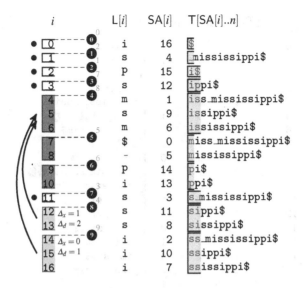

i		$L[i]$	$SA[i]$	$T[SA[i]..n]$
0		i	16	$
1		s	4	_mississippi$
2		p	15	i$
3		s	12	ippi$
4		m	1	iss_mississippi$
5		s	9	issippi$
6		m	6	ississippi$
7		$	0	miss_mississippi$
8		-	5	mississippi$
9		p	14	pi$
10		i	13	ppi$
11		s	3	s_mississippi$
12	$\Delta_x = 1$	s	11	sippi$
13	$\Delta_d = 2$	s	8	sissippi$
14	$\Delta_x = 0$	i	2	ss_mississippi$
15	$\Delta_d = 1$	i	10	ssippi$
16		i	7	ssissippi$

Fig. 1. Forwards BWT for T = "miss_mississippi". The suffix blocks formed when $b = 3$ are shown on the left, and the block prefix strings (BPS) are shaded on the right. The block at $i = 14, 15, 16$ is reducible to the block at $i = 4, 5, 6$; and the block at $i = 12, 13$ is reducible in two steps to locations $i = 5, 6$.

Gog et al. [6] refine the LOF-SA to make a new structure they call the RoSA, or Reduced On-Disk Suffix Array, requiring around $3|T|$ bytes. Key changes include:

– replacing the in-memory trie by a condensed BWT-based index structure over a set of reversed substrings, giving rise to smaller query-time memory requirements and significantly better worst-case space performance;
– handling blocks identified as being *reducible* – in that they can be translated on to contiguous sequences of suffix pointers contained within other blocks – via pointer reductions rather repeating them, giving rise to substantial disk space savings, in a manner also noted by Mäkinen and Navarro [10].

Figure 1 shows the 10 RoSA blocks created for the string T = "miss_mississippi" with $b = 3$. There are two *reducible* blocks, signified by all of the suffixes in the block having the same BWT character $L[i]$; three *irreducible* blocks; and five *singleton* blocks.

Searching: In the RoSA each pattern P is first searched for in a *condensed BWT* built over the reversals of the block prefix strings (BPS's), and all their prefixes. For example, in Figure 1 there are 10 BPS's: "$", "_", "i$", "ip", "is", "m", "p", "s_", "si", and "ss". If a pattern P commences with any of these strings, or any prefix of any of these strings, the corresponding suffix block(s) must be retrieved so that the remainder of the pattern can be checked. On the other hand, if P does not commence with one of the BPS's, then it is certain that P does not appear in T. The set of BPS reversals and prefixes of reversals covers the strings: "$", "$i", "_", "_s", "i", "is", "m", "p", "pi", "s", "si", and "ss". Figure 2 shows the BWT and suffixes generated from the

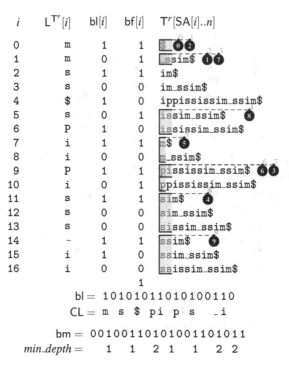

i	$\mathsf{L}^{T^r}[i]$	$\mathsf{bl}[i]$	$\mathsf{bf}[i]$	$T^r[SA[i]..n]$
0	m	1	1	$\$$ ❶❷
1	m	0	1	⌞ssim$\$$ ❶❼
2	s	1	1	im$\$$
3	s	0	0	im⌞ssim$\$$
4	$\$$	1	0	ippississim⌞ssim$\$$
5	s	0	1	issim⌞ssim$\$$ ❽
6	P	1	0	issississim⌞ssim$\$$
7	i	1	1	m$\$$ ❺
8	i	0	0	m⌞ssim$\$$
9	P	1	1	pississim⌞ssim$\$$ ❻❸
10	i	0	1	ppississim⌞ssim$\$$
11	s	1	1	sim$\$$ ❹
12	s	0	0	sim⌞ssim$\$$
13	s	0	0	sissim⌞ssim$\$$
14	–	1	1	ssim$\$$ ❾
15	i	1	0	ssim⌞ssim$\$$
16	i	0	0	ssississim⌞ssim$\$$
			1	

$$\mathsf{bl} = 10101011010100110$$
$$\mathsf{CL} = \text{m s \$ pi p s ⌞i}$$

$$\mathsf{bm} = 0010011010101001101011$$
$$min\_depth = \quad 1 \quad 1 \quad 2 \; 1 \quad 1 \quad 2 \; 2$$

Fig. 2. The reverse BWT of the example string, together with the structures that represent it. Only bitvectors bf, bl, and bm, plus arrays *min_depth*, CL, and CC (cumulative symbol frequency counts derived from CL, not shown), plus a dense array of pointers mapping backward block identifiers to disk addresses, are required during operation.

reverse of the example string, and the relationship between the forwards block identifiers (circular black labels), and the backwards block identifiers derived from the set of reversed BPS's (grey superscripts). The mapping between the original BWT sequence and the condensed sequence is recorded via three bitvectors bf, bl, and bm: bf contains a 1-bit at the beginning of every interval that is required during a search for any of the BPS reversals, plus a 1-bit at the first suffix after any of those intervals; bl is the inverse mapping, and has a 1-bit for every BWT symbol that is required while those same intervals are being accessed; and bm records the interleaving of the block identifiers and the 1-bits of bf, with 0-bits in bm indicating block identifiers (black circles), and 1-bits indicating corresponding 1-bits in bf.

Gog et al. [6] show how these structures can be used to support pattern search, and give comprehensive results including for a 62.5 GB text, with the complete suffix array index for that file requiring just 134 GB, or an average of 17.4 bits per pointer (with a total disk requirement of 197 GB = 2.9|T|); and the in-memory index taking 1.4 GB. Searching for infrequent patterns (*count* queries) of length $10 \le |P| \le 100$ requires two disk accesses and around 40 msec elapsed time on a MacBook Pro with mechanical disk; and under 2 msec on a MacBook Air with solid-state disk.

Fig. 3. The result of parsing the example string into factors. Each of the factors is a backwards block identifier, shown in Figure 2 as a grey superscript.

3   Reducing Disk Space

As with all suffix array structures, the text T is an integral part of the RoSA. In this section we describe how the condensed BWT can be used to store T as a set of factors. In particular, the set of block prefix strings can be interpreted as a phrase-book and used in a dictionary-based compression regime. Issues that must be addressed include:

- the parsing strategy used to convert T into a sequence of factors drawn from the set of available strings;
- the encoding used to store the factor identifiers, including how random access decoding is supported; and
- how the phrase-book is represented, and how decoding to be carried out using it.

Greedy Parsing: Every suffix of T has as a prefix a member of the set of BPS's which can be used as a factor. To identify a subset of those prefixes in a left-to-right greedy manner, the BWT depicted in Figure 2 is used, following the same backwards search process as is used for pattern matching. In addition, because the terminator symbol "$" is unique in T, it is a singleton suffix block and hence one of the BPS's, guaranteeing that a left-to-right parse can be properly closed.

For example, suppose that two factors "m" and "is" have been identified, and that "s_mississippi$" remains. Starting from the initial interval $(0, 16)$, the (lb, rb) range is narrowed to $(11, 16)$ as the next character, "s", is incorporated. This interval is greater than b, the blocksize, and so a BPS has not yet been identified. The "_" character narrows the interval to $(1, 1)$, corresponding to the reversed string "_s" and forwards string "s_". Now the range is smaller than b, and the backward identifier 3 is emitted (the grey superscript in Figure 2). Continuing the same process identifies the full set of factors shown in Figure 3. A full description of pattern searching appears in Gog et al. [6].

Encoded Representation: The factorization process generates a sequence T' of length $n' \leq n$, over an alphabet of $\sigma' \geq \sigma$ symbols. One obvious way of representing T' is to use fixed-width binary codes of $\log \sigma'$ bits each.

That approach is effective if the majority of BPS's are used as factors. If only a small subset occur in T', an alternative is to extract a subalphabet, and to use (potentially) shorter binary codes. We refer to this second approach as being a *dense alphabet* representation and the original as being *sparse*. The dense approach has the potential to generate superior compression, but requires an in-memory rank/select bitvector mapping of size σ' to convert between the dense domain and the set of BPS's.

```
00    decode_factor(bwd_id)
01       zero_pos ← select(bm, bwd_id, "0")
02       ones ← zero_pos − bwd_id
03       d ← min_depth[rank(bm, zero_pos, "1")]
04       if ones = 0 then
05          d ← d + zero_pos
06       else
07          d ← d + zero_pos − (select(bm, ones − 1, "1") + 1)
08       lb ← select(bf, ones, "1")
09       for i ← d − 1 downto 0 do
10          c ← first_row_character(lb)
11          factor[i] ← c
12          c_rank ← rank(bf, lb, "1") − CC[c]
13          c_pos ← select(CL, c_rank, c)
14          lb ← select(bl, c_pos, "1")
15       return factor[0..d − 1]
```

Fig. 4. Decoding a factor using the condensed BWT, given its backwards identifier. Function *first_row_character*(*lb*) returns the first character of the *lb* th suffix in the reversed BWT.

A third option is to use an entropy coder, and represent the factors using a Huffman or similar mechanism. The drawback of this approach is that to provide random access into the compressed bitstream an auxiliary structure storing indexed entry points is required; fixed-width binary codewords are much more useful in this regard.

Decoding: Figure 4 describes the process of converting a backwards identifier, extracted from T', back into a string over σ. Lines 01–08 determine the length of the factor, which is the depth in the condensed BWT of the corresponding BPS; and the start of the corresponding interval, *lb*. The loop from line 09 to 14 then extracts one character at a time, from left-to-right in the condensed BWT, and hence from right-to-left in P. The following lemma summarizes the decoding process.

Lemma 1: Function decode_factor() *correctly regenerates a factor from its backward identifier. A factor of d characters is computed in* $O(d \log \sigma)$ *time.*

Proof: Function *decode_factor*() executes in two distinct phases. The first eight lines access bitvector bm multiple times, with the goal of determining the length of the factor (variable d) and the left bound (variable *lb*) of the interval associated with the factor in the list of suffixes of the reversed text, T^r. The second group, starting at step 09, iteratively determine the symbols associated with that factor.

Consider the first block, steps 01 to 08. By construction, bitvector bm indicates the relative positions in bf at which the backward blocks occur, with each group ending in a "1" bit representing in unary (that is, by counting the "0"s) the number of backward block identifiers associated with the corresponding "1" bit in bf. For example, in Figure 2, bm commences with "0010011" to record that two backward blocks (hence, block number 0 and number 1) are associated with the first "1" bit in bf, the one that happens to be in bf[0]; that two more backward blocks are associated with the second "1" bit in bf, being the one in bf[1]; and that no backward blocks are associated with the third "1" bit in bf, the one in bf[2]. With this structure for bm, the *select* at step 01 identifies the location in bm of the "0" that corresponds to the block number supplied as argument *bwd_id*; then the difference calculation at step 02 determines how many "1"s there are through until that point; and then at step 08 the location in bf of that *ones*'th "1"-bit is determined using the *select* operator, and taken as the initial value of *lb*.

Because multiple backward blocks can have the same *lb*, the depth *d* of the block, which is the length of that block's corresponding BPS in the forwards suffix array (see Figure 1), serves as a secondary identification. Array *min_depth* stores an integer value for each "01" pair that arises in bm, showing the minimum depth of any of the blocks – of which there must be one, because of the "0", and might be more – that correspond to that "1"-bit in bm, and hence share a common "1" in bf. Backward block identifiers for the blocks associated with this "1"-bit in bf are assigned in increasing order from that minimum depth, noting that the block formation process is such that there cannot be any subsequent gaps in the sequence of block depths, and for any *lb* entry point, block depths form a contiguous range.

Once the minimum depth for a set of backward blocks has been established and assigned to *d* at step 03, the relative depth of the backward block indicated by *bwd_id* is computed at step 05, in the case that it is associated with the first "1"-bit in bf, or at step 07, in the case when the backward block corresponds to an entry point beyond the first "1"-bit in bf. That is, step 07 increments *d* by the distance between the *bwd_id*'th "0" location in bm and the most recently preceding "1" bit. By step 08, both *lb* and *d* have been assigned values, and in combination describe a single prefix in the sorted list of suffixes of the reverse string. Note that for symmetry with *rank* we regard *select* as starting at 0. For example, in Figure 2, *select*(bm, 3, "1") returns 8.

Now the decoding loop from step 09 commences. As an invariant, *lb* is the index in the sorted list of suffixes of T^r of the first *i* symbols of the factor. That invariant is established via the computation already described; and the loop exits once *i* reaches zero and no further symbols remain.

For a given *lb* value, the symbol that is required is the *lb*'th one in the list of sorted symbols of T. But *lb* is an index in the original domain, and CC stores the cumulative character frequencies in the condensed domain. Hence, the first action of function *first_row_character*() is to apply the full-to-condensed mapping that is encapsulated in bf, and compute *rank*(bf, *lb*, "1"). The offset that is generated is then binary searched for in the CC array, and the corresponding character *c* returned and stored in *factor*[*i*].

Steps 12–14 then adjust *lb*, in preparation for the next iteration. It must be updated to address the suffix in the reversed domain of the current string. For example, in Figure 2, if *bwd_id* = 8 when function *decode_factor*() is called, then *lb* is initially 11 to indicate the suffix "sim$", and *d* is 2, to indicate that the first two characters of that suffix are to be extracted and reversed. The first row character from row 11 when *i* = 1 is "s", which is stored in *factor*[1]. Now *lb* must be shifted so that it addresses the one-symbol-shorter suffix "im$". That adjustment is done in three operations: step 12 determines which of the *c* symbols in the condensed BWT string the one in question is, by first using bf to map into the condensed domain, and then offsetting by CC[*c*]; step 13 interrogates CL to determine where that same symbol appears within it; and then step 14 converts the rank of that symbol back into the uncondensed domain, to get the new *lb* value.

Table 1. Operations required during function *decode_factor()*

Line	Structure	Operation	Implementation	Number	Cost each
01	bm	*select*	compressed bitvector	1	$O(1)$
03	bm	*rank*	compressed bitvector	1	$O(1)$
03	*min_depth*	*access*	array	1	$O(1)$
07	bm	*select*	compressed bitvector	1	$O(1)$
08	bf	*select*	compressed bitvector	1	$O(1)$
10	bf	*rank*	compressed bitvector	d	$O(1)$
10	CC	*search*	sorted array	d	$O(\log \sigma)$
12	bf	*rank*	compressed bitvector	d	$O(1)$
13	CL	*rank*	Huffman wavelet tree	d	$O(H_0(\text{CL})) = O(\log \sigma)$
14	bl	*select*	compressed bitvector	d	$O(1)$

In terms of the example, starting with $lb = 11$ and $d = 2$, the three assignments compute $c\_rank = 0$ (that "s" is the first of the "s"s); $c\_pos = 1$ (that first "s" is the second symbol in CL); and then $lb = 2$ (that second symbol in CL corresponds to the third symbol in the uncondensed domain). Continuing the example, when $lb = 2$, the symbol "i" is prefixed to the previous "s", to make the complete factor "is". The loop then ends, and the decoded factor "is" derived from $bwd\_id = 8$ is made available to the calling environment.

To establish the execution cost of the decoding process, Table 1 summarizes the low-level operations required by *decode_factor()*. Note that function *first_row_character()* involves a *rank* on the bf bitvector, and then a binary search over the CC cumulative count array for the condensed BWT string, in order to identify the first character of the corresponding suffix. Summing the costs shown in the table yields the required bound on execution time.   □

If just *existence* and *count* queries are to be processed, the only additional space requirement compared to the original RoSA is that of providing *select* operations on two of the structures, CL and bl. But if *locate* and *context* queries are also anticipated, then a mapping that converts factor numbers to byte addresses in T is necessary.

Suffix Pointers: In the original RoSA, each suffix pointer is a byte address in T. With T replaced by a sequence of factors, there are now two options for the suffix pointers:

- retain them as byte offsets, and provide a mapping that converts byte addresses to ⟨factor number, offset within factor⟩ pairs.
- store the factor offset in which each suffix string commences; and after decoding, search within the decompressed factor to identify the commencing byte position of the pattern.

The second option requires that a mapping from factor numbers to byte offsets be maintained so that *locate* queries can be handled, but has the advantage of reducing the space taken by the suffix pointers, since, as is shown in Section 4, the sequence of factors is less than $1/10$ the length of the text T. That is, in the second option there is also a saving of around 3 bits per suffix pointer.

Table 2. Sample texts used in experiments. The values listed in the fourth column are generated by executing xz --best and expressing the output size as a fraction of the input size. The ROSA sizes are from Table 5 of Gog et al. [6], expressed as multiples of $|T|$, computed with $b = 4,096$.

File	Type of data	Length in characters	xz $(/|T|)$	Original ROSA $(/|T|)$		
				Memory	Disk	Total
WEB-4G	HTML/Web	4.19×10^9	0.071	0.025	1.961	2.986
WEB-64G	HTML/Web	6.87×10^{10}	0.076	0.022	1.900	2.922
DBLP-1G	XML/Bibliographic	1.08×10^9	0.112	0.020	2.126	3.146
DNA-3G	Text/Genomic	3.10×10^9	0.206	0.116	4.704	5.820

Approximate Suffixes: If factor rather than byte addresses are being stored as suffix pointers, then localized string search is required over the decoded factor(s) in order to find the starting point of the pattern. A further enhancement is then possible, of storing *approximate suffix pointers* that are restricted to certain quantized values – if the only addresses that may be stored as suffix pointers are the multiples of some fixed value R, then around $\log R$ bits can be saved in each suffix pointer. Access to T in this arrangement consists of a sequence of steps: first, identification of the suffix pointer that corresponds to the pattern P; second, retrieval of $m + R - 1$ factors from that starting location in T'; third, decoding those factors using Figure 4; and, finally, use of KMP or equivalent to search the decoded string for the pattern P. If a match is found and locations or contexts in T are required, an in-memory mapping is used to convert the ⟨factor number, byte offset⟩ coordinates back into a byte address in T. This mapping has a non-trivial cost, especially when R is small. For larger values of R the cost is amortized, and the mapping does not dominate the size of the index. The mapping could also be interspersed in the compressed text T' stored on disk, in which case it would not affect the query-time memory footprint. Either way, it gets smaller as R is increased from 1, further allowing space to be traded against execution time.

4   Experimentation

Text, Patterns, and Hardware: Table 2 summarizes the data files used in the experimentation, and lists previously reported sizes for the original ROSA when applied to them. The three English-plus-markup data files require an overall retrieval system that occupies around $3|T|$, and can be accessed via an index that requires less than 2.5% of the size of the raw text. The DNA data is more expensive both in memory and on disk.

Query pattern sets were generated by identifying all substrings of a given length and a given approximate frequency across the text. Once the set of substrings that met the two criteria was identified, 1,000 of them were chosen at random to make each test set. Using this methodology we are able to execute tests in which both the pattern length m is known and, independently, the occurrence frequency k is known.

The experimental machine was a MacBook Air with 1.8 GHz Intel Core i7 processor, 4 GB RAM, a 250 GB solid-state disk, and running Mac OS X 10.7.3.

Compressing the Text: Table 3 demonstrates that the factorization process yields effective compression, and reduces the space required by T by more than 75%. Comparing

Table 3. Representing T via the BPS's. In all cases the RoSA blocksize is $b = 4,096$.

File	Length in factors	Av. factor in characters	Sparse		Dense		H_0 factors
			range	$(/|\mathsf{T}|)$	range	$(/|\mathsf{T}|)$	$(/|\mathsf{T}|)$
WEB-4G	2.43×10^8	17.27	1.54×10^7	0.174	0.45×10^7	0.167	0.131
WEB-64G	6.71×10^{10}	24.70	2.19×10^8	0.142	4.74×10^7	0.132	0.091
DBLP-1G	8.99×10^7	12.03	2.89×10^6	0.229	1.35×10^6	0.218	0.190
DNA-3G	2.56×10^8	12.13	5.86×10^7	0.268	0.21×10^7	0.216	0.166

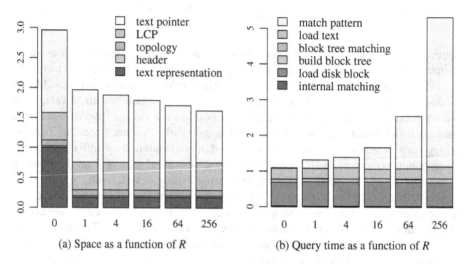

(a) Space as a function of R (b) Query time as a function of R

Fig. 5. Space and speed tradeoffs for *count* queries using WEB-4G, with $b = 4,096$, and patterns of length $m = 40$ and $k = 100$ occurrences. Space is measured in units of $|\mathsf{T}|$, and execution time as elapsed query time in msec per query, averaged over 1,000 distinct queries.

the sparse- and dense-alphabet versions shows that for the three English files there is only a modest additional gain achieved by the use of a subalphabet; whereas on the DNA file the gain is substantial. The final column in Table 3 shows the H_0 entropy of the factors, and represents the compression that could be expected using a Huffman code. The further gains do not justify the increased complexity that would be required.

Trading Space for Speed: Figure 5 shows how space and search time vary as a function of R, the quantization factor for suffix pointers. The original RoSA is shown as $R = 0$. When $R = 1$, significant space is saved as a result of using the factored text (the green bar at the bottom of the left graph); with only slightly increased querying time. Thereafter, each increase in R further reduces the cost of the suffix pointers (the grey top component in the left-hand graph), at the cost of increased decoding and KMP scanning time (the grey top component in the right-hand graph). Taking $R = 16$ yields a useful compromise between space and speed.

Query Types: Table 4 shows the relative cost of *count* and *locate* queries, with time measured in elapsed milliseconds per query (msec), that is, including the cost of

Table 4. Execution cost (elapsed query time in msec per query) for *count* and *locate* queries using different pattern matching structures and different numbers of answer occurrences k. All results are for $b = 4,096$ and $|P| = 40$ and are the average over 1,000 distinct queries.

Structure	File	Memory ($\%|T|$)	Disk ($/|T|$)	*count* queries $k = 10$	$k = 100$	*locate* queries $k = 10$	$k = 100$
RoSA, $R = 0$	WEB-4G	2.5	2.96	1.0	1.1	1.0	1.1
RoSA, $R = 16$	WEB-4G	3.0	1.80	1.6	1.7	7.0	46.8
RoSA, $R = 64$	WEB-4G	2.5	1.71	2.4	2.5	13.1	101.9
FM-INDEX, *sampling* = 16	WEB-4G	44.4	—	0.1	0.1	0.3	0.8
FM-INDEX, *sampling* = 64	WEB-4G	22.2	—	0.1	0.1	0.6	3.3
RoSA, $R = 0$	WEB-64G	2.3	2.89	1.1	1.1	1.1	1.2
RoSA, $R = 16$	WEB-64G	2.6	1.75	2.8	2.2	11.5	96.7
RoSA, $R = 64$	WEB-64G	2.4	1.68	4.5	3.9	33.4	221.7
FM-INDEX, *sampling* = 64	WEB-64G	21.6	—	120.3	116.9	268.5	644.7

accessing secondary storage. In the $R = 0$ original RoSA both types of query are very fast, taking around a msec on a machine with SSD drive. When T has been compressed, *count* queries are still fast, but *locate* queries require factor decompression plus KMP searching, neither of which is needed when $R = 0$. The added searching cost is offset by a saving of more than $1.0|T|$ disk space between $R = 0$ and $R = 16$. The $R = 16$ enhanced RoSA structures store the text T plus its suffix array in just $1.8|T|$.

Compared to the FM-Index and the FEMTO: Table 4 also gives query timings for the FM-INDEX [5], using the best available implementation [7]. The FM-INDEX has the great advantage of being smaller than the text T, even when sampled access points are provided to allow *locate* and *context* queries; but requires that all of its data be memory resident. When that is not possible, query times increase dramatically, as shown by the last row of Table 4. The same experiment on a MacBook Pro with a mechanical disk resulted in a query time of more than two seconds, whereas the RoSA performs just two disk accesses and requires 41 msec on a MacBook Pro. That is, regardless of whether mechanical disk or SSD is used, if the index can be maintained in memory, the FM-INDEX should be preferred; but once the size of the FM-INDEX index exceeds the available memory, the RoSA is faster.

The work of Ferguson [3] confirms what happens when an FM-INDEX is used in external memory. Using SSD storage, and a 43 GB collection of English text, the FEMTO system required approximately 90 msec to execute a *count* query for a single $m = 28$-byte test pattern of unknown collection frequency, and a further 40 msec to identify 10 of the locations at which that three-word string appeared. When executed using disk storage, the same query took more than 3 seconds. These times are considerably longer than those shown in Table 4, but are achieved with between $0.5|T|$ and $1.0|T|$ of space, less than the $1.8|T|$ required by the enhanced RoSA.

Ferguson also explored batching of queries, and demonstrated a substantial increased in query throughput, at the cost of corresponding increases in query latency, by "elevator sorting" storage requests.

5 Summary

We have demonstrated that the approximately $3|T|$ disk space requirement of the ROSA for English text can be reduced to less than $2|T|$. Moreover, the techniques used – factorizing T, and quantizing suffix pointers – are implemented without the need for an explicit in-memory phrase-book, a significant achievement. Search times increase somewhat, but remain better than other external-memory approaches. In future work we plan to turn our attention to the LCP component (the pink section in Figure 5(a), second from the top, and larger than the compressed text), and to fast sequential search mechanisms that can avoid fully decoding factors while carrying out the scanning phase.

Acknowledgments. This work was funded by the Australian Research Council. We also thank Giovanni Manzini for his input.

The ROSA software is available at `https://github.com/simongog/RoSA`.

References

1. Baeza-Yates, R.A., Barbosa, E.F., Ziviani, N.: Hierarchies of indices for text searching. Inf. Systems 21(6), 497–514 (1996)
2. Colussi, L., De Col, A.: A time and space efficient data structure for string searching on large texts. Inf. Processing Letters 58(5), 217–222 (1996)
3. Ferguson, M.P.: FEMTO: Fast search of large sequence collections. In: Kärkkäinen, J., Stoye, J. (eds.) CPM 2012. LNCS, vol. 7354, pp. 208–219. Springer, Heidelberg (2012)
4. Ferragina, P., Grossi, R.: The string B-tree: A new data structure for search in external memory and its applications. J. ACM 46(2), 236–280 (1999)
5. Ferragina, P., Manzini, G.: Indexing compressed text. J. ACM 52(4), 552–581 (2005)
6. Gog, S., Moffat, A., Culpepper, J.S., Turpin, A., Wirth, A.: Large-scale pattern search using reduced-space on-disk suffix arrays. IEEE Trans. Knowledge and Data Engineering (to appear)
7. Gog, S., Petri, M.: Optimized succinct data structures for massive data. Software Practice & Experience (to appear, 2013), `http://dx.doi.org/10.1002/spe.2198`
8. González, R., Navarro, G.: A compressed text index on secondary memory. J. Combinatorial Mathematics and Combinatorial Comp. 71, 127–154 (2009)
9. Kärkkäinen, J., Rao, S.S.: Full-text indexes in external memory. In: Meyer, U., Sanders, P., Sibeyn, J.F. (eds.) Algorithms for Memory Hierarchies. LNCS, vol. 2625, pp. 149–170. Springer, Heidelberg (2003)
10. Mäkinen, V., Navarro, G.: Compressed compact suffix arrays. In: Sahinalp, S.C., Muthukrishnan, S.M., Dogrusoz, U. (eds.) CPM 2004. LNCS, vol. 3109, pp. 420–433. Springer, Heidelberg (2004)
11. Manber, U., Myers, G.W.: Suffix arrays: a new method for on-line string searches. SIAM J. Comp. 22(5), 935–948 (1993)
12. Moffat, A., Puglisi, S.J., Sinha, R.: Reducing space requirements for disk resident suffix arrays. In: Zhou, X., Yokota, H., Deng, K., Liu, Q. (eds.) DASFAA 2009. LNCS, vol. 5463, pp. 730–744. Springer, Heidelberg (2009)
13. Sinha, R., Puglisi, S.J., Moffat, A., Turpin, A.: Improving suffix array locality for fast pattern matching on disk. In: Wang, J.T.-L. (ed.) Proc. ACM SIGMOD Int. Conf. Management of Data, pp. 661–672 (2008)

You Are What You Eat:
Learning User Tastes for Rating Prediction

Morgan Harvey[1], Bernd Ludwig[2], and David Elsweiler[2]

[1] Faculty of Informatics, University of Lugano, Lugano, Switzerland
[2] Inst. for Info. and Media, Lang. and Culture, University of Regensburg, Germany
morgan@derharvey.de, bernd.ludwig@ur.de, david@elsweiler.co.uk

Abstract. Poor nutrition is one of the major causes of ill-health and death in the western world and is caused by a variety of factors including lack of nutritional understanding and preponderance towards eating convenience foods. We wish to build systems which can recommend nutritious meal plans to users, however a crucial pre-requisite is to be able to recommend recipes that people will like. In this work we investigate key factors contributing to how recipes are rated by analysing the results of a longitudinal study (n=124) in order to understand how best to approach the recommendation problem. We identify a number of important contextual factors which can influence the choice of rating. Based on this analysis, we construct several recipe recommendation models that are able to leverage understanding of user's likes and dislikes in terms of ingredients and combinations of ingredients and in terms of nutritional content. Via experiment over our dataset we are able to show that these models can significantly outperform a number of competitive baselines.

1 Introduction

In the developed world people have the luxury of an abundance of choice with regard to the food they eat. While huge choice offers many advantages, making the decision of what to eat is not always straightforward, is influenced by several personal and social factors [9] and can be complex to the point of being overwhelming [12]. Therefore, many people would benefit from assistance that allows them to strike a balance between a diet that is healthy and will keep them well and one that is appealing and they will want to eat. After all, it is no good providing users with healthy diet plans if they do not cook and eat the recipes therein, but instead choose unhealthy meals which are more appealing to them.

This is a problem for which recommender systems (RS) are ideally suited: If systems can predict recipes that the user would actually *like to eat*, this could be combined within a system modelling expert nutritional knowledge to generate appealing meal recommendations that are also healthy and nutritious. A prerequisite, therefore, is an understanding of the factors that influence the decision of whether a recommended meal will be eaten and prepared or not. In this work we investigate these factors by analysing the results of a long-term user study, using the insights obtained to build new RS which are able to significantly outperform the current state-of-the-art in this field.

O. Kurland, M. Lewenstein, and E. Porat (Eds.): SPIRE 2013, LNCS 8214, pp. 153–164, 2013.

2   Related Work

RS provide suggestions, in the form of items, that are predicted to offer utility to the user. Such systems are particularly beneficial in situations where there is an overwhelming choice of alternatives and/or where the user lacks sufficient personal experience, competence or time to evaluate potential options [10]. Correspondingly, recommendations are usually made based on knowledge of the user's needs, preferences, and past behaviour. Many RS only use past ratings in order to predict ratings for previously unseen combinations of user and item. A common approach to generating recommendations is to mimic the natural human behaviour of making decisions based on recommendations from peers. More modern approaches [5] attempt to learn a model of how ratings are generated by breaking the rating process down into a number of components or "biases" which contribute to the final rating. In the case of recommending recipes there are many content-related features that could be used to base predictions on, e.g., cooking time, ingredients, nutritional properties, classification of dish, skills required. The open questions are: which content is useful and how can you best make use of this content in recommendation models?

While food recommendation is not frequently studied, there is a small body of appropriate related work. Early attempts to design automated systems using case-based planning to recommend meals include CHEF [4] and JULIA [7]. Hybrid recommenders have been presented [13] for recommending recipes and systems have been proposed based on grouping of users [14]. More recent efforts try to understand user's tastes, improving recommendations by breaking recipes down into individual ingredients, which has been demonstrated to work well [2,3]. This work has shown that, in the case of recipes, new approaches to the RS problem are necessary. We hypothesise that the process of rating a recipe is complex and several factors will combine to determine the rating assigned, beyond purely the user's tastes and that these tastes must be carefully modelled. Factors such as how well the preparation steps are described and perhaps the nutritional properties of the dish and the availability of ingredients could have a bearing on the user's opinion of the recommendation [6]. We believe that by building recommender algorithms that incorporate or exploit these kinds of aspects we will be better able to accurately predict ratings. However we also believe that it is important that such factors can be automatically ascertained from ratings data rather than relying on the users themselves. The amount of information expected from users is therefore minimised. Below we describe how data was collected and analysed to understand how content and contextual factors may influence the way a recipe is rated.

3   Data Collection

To collect data we developed a simple food rating system, which selected recipes from a pool of 912 recipes sourced from a popular German recipe web site. While there is quite a strong emphasis on German food (which is beneficial as most users were German), the web site also contains a large number of recipes from all of the major world cuisines. Users were given a personalised URL and when this was accessed, they were presented with a randomly selected recipe. The system did not attempt to perform any recommendation or try to match recipes to a user's tastes. The user was then asked to provide a rating for the recipe in context - either as a main meal or breakfast for the following day - by clearly stating which meal the user should have in mind when rating, e.g. Please rate this recipe as a breakfast for tomorrow. Recipe meta-data was used to determine which meals should be recommended for which time period. This is important because, in contrast to previous data collection methods, the user is not only rating the recipe with respect to how appealing it is, but also how suitable the recipe is given a specific context. In addition to collecting ratings, the web interface offered the user the chance to explain his rating by clicking appropriate check boxes representing reason. These check boxes were grouped into reasons to do with personal preferences, reasons related to the healthiness of the recipe and reasons related to the preparation of the recipe. The listed explanations were generated through a small user study, whereby 11 users rated recipes and explained their decisions in the context of an interview. The web interface also provided a free-text box for reasons not covered by the checkboxes [1].

After publicising the system on the Internet, through mailing lists and twitter, 124 users from 4 countries provided 4,472 ratings over a period of 9 months. We argue that although this is a relatively small and sparse dataset, it is an improvement on previous recipe ratings data collection methods, which have used mechanical turk, where there are no validity controls and users are incentivised to rate as many recipes a possible as they are being paid [2,3] and surveys where participants rate large numbers of recipes or ingredients in a single session.

Our dataset also differs from previous work in terms of matrix density. The number of ratings per user is Zipfian (median = 7, mean = 29.93 max = 395 min =1; 18 users have 1, 52 have 10+). Whereas previous food recommender papers report user-ratings densities of 22%-35% [2,3], our dataset exhibits a more realistic density of 3.95% and a median 3 ratings per recipe (mean = 4.04, max=14, min=2), more in line with collections such as movielens and netflix. Our dataset is, therefore, not only realistic, but also a challenging platform for experimentation as it is both sparse and variant in terms of ratings (sd = 1.43).

4   Exploratory Analysis

To learn about the decision process undertaken when users rate recipes, as well as the factors that influence this process, we statistically analysed the reasons provided by the users when they rated. The most common reasons for negatively rating a recipe were that the recipe contained a particular disliked ingredient, the combination of ingredients did not appeal, or the recipe would take too long to prepare and cook. The most common positive factors included ease or quickness of preparation, the type of dish or the recipe being novel or interesting.

---

[1] Screenshot of the interface - http://tinypic.com/r/1zx4p77/4

Health related reasons, such as the recipe containing too many calories, the recipe being perceived as unhealthy, or positive factors like the recipe being balanced or easily digestible were clicked less often overall. However, these were clicked very frequently for a particular subset of users; those who ever chose a health reason did so, on average, for 16.3% of the recipes they rated.

We trained a number of linear models to understand how relationships between factors contribute to a final rating. The final model (adj. R^2=0.329) shows that 17 factors were significant. Ingredient factors, such as the presence of particular ingredients or combination of ingredients and whether meat was in the recipe had particularly strong predictive power. Furthermore, the data show that ingredient factors can have both a positive and negative influence on ratings and that the combination of ingredients can be important, neither of which are considered by current models.

Although the health factors did not add significantly to the predictive power of the models, we wanted to understand if they might help predict ratings on a per-user basis. We looked at the correlation between calorie and fat content of recipes and the ratings provided by two groups of users, those who had clicked on a health related factor once or more (Care-about-Health, n= 54, 3130 ratings), and those who never clicked on a health reason (Don't-Care-About Health, n=70, 1342 ratings)[2]. Figure 1 shows clear differences between the rating behaviour exhibited in these groups. There is a strong trend that for the Care-about-Health group, the higher the fat (R^2=0.88, p=0.012) or calorific content (R^2=0.87,p=0.022) of the recipe, the lower the rating. However, this trend is not present in the second group. If anything, there seems to be a slight tendency toward the reverse trend, whereby recipes higher in fat (R^2=0.23,p=0.643) and calories (R^2=0.73, p=0.064) are assigned a higher rating. This analysis suggests that accounting for nutritional factors in recommendation models will allow more accurate predictions to be generated. To summarise, these analyses demonstrate

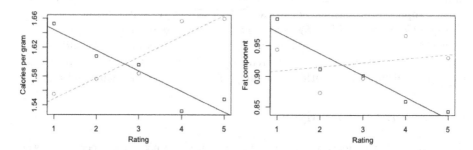

Fig. 1. Influence of Calorific Content and Fat on Ratings

the complexity of rating decision process. Even with 17 significant explanatory variables, the best model is still only able to return an adj. R^2 value of 0.329. Nevertheless, they hint that several factors could be exploited in recipe recommendation algorithms to improve accuracy. In the following section we describe

---

[2] Nutritional content of recipes was calculated using the system as described in [8].

models that exploit the factors and trends uncovered. As a starting point, we focus on building powerful ingredient based models, as these were shown to be important and have been emphasised in previous work. We then extend these ingredient models to take account of nutritional aspects in terms of fat and energy, which aligns well with our long-term research aims.

5 Recommendation Models

Before describing the new models, we first introduce appropriate notation in Table 1. Note for the purposes of this discussion recipes may also be referred to as "items" and ingredients as "features", these terms will be used interchangeably.

Table 1. List of notation for recommender models

Symbol	Description	Symbol	Description
$d \in \mathbf{d}$	set of items (recipes)	Φ	item feature weights $\mathbf{d} \times \mathbf{f}$
$u \in \mathbf{u}$	set of users	$\phi_{d,f}$	weight of feature f in item d
$f \in \mathbf{f}$	set of features (ingredients)	Ψ	user feature weights $\mathbf{u} \times \mathbf{f}$
R	ratings matrix $\mathbf{u} \times \mathbf{d}$	$\psi_{u,f}$	weight of feature f for user u
b_u	bias due to user u	Ψ^+	matrix of positive user features
b_d	bias due to item (recipe) d	Ψ^-	matrix of negative user features
$r_{u,d}$	rating for item d by user u	IUF_f	inverse user frequency of feature f
IDF_f	inverse item frequency of f		

Ingredients contained in recipes are like words in documents and can be referred to as features. Based on this assumption, we can build an item-feature matrix Φ which can be either binary, indicating the presence or absence of an ingredient in a recipe, or the relative weight of each ingredient in the recipe. The weight of feature f in item d is $\phi_{d,f}$. To compute a similarity between users and items, we can construct a similar feature matrix for users. Such a matrix Ψ (where $\psi_{u,f}$ is the weight of feature f for user u) can be constructed by considering the ingredients contained within the recipes rated by the user. Concretely: $\psi_{u,f} = \sum_{d \in D} \phi_{d,f} \, I\{r_{u,d} > 0\} \, \pi$, where π is an additional weighting factor that may be item, user or feature-dependent (in the un-weighted case this defaults to 1) and $I\{\}$ is the indicator function which is 1 when the condition within the braces - in this case that user u has rated item d - is satisfied.

From our analysis we know that the contextual factors which affect ratings can be both positive and negative and vary in their influence. In the case of ingredients, it was shown that users have a set of ingredients (and combinations thereof) that they like as well as a set of those that they do not like. Previous work has attempted to incorporate this observation into the modelling process by weighting ingredient-features by the rating assigned to the recipes containing that ingredient [3]. In one model, ratings are exploited by assigning weighting to ingredients based on their parent recipe's rating (i.e. π is set to $r_{u,d}$). The problem with this approach is that it implicitly assigns some positive rating to ingredients which the user dislikes, particularly compared to ingredients which they have not yet rated.

We take a different approach by using two separate user-feature matrices Ψ^+ and Ψ^- containing weighted values for ingredients the users like and those that they do not like. Ψ^+ is derived from recipes to which the user assigned a rating of 4 or 5 and Ψ^- from those that the user assigned a score of 1 or 2. Here we utilise the weighting factor π; for Ψ^+ we can assign a weighting of 1 for those rated 4 and a weighting of 2 for those rated 5, for Ψ^- recipes rated 1 receive a weighting of 2 and those rated 2, a weight of 1. This preserves the idea that a rating of 5 indicates a stronger positive preference than a rating of 4 whereas a rating of 1 should be more strongly negative than a rating of 2.

5.1 Predicting Ratings

We now need a metric to determine how similar (or dissimilar) an item and a user are. We use a variation on TF-IDF weighting [11] as this will give high weights to ingredients that are frequently rated positively by the user of interest, but not generally by all users. Similarity between two items can be computed using the cosine similarity metric, resulting in a vector space (VS) model:

$$sim_{VS}(u, d) = \sum_{f \in \mathbf{f}} \frac{(\psi_{u,f}\, IUF_f)(\phi_{d,f}\, IDF_f)}{\sum_{\mathbf{f}} \sqrt{(\psi_{u,f} IUF_f)^2} \sum_{\mathbf{f}} \sqrt{(\phi_{d,f} IDF_f)^2}} \tag{1}$$

When analysing the ratings matrix we noted that it was rather sparse, particularly on a per-user basis with many users having only rated a small number of recipes. This introduces problems for the basic TF-IDF model as a large number of users will have very sparse feature vectors. Our analyses also show that people like types of ingredients and specific combinations of them and therefore performance may be improved by trying to learn which ingredients are similar through their co-occurrence in recipes. This can be achieved implicitly by the use of dimensionality reduction techniques on the feature matrix.

We can therefore apply a Singular Value Decomposition (SVD) to the feature matrices in a similar fashion to previous work in information retrieval [1] where this method has been used successfully to improve accuracy for document retrieval. SVD is commonly used to reduce the amount of noise within matrices and can uncover relationships between variables that are not obvious from the explicit first-order co-occurence data. The reader is referred to [1] for a more detailed treatment of the subject. Given a reduced-dimensionality representation of the original feature matrices, a similarity metric between two items is simply the cosine of the angle between their vectors over the new feature space.

5.2 User and Item (Recipe) Biases

As noted in the related work section, many modern RS estimate ratings based on a number of biases. Two sets of biases which have been shown to have a large impact on the rating ultimately given are dependent on each individual user and on the item (in this case recipe) being rated. For example, some users may naturally rate items higher than others and some may naturally choose from a lower baseline score. Similarly some items are intrinsically better than others and are therefore likely to be rated higher by all users. By calculating these biases as part of our model, we can effectively remove these eccentricities from the ratings. This gives the ingredient similarity measures the freedom to deal purely with the variations caused by each user's tastes. The bias due to user u is denoted b_u and the bias for recipe/item d is denoted b_d.

These biases can be calculated by means of iterating fixed-point gradient descent optimisation routine based on the training ratings until convergence is observed via the following update rules:

$$\hat{b_u} = b_u - \lambda(eb_d - \alpha b_u) \tag{2}$$

$$\hat{b_d} = b_d - \lambda(eb_u - \alpha b_d) \tag{3}$$

where $\hat{b_u}$ and $\hat{b_d}$ are the updated values for the parameters, λ is a fixed scalar parameter which determines the learning rate of the optimiser, α is a regularisation parameter to prevent over-fitting and e is the error of the following simple model estimate $\hat{r}_{u,d} = \mu + b_u + b_d$ where μ is the mean training rating.

5.3 Including Nutritional Information

Our analysis indicate that there is a notable split between users who appear to care about the healthiness of a dish and those to whom this factor is perhaps not so important. We know, for example, that those users for whom nutrition is important will, in the mean, rate items with high levels of fat and calories lower than other recipes. This information could be used to introduce an additional bias into the model in order to improve prediction performance. To model this bias, we first split the recipes into "bins" based on their calorific and fat content. Bins were chosen by calculating the q quantiles of the calories and fat respectively and assigning each recipe to its corresponding bin. We separated users into "healthy" and "unhealthy" groups based on their use of the *calories* and *healthy* checkboxes in the training ratings. For each of the two groups a vector of biases was computed for all bins over both the calories and the fat content, where the biases are simply the expected mean-normalised change in the rating for rated items within the bins. These biases are then included as additional explanatory variables in the linear model. Due to the splitting of users it is necessary to calculate two separate models, one for each user group, since the coefficients for both the calories and fat biases will be different for the two distinct groups.

To predict a rating $\hat{r}$ for user u given a recipe (or item) d we can learn a linear weighted model based on the output from the similarity metrics over both positive and negative feature matrices and the biases:

$$\hat{r}_{u,d} = \theta_0 + \theta_1 sim^+(u,d) + \theta_2 sim^-(u,d)$$
$$+ \theta_3 b_c(u,d) + \theta_4 b_f(u,r) + \theta_5(b_u + b_d)$$

where $b_c(u, d)$ and $b_f(u, d)$ are the predicted calorie and fat biases for user u and recipe d (based on the calorie and fat bins d belongs to). The terms in this linear equation can actually be seen as the factors that combine to bias the rating in either a positive or negative direction, thus perturbing the rating from some baseline "standard" rating. θ_0 can be seen as approximating a standard or average rating, θ_1 is the factor biasing the rating in a positive direction and θ_2 biases in the opposite direction. θ_3 and θ_4 represent the biases due to nutritional content and θ_5 encodes the influence of the user and item-specific biases. These weights can be optimised using a large number of numerical optimisation procedures including gradient descent, neural networks and generalised linear models. Due to its stability and relative simplicity we use the latter method in this work.

6   Experimental Results

To test the performance of our models for recipe recommendation we must ascertain how well they are able to predict ratings for unknown pairs of users and recipes. To do so we randomly separated our dataset into 5 equal partitions and conducted split-fold testing where for each test 4 of the partitions is used for training the models and the remaining partition is used to test performance. resulting in a total of 3,624 training ratings and 848 test ratings.

The prediction problem is best described by saying that we would like to "fill in" the sparse ratings matrix, extrapolating (or predicting) a rating $\hat{r}_i$ for every possible user-item pair from the limited data available. More practically we wish to define some function or model which will minimise the root mean squared prediction error over the test data $RMSE = \sqrt{\frac{1}{N_{test}} \sum_{i=1}^{N_{test}} (r_i - \hat{r}_i)^2}$. The RMSE is commonly used in statistics for measuring the difference between the set of values predicted by a model and the values actually observed from the system being modelled. We also report the Mean Absolute Error (MAE) which is simply the mean absolute difference between the predicted rating and the actual rating, over the whole test set. We report both metrics as they provide different information regarding the performance of predictions: the RMSE penalises large errors much more than small errors while the MAE penalises all errors equally relative to their size.

6.1   Models and Parameters

We compare the performance of our models against 3 baselines from the CF literature, including the state-of-the-art recipe recommendation model:

mean-r naïve baseline, returns the mean rating as an estimate for all u, d pairs.
CF nearest-neighbour method, Pearson correlation coefficient similarity metric.
CB best-performing content-based algorithm by Freyne et al [3].

In this section we evaluate the performance of the following 4 recipe recommendation models as described in Section 5:

VS weighted model with VS similarity measure
VS+n weighted model, VS similarity measure, nutritional biases
VS+n+b weighted model, VS similarity measure, all biases
SVD weighted model, SVD-based similarity measure
SVD+n weighted model, SVD-based similarity measure, nutritional biases
SVD+n+b weighted model, SVD-based similarity measure, all biases

For CF we use a maximum of 10 neighbours ignoring those with low similarity
(<0.2). Both SVD models were trained over 100 dimensions. For the +n models
q was set to 20 quantiles. The user and item optimiser converges as would be
expected, with major gains being made over the earlier iterations and becoming
smaller as the optimal values are reached, completely flattening out near the end.
This hints that the algorithm has fully converged by this point. The learning rate
λ was set to 0.001 and the regularisation parameter α was set to 0.05 as this
resulted in the fastest convergence times and best held-out likelihood.

6.2   Average Performance

Table 2. Best results from each model. % indicate improvement over mean baseline.
* indicates statistically significant improvement over mean, † over CB model.

Model	Prediction error		Improvement	
	MAE	RMSE	MAE	RMSE
mean	1.180	1.383	-	-
CF	1.175	1.379	0.42%	0.28%
CB	1.154	1.347	2.2%	2.6%
VS	1.115 *	1.308 *	5.5%	5.4%
VS+n	1.109 *	1.299 *	6%	6.1%
VS+n+b	1.079 * †	1.269 * †	8.6%	8.2%
SVD	1.095 * †	1.296 * †	7.2%	6.3%
SVD+n	1.086 * †	1.289 * †	8%	6.8%
SVD+n+b	1.072 * †	1.256 * †	9.2%	9.2%

Table 2 shows the average performance figures yielded by the models. Significance is determined based on the p-value returned by a paired Student's-t test.
Exact p-values were: SVD-CB = 0.02, SVD+n-CB = 0.011, mean-CB = 0.39.
The p-values comparing the mean with all of the models presented in this paper
were ≪ 0.01. The results indicate that all the content-based recommenders are
able to outperform both the mean rating and the neighbourhood-based algorithm, which returns particularly poor figures for this dataset. This is likely due
to the sparsity of the data making it difficult for the algorithm to find suitable
neighbours from which to derive its estimates. Among the content-based methods it is clear that the VS method outperforms the CB method and that the
SVD method in turn outperforms VS.

Addition of the nutritional information into the model improves performance for both the VS and SVD variants, however in neither case is this improvement significant. The addition of the individual user and item biases is, however, significant and increases the performance of both the VS and SVD-based models. In fact, the performance gain is such that the VS model with the biases is even able to beat both of the SVD models without the biases. As would be expected, the performance of the SVD algorithms are somewhat dependent on the number of dimensions. Performance with a small number of dimensions (i.e. 10) is poor, but increases consistently until it reaches an informational saturation point at approximately 100 dimensions, after which performance gain is asymptotic. The performance of all of the trained models increases with the proportion of training data, however it appears that the newer models are better able to exploit the extreme case where 90% of the data is used for training. The errors returned at the other extreme (i.e. where only 50% of the data is kept for training) suggest that the SVD-based model is able to cope better in the case of sparse data than the VS-based one.

6.3   Standard Deviation of Errors

The RMSE and MAE provide useful information regarding the performance, and more specifically the expected error, of a given prediction algorithm. However, users do not want excessively large errors as this can rapidly destroy their trust in the system and therefore the standard deviation of the errors is also important.

The most variant errors are returned by the CF algorithm followed by the mean, with these returning 0.716 and 0.715 respectively. The content-based recommenders perform better: CB = 0.696, VS = 0.669, SVD = 0.665, VS+n = 0.66 and SVD+n= 0.659. By adding in the user and item biases the standard deviation is further reduced to 0.647 for SVD+n+b and 0.649 for VS+n+b. These results illustrate further that differences in performance suggested by the RMSE and MAE results are likely to make a tangible difference to the accuracy of the recommender. There is little difference between the performance of the 3 baselines, however there is a large step-up in the performance of the models outlined in this work. The fact that the improvements (over the baseline) for the RMSE scores are larger than for MAE also suggests that the models presented in this paper make fewer large errors. As discussed previously, this is advantageous as making large errors can have serious implications with regard to the trust of the user in the RS. The much lower standard deviations for the models that incorporate the user and item biases also illustrate how much extra prediction power and flexibility these are able to provide.

7   Discussion of Results and Conclusion

In this paper we have investigated the decisional process involved in rating recommended recipes. We described a large naturalistic data collection method with 124 users in which recipes were rated in context over a period of 9 months. This resulted in a realistic dataset for testing RS for this specific problem which approximates well the kind of data that would be generated in a real recipe recommendation system, especially compared to recipe datasets used in the past. Analyses of the dataset underlined the complexity of the recipe rating process with 17 factors having a significant influence on the rating in the best linear model. Yet, this model is only able to explain about a third of the variance in the rating. However, based on insights obtained from analyses performed, we developed new models and showed empirically that these models offer performance improvement over strong baselines.

The results justify choices made in the modelling process; the new models offer improved performance both in terms of reducing error and the variance of the error. The results show that the ingredients contained within recipes are important and that this data can be better exploited by using models that account for positive and negative weighting and by applying dimensionality reduction techniques. Furthermore it is clear from the results obtained that training separate bias parameters for each individual user and item is extremely beneficial, particularly given the low cost in terms of increased model estimation time.

Including nutritional information in the model was also shown to be beneficial. The models incorporating calorie and fat data offered improved performance, although the differences were not significant. It is our intuition that as our dataset grows, the results of models exploiting nutritional information will improve. It should also be noted that our current method of incorporating recipe nutritional information in the rating prediction is quite simple and could certainly be improved. More sophisticated models might, for example, use a continuous function to estimate weights for calories or fat values rather than simply using bins. Moreover, future models may learn weights on a per user basis rather than relying on pre-defined groups as we do now.

The presented work represents a single component in a much larger project aimed at building RS that can promote healthier dietary choices. Our short term goals include continuing the work here to build models that better predict user food preferences using the ideas suggested above or, for example, by incorporating other content factors. We are continuing to collect data and hope to investigate how performance of models change as the collection size increases. For example, will the CF approaches eventually match content approaches when the collection achieves a certain density? We acknowledge that, in contrast to our long-term goals, the nutrition-aware models could be improving performance by offering unhealthy choices to users who prefer such recipes. It would be interesting to look at how these models influence the error for the two user groups. Further analysis might also provide an understanding of where nutritional content plays a role i.e. at what level of fat content do users start to rate differently? Our initial analyses in Section 4 suggest that there is also scope for further performance improvement by developing models that take other content-related factors into account.

In the longer term we plan to move beyond recommending recipes in isolation to recommending full dietary plans. This would involve recommending sequences of recipes under a number of constraints such as the daily recommended intake suggested by the WHO, and user activity patterns. Achieving this will present several algorithmic and usability challenges and will necessitate the development of more sophisticated models. We are also interested in understanding how using this kind of RS can influence user behaviour and knowledge of nutritional principles. Will users learn about nutrition over time from the suggestions made or will they instead simply rely on the system to provide them with meal plans without truly understanding what they should be eating to keep healthy?

It is worth noting that the models presented in this paper could also be used for other RS problems where contextual data is available for items that could be used as features for the similarity matrices. For example, the models could be adapted to recommend movies by using features such as directors, actors and genres instead of ingredients.

References

1. Deerwester, S., Dumais, S., Landauer, T., Furnas, G., Harshman, R.: Indexing by lsa. J. of the Am. Soc. of Inf. Sci. 41(6), 391–407 (1990)
2. Freyne, J., Berkovsky, S.: Intelligent food planning: personalized recipe recommendation. In: 15th Int. Conf. on Intelligent User Interfaces, IUI 2010, pp. 321–324. ACM, New York (2010)
3. Freyne, J., Berkovsky, S., Smith, G.: Recipe recommendation: Accuracy and reasoning. In: Konstan, J.A., Conejo, R., Marzo, J.L., Oliver, N. (eds.) UMAP 2011. LNCS, vol. 6787, pp. 99–110. Springer, Heidelberg (2011)
4. Hammond, K.: Chef: A model of case-based planning. In: Proceedings of the National Conference on AI (1986)
5. Harvey, M., Carman, M.J., Ruthven, I., Crestani, F.: Bayesian latent variable models for collaborative item rating prediction. In: Proc. CIKM 2011, pp. 699–708. ACM (2011)
6. Harvey, M., Ludwig, B., Elsweiler, D.: Learning user tastes: a first step to generating healthy meal plans? In: ACM RecSys 2012 LifeStyle Workshop (2012)
7. Hinrichs, T.: Strategies for adaptation and recovery in a design problem solver. In: Proceedings of the Workshop on Case-Based Reasoning (1989)
8. Mueller, M., Harvey, M., Elsweiler, D., Mika, S.: Ingredient matching to determine the nutr. properties of internet-sourced recipes. In: Pervasive Health (2012)
9. Nestle, M., Wing, R., Birch, L., DiSogra, L., Drewnowski, A., Middleton, S., Sigman-Grant, M., Sobal, J., Winston, M., Economos, C.: Behavioral and social influences on food choice. Nutrition Reviews 56(5), 50–64 (1998)
10. Ricci, F., Rokach, L., Shapira, B., Kantor, P.B. (eds.): Rec. Systems Handbook. Springer (2011)
11. Salton, G., Buckley, C.: Weighting approaches in automatic text retrieval. IP and M 24(5), 513–523 (1988)
12. Scheibehenne, B., Greifeneder, R., Todd, P.M.: Can there ever be too many options? A meta-analytic review of choice overload. J. of Consumer Rsrch. 37, 409–425 (2010)
13. Sobecki, J., Babiak, E., Słanina, M.: Application of hybrid recommendation in web-based cooking assistant. In: Gabrys, B., Howlett, R.J., Jain, L.C. (eds.) KES 2006. LNCS (LNAI), vol. 4253, pp. 797–804. Springer, Heidelberg (2006)
14. Svensson, M., Laaksolahti, J., Höök, K., Waern, A.: A recipe based on-line food store. In: 5th Int. Conf. on Intelligent User Interfaces, IUI 2000, pp. 260–263. ACM, New York (2000)

Discovering Dense Subgraphs in Parallel for Compressing Web and Social Networks[*],[**]

Cecilia Hernández[1,2] and Mauricio Marín[3]

[1] Dept. of Computer Science, University of Concepción, Chile
[2] Dept. of Computer Science, University of Chile, Chile
[3] Yahoo Research, Santiago
chernand@dcc.uchile, mmarin@yahoo.com

Abstract. Mining and analyzing graphs are challenging tasks, especially with today's fast-growing graphs such as Web and social networks. In the case of Web and social networks an effective approach have been using compressed representations that enable basic navigation over the compressed structure. In this paper, we first present a parallel algorithm for reducing the number of edges of Web graphs adding virtual nodes over a cluster using BSP (Bulk Synchronous Processing) model. Applying another compression technique on edge-reduced Web graphs we achieve the best state-of-the-art space/time tradeoff for accessing out/in-neighbors. Second, we present a scalable parallel algorithm over BSP for extracting dense subgraphs and represent them with compact data structures. Our algorithm uses summarized information for implementing dynamic load balance avoiding idle time on processors. We show that our algorithms are scalable and keep compression efficiency.

Keywords: Parallel algorithms, Compressed Web and social graphs.

1 Introduction

Massive graphs appear in a wide range of domains including the Web, social networks, RDF graphs, protein networks and many more. For instance, the Web graph on a recent estimation has more than 7.8 billion pages with more than 200 billions of edges (mentioned in previous work [1]).

In the last decade, many graph algorithms have been proposed to address some of the problems associated with large graphs. Different approaches have been used to manage large graphs. One approach consists of representing graphs in compressed form while being able to resolve queries of interest without decompression. Although these compressed structures are usually slower than uncompressed representations, they are faster than having to access the disk. Many of these compressed structures target Web graphs, some support out-neighbor queries [2,3], that is retrieving the outgoing links of a node x, and some also

---

[*] Partially funded by Millennium Nucleus Information and Coordination in Networks ICM/FIC P10-024F.
[**] Partially funded by FONDEF IDeA CA12i10314.

O. Kurland, M. Lewenstein, and E. Porat (Eds.): SPIRE 2013, LNCS 8214, pp. 165–173, 2013.
© Springer International Publishing Switzerland 2013

support in-neighbor queries, that is incoming links of a node x [4]. Another approach is the use of distributed systems where distributed memory is aggregated to process the graph. Distributed memory is useful when data is larger than the memory available on a commodity machine. Pregel [5] is a graph system that works on BSP model, Pegasus [6] is a graph mining library over Hadoop, which is the free implementation of MapReduce [7]. Pace [8] discusses important differences between BSP and MapReduce and shows that iterative algorithms are more efficient using BSP than MapReduce.

The main contributions of this paper are:

- A scalable BSP parallel algorithm for reducing the number of edges of Web graphs by finding dense subgraphs and adding virtual nodes. This algorithm is based on DSM (Dense Subgraph Mining) algorithm, which used with virtual nodes, BFS ordering and K2tree [9] achieves the best compression on Web graphs [1].
- A scalable parallel DSM algorithm for extracting dense subgraphs. The algorithm exploits locality of adjacency lists and uses dynamic load balanced for maximizing processor utilization avoiding idle times. Representing these dense subgraphs with compact data structures [10] combined with an improved version of MPk [11] provides the best space/time tradeoffs for social networks [10].

2   Related Work

Compressing the Web has been an active research area for some time. Some of the earlier proposals include basic navigation, which is reduced at retrieving out-neighbors [2,3], and others that include retrieving out/in-neighbors [4,11,10]. Compression techniques for Web graphs use different patterns, such as locality and similarity of adjacency lists [2], the sparse nature of the adjacency matrix [4], label ordering [3,2], edge reduction [12], and dense subgraphs [10,1]. In social networks, successful representations use clique-like structures [13,11] and more dense subgraph patterns, such as cliques, bicliques and other patterns that combine cliques and bicliques [10]. Some of these structures [11,10,1] use compact data structures based on bit vectors and symbol sequences. Compact data structures use space efficiently and their basic operations are rank/select/access.

Discovering dense subgraphs in large graphs is a challenging problem in data analysis and has a wide-range of applications, including community mining, spam detection, and social analysis. The general problem has many variants such as finding and enumerating cliques [14] and detecting dense subgraphs or communities [15,16]. Although, there are some differences in the terminology defining a dense subgraph, all works consider the density as measuring the number of edges in relation with the number of nodes in such structures.

In recent years, parallel and distributed data management has gained attention due to the success of MapReduce [7] and Hadoop. MapReduce is simple to use and provides high throughput. Pregel [5] aims processing graphs and it is based on vertex computation using BSP. However, MapReduce and Pregel require hundreds or thousands of machines in order to process large graphs. For instance, Pegasus [6] and Pregel focus on large graph querying and mining, Pegasus is built on top of Hadoop and Pregel is built using BSP. Pregel improves upon MapReduce by passing computation results instead of graph structures among processors.

3 Our Approach

We represent a web graph as a directed graph $G = (V, E)$ where V is a set of vertices (pages) and $E \subseteq V \times V$ is a set of edges (hyperlinks). For an edge $e=(u,v)$, we call u the *source* and v the *center* of e. We find patterns given by the following definition.

Definition 1. A *dense subgraph* $H(S,C)$ of $G = (V, E)$ is a graph $G'(S \cup C, S \times C)$, where $S, C \subseteq V$.

Note that this definition includes cliques ($S = C$) and bicliques ($S \cap C = \emptyset$). Our goal is to represent the $|S| \cdot |C|$ edges of a dense subgraph $H(S,C)$ in space proportional to $|S| + |C| - |S \cap C|$. Thus, the bigger the dense subgraphs we detect, the more space we save at representing their edges.

The parallel algorithms presented here are based on a sequential algorithm for discovering dense subgraphs, DSM (Dense Subgraph Mining) [1]. DSM consists of 2-step clustering and 2-step mining. The clustering algorithm computes $|R|$ hash values for each adjacency list conforming a matrix of hash values of dimension $|R \cdot V|$ (Step 1). The matrix is sorted by columns where each cluster is formed by similar rows (Step 2). The mining phase takes the adjacency lists related to hash rows of each cluster and sorts edges by frequency (Step 3). Then, each adjacency list of the cluster is inserted into a prefix tree, discarding edges of frequency 1. Each node v in the prefix tree has a label (consisting of the node id), and it represents the sequence $l(v)$ of labels from the root to the node. Such node v stores also the range of graph nodes whose list start with $l(v)$ (Step 4). Figure 1 shows an example.

Our first parallel algorithm (Algorithm 1 in Table 1) uses DSM for reducing edges by a factor between 5 and 10, adding a small percentage of virtual nodes

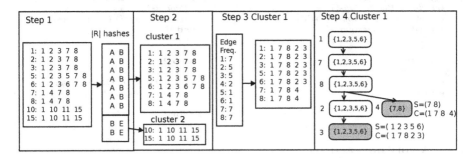

Fig. 1. Example of the Dense Subgraph Mining (DSM) algorithm

(around 10 and 15 %). Then, we apply BFS ordering and k2tree over the edge-reduced Web graphs. Our second parallel algorithm (Algorithm 2 in Table 1) uses DSM for extracting dense subgraphs and represent them with compact data structures. Such representation is based on the following components.

Let $\mathcal{H} = \{H_1, \ldots, H_N\}$ be the dense subgraph collection found in the graph, based on Definition 1. We represent $\mathcal{H}$ as a sequence of integers X with a corresponding bitmap B. Sequence $X = X_1 : X_2 : \ldots : X_N$ represents the sequence of dense subgraphs and bitmap $B = B_1 : B_2 : \ldots B_N$ is used to mark the separation between each subgraph. We now describe how a given X_r and B_r represent the dense subgraph $H_r = H(S_r, C_r)$.

We define X_r and B_r based on the overlapping between the sets S and C. Sequence X_r will have three components: L, M, and R, written one after the other in this order. Component L lists the elements of $S - C$. Component M lists the elements of $S \cap C$. Finally, component R lists the elements of $C - S$. Bitmap $B_r = 10^{|L|}10^{|M|}10^{|R|}$ gives alignment information to determine the limits of the components. In this way, we avoid repeating nodes in the intersection, and have sufficient information to determine all the edges of the dense subgraph.

Table 1. Algorithms DSM with virtual nodes (Algorithm 1), and DSM for extracting dense subgraphs (Algorithm 2)

Algorithm 1	Algorithm 2
Input: G_p, ES, T	**Input:** G_p, $esArray$, T, $threshold$.
Output: Reduced $RG(|V + VN|, E2)$ graph	**Output:** Dense subgraph collection
Each processor reads its data partition	Each processor reads its data partition
{**Step 0**}	$ES = esArray.first()$
for $(i \leftarrow 0 \text{ to } T - 1)$ **do**	{**Step 0**}
$\quad clusters = FindClusters()$	**for** $(i \leftarrow 0 \text{ to } T)$ **do**
$\quad$ **for** $(c \in clusters)$ **do**	$\quad clusters = FindClusters()$
$\quad\quad Sets(S, C) = FindDenseSubs(c, ES)$	$\quad$ **for** $(c \in clusters)$ **do**
$\quad\quad localVnodes = DefineSets(S, C)$	$\quad\quad Sets(S, C) = FindDenseSubs(c, ES)$
$\quad\quad Replace(G_p, Sets(S, C), localVnodes)$	$\quad\quad numDSs = |Sets(S, C)|$
$\quad\quad AddVnodes(G_p, Sets(S, C), localVnodes)$	$\quad\quad WriteToDisk(Sets(S, C))$
$\quad$ **end for**	$\quad$ **end for**
end for	**if** $(i == period())$ **then**
$sendLocalVnodeMsg()$	$\quad sendLoadMsg()$
$sync()$	$\quad sync()$
{**Step 1**}	$\quad$ {**Step 1**}
if $(proc == 0)$ **then**	$\quad$ **if** $(proc == 0)$ **then**
$\quad lvnodes = RecibeMsgs()$	$\quad\quad ProcessLoad()$
$\quad gvnodes = ProcVNodeGlobal(lvnodes)$	$\quad\quad sendDistInfo()$ (to all procs)
$\quad sendGlobalVnodes(gvnodes)$	$\quad$ **end if**
end if	$\quad sync()$
$sync()$	**end if**
{**Step 2**}	{**Step 2**}
$gvnodes = RecieveMsgs()$	$sendData()$
$replaceVnodes(G_p, lvnodes, gvnodes)$	**if** $(numDSs < threshold)$ **then**
return RG	$\quad ES = esArray.next$
	end if
	end for

3.1 Algorithms and Analysis

The BSP model provides an efficient parallel distributed memory model that considers relevant parameters of a real parallel computer system. A BSP computer is defined by P processors with local memory, connected via a point-to-point communication link. BSP algorithms proceed in supersteps in each of which processors receive input data, perform asynchronous computation over its data and communicate output at the end. Supersteps are synchronized at the end using barriers. An algorithm designed in BSP is measured by three main features: *computation*, *communication*, and *synchronization* costs. The cost model is given by $W + Hg + L$, where W is the maximum cost of computation on a processor, H is the maximum input/output communicated among processors, g is the latency and L is the synchronization cost.

Algorithm 1 in Table 1-(left) describes our parallel DSM for reducing edges and adding virtual nodes. During **Step 0** each processor processes G_p in parallel locally. Each iteration finds all clusters on G_p and on each cluster the mining algorithm discovers dense subgraphs of the type $\mathcal{H}$ with components (S,C) of size at least ES. For each subgraph, we create local virtual node ids (*localVnodes*) to separate sets (S,C).

In **Step 0** all processors sends a tuple with $(lvnodeInit, numberLVnodes)$ to processor 0. Thus, **Step 0** is $O(O(T\frac{|E|}{P} \log \frac{|E|}{P})$, where $|E|$ is the number of edges in G, P the number of available processors, and T the number of iterations. In **Step 1** processor 0 relabels local virtual nodes to global ids and sends that information to all processors. Relabeling is done by changing the $gVnodeInit$ based on the number of virtual nodes found in previous processed processor tuple, that is $gVnodeInit_i = vninit$ and $gVnodeInit_{i+1} = \sum numberLVnodes_i$. We work with global virtual node ids instead of locals to minimize mapping space. Thus, **Step 1** is $O(Pg + L)$. In **Step 2** all processors receive tuples with global virtual node ids and each processor replaces local virtual node ids for global ones. Then, this step is $O(|V + VN|)$ which indicates that the algorithm scales up efficiently since the amount of required communication is much smaller than the amount of computation performed by processors on local data. Therefore, the total cost is $O(T(\frac{|E|}{P} \log \frac{|E|}{P}) + Pg + L + |V + VN|)$.

Algorithm 2 in Table 1-(right) describes our parallel algorithm for extracting dense subgraphs using dynamic load balance. This is an iterative algorithm, where each iteration has several steps. In **Step 0** each processor computes clustering and mining and extracts dense subgraphs and sends periodically its workload information to processor 0. Processor workload tuple is given by ES and $numDSs$, where ES is the current size of the dense subgraphs that are mined and $numDSs$ is the number of subgraphs at the current iteration. The clustering is $O(\frac{|E|}{P} \log \frac{|E|}{P})$ and all processors send local workload tuples to processor 0 in $O(Pg + L)$ periodically. Function $period()$ determines how often processors send their load. In **Step 1** processor 0 receives local load from all processors, computes a global load tuple containing $(minP, maxP, minES, maxES, minDSs, maxDSs)$, and decides whether load balance is performed and the amount of data to move. If it decides to apply load balance, it sends global load balance tuple to all processors. In **Step 2** each processor receives the global load tuple and the heavier processor sends a portion of its data to the lighter processor.

Step 0 is computed T times and each processor sends workload tuples to processor 0 T_p times. During *Step 1* processor 0 computes workload tuples and decide whether heavier processors will send data to lighter processors, which is $O(Pg + L)$. Applying load balance depends on the distance between $(maxES, minES)$ and $(minDSs, maxDSs)$ among processors, and it can happen T_d times. This step is computed in $O(Mg + L)$, where M is a portion of G_p to move. The total cost is $O(T(\frac{|E|}{P} \log \frac{|E|}{P}) + T_p(Pg + L + P) + T_d((P + M)g + L))$.

4   Experimental Evaluation

We perform different experiments over Web and social graphs described in Table 2. [1]. We use the natural order for input graphs in all our experiments. We implemented parallel algorithms using C++ and BSP over a cluster with at most 64 processors. Each processor is an Intel 2.66 GHz, with 24 GB of RAM and 8 MB of cache. We partition input graphs among processors by equal number of edges contained by complete list of out-neighbors. This partition scheme gave us more balanced processor work load.

We study the performance of our parallel DSM with virtual nodes and extracting dense subgraphs using dynamic load balance. We analyze the effect of using different number of processors in terms of compression efficiency, running times, and speedup. We also compute the Edge ratio (ER). ER is the total number of edges (belonging to dense subgraphs) extracted in parallel versus the total number of edges in dense subgraphs extracted with the sequential algorithm.

Table 2. Number nodes, edges and size in MBs of graphs. A1S stands for the speedup(S) for 8 and 64 processors when using DSM in Algorithm 1, and A2S when using DSM in Algorithm 2. ER (Edge ratio) is the number of edges (belonging to dense subgraphs) extracted in parallel versus the ones extracted sequentially.

Data Set	Nodes	Edges	MB	A1S (8)	A1S (64)	A2S (8)	ER	A2S (64)	ER
eu-2005	862,664	19,235,140	77	4.95	20.88	14.46	0.99	58.6	0.99
indochina	7,414,866	194,109,311	765	2.18	23.85	5.39	0.96	52.4	0.99
uk-2002	18,520,486	298,113,762	1,200	10.10	68.18	11.21	0.90	102.87	0.97
arabic-2005	22,744,080	639,999,458	2,500	10.52	55.40	6.95	0.92	66.80	0.96
dblp-2011	986,324	6,707,236	30	-	-	29.08	0.76	117	0.87
LJSNAP	4,847,571	68,993,773	280	-	-	8.63	0.45	55.34	0.65

Figure 2 shows parallel running times and compression performance (bpe) using different numbers of processors for different Web graphs. We include running times for computing DSM-ESx-T10 (where $ES = x$ for finding dense subgraphs of at least size x, and $T = 10$ i.e. 10 iterations); and the running time for achieving the complete compression structure, which consists of two parts; DSM-ESx-T10 builds a graph with fewer edges and virtual nodes (RG); and K2treeBFS applies BFS and k2tree over RG. As observed, the running time improves greatly without affecting compression. These results suggest that there

---
[1] Data sets available at: law.dsi.unimi.it and snap.stanford.edu/data

is a great amount of locality of reference in adjacency lists. Figure 2 shows that the cost of applying k2treeBFS, which is sequential, has more impact on larger graphs. This is seen by the distance between the two running time plots visible on Arabic data set.

Table 2 shows the speedup achieved using 8 and 64 processors (A1S (8) and A1S (64)) using DSM with virtual nodes (k2tree not included). We observe that the speedup is higher for larger graphs, which suggest that such graphs take more advantage of memory aggregation in the cluster system.

We evaluate our second parallel algorithm (extracting dense subgraphs with DSM) measuring running times, speedup and the Edge ratio (ER). We use 100 iterations for extracting dense subgraphs and dblp-2011 and 200 iterations for LJSNAP (LiveJournal). Figure 3 shows the running time for DSM with dense subgraph extraction, considering only time for extraction, complete compression time (including mpk), and the compression achieved for social networks using $\mathcal{H}$ and $\mathcal{R}$ with mpk [11]. This figure also shows that the sequential part of the compression construction slows down the compression time. Table 2 shows the speedup for 8 and 64 processors (A2S (8) and A2S (64)). We also measure ER, which is the total number of edges extracted in dense subgraphs using our parallel algorithm versus the total number of edges extracted belonging to dense subgraphs using the sequential algorithm. We extract in parallel more than 90% edges belonging to dense subgraphs achieving good speedups on Web graphs. However, it is less effective on social graphs where ER is lower as seen in Table2.

5   Conclusions

This paper proposes two parallel algorithms for DSM, a sequential algorithm for discovering dense subgraphs [1] for compressing Web and social graphs. Our first parallel algorithm uses DSM with virtual nodes for reducing the number of edges. This algorithm exploits locality of reference of adjacency lists. Applying BFS ordering and k2tree over parallel edge-reduced Web graphs does not degrade compression efficiency. Our second parallel algorithm extracts dense subgraphs

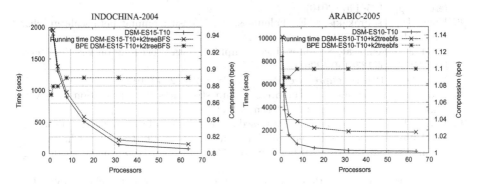

Fig. 2. Parallel running time with corresponding compression for Web graphs

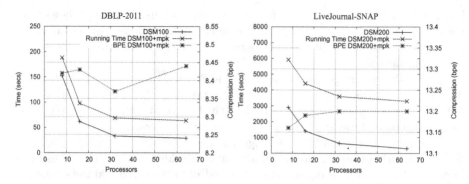

Fig. 3. Parallel running time for extracting dense subgraphs and bpe for social graphs

in parallel using dynamic load balance. Both algorithms provide good speedup and compression efficiency. However, since both algorithms are used with other sequential compression techniques such as *k2tree* [9] and *mpk* [11], they limit our compression speed.

References

1. Hernández, C., Navarro, G.: Compressed representations for web and social graphs. To appear in Knowledge and Information Systems (2013), http://link.springer.com/article/10.1007/s10115-013-0648-4
2. Boldi, P., Rosa, M., Santini, M., Vigna, S.: Layered label propagation: A multiresolution coordinate-free ordering for compressing social networks. In: WWW, pp. 587–596 (2011)
3. Apostolico, A., Drovandi, G.: Graph compression by bfs. Algorithms 2(3), 1031–1044 (2009)
4. Brisaboa, N.R., Ladra, S., Navarro, G.: k^2-Trees for compact web graph representation. In: Karlgren, J., Tarhio, J., Hyyrö, H. (eds.) SPIRE 2009. LNCS, vol. 5721, pp. 18–30. Springer, Heidelberg (2009)
5. Malewicz, G., Austern, M.H., Bik, A.J.C., Dehnert, J.C., Horn, I., Leiser, N., Czajkowski, G.: Pregel: A system for large-scale graph processing. In: SIGMOD Conference, pp. 135–146 (2010)
6. Kang, U., Tsourakakis, C.E., Faloutsos, C.: Pegasus: mining peta-scale graphs. Knowl. Inf. Syst. 27(2), 303–325 (2011)
7. Dean, J., Ghemawat, S.: Mapreduce: Simplified data processing on large clusters. In: OSDI, pp. 137–150 (2004)
8. Pace, M.F.: Bsp vs mapreduce. Procedia CS 9, 246–255 (2012)
9. Ladra, S.: Algorithms and compressed data structures for information retrieval. Ph.D. Thesis, University of A. Coruña (2011)
10. Hernández, C., Navarro, G.: Compressed representation of web and social networks via dense subgraphs. In: Calderón-Benavides, L., González-Caro, C., Chávez, E., Ziviani, N. (eds.) SPIRE 2012. LNCS, vol. 7608, pp. 264–276. Springer, Heidelberg (2012)
11. Claude, F., Ladra, S.: Practical representations for web and social graphs. In: CIKM, pp. 1185–1190 (2011)

12. Buehrer, G., Chellapilla, K.: A scalable pattern mining approach to web graph compression with communities. In: WSDM, pp. 95–106 (2008)
13. Maserrat, H., Pei, J.: Neighbor query friendly compression of social networks. In: KDD, pp. 533–542 (2010)
14. Schmidt, M.C., Samatova, N.F., Thomas, K., Park, B.-H.: A scalable, parallel algorithm for maximal clique enumeration. J. Parallel Distrib. Comput. 69(4), 417–428 (2009)
15. Kumar, R., Raghavan, P., Rajagopalan, S., Tomkins, A.: Trawling the Web for emerging cyber-communities. Computer Networks 31(11-16), 1481–1493 (1999)
16. Dourisboure, Y., Geraci, F., Pellegrini, M.: Extraction and classification of dense communities in the web. In: WWW, pp. 461–470 (2007)

Faster Lyndon Factorization Algorithms for SLP and LZ78 Compressed Text

Tomohiro I[1,2], Yuto Nakashima[1], Shunsuke Inenaga[1],
Hideo Bannai[1], and Masayuki Takeda[1]

[1] Department of Informatics, Kyushu University, Japan
{tomohiro.i,yuto.nakashima,inenaga,bannai,takeda}@inf.kyushu-u.ac.jp
[2] Japan Society for the Promotion of Science (JSPS)

Abstract. We present two efficient algorithms which, given a compressed representation of a string w of length N, compute the Lyndon factorization of w. Given a straight line program (SLP) $\mathcal{S}$ of size n and height h that describes w, the first algorithm runs in $O(nh(n + \log N \log n))$ time and $O(n^2)$ space. Given the Lempel-Ziv 78 encoding of size s for w, the second algorithm runs in $O(s \log s)$ time and space.

1 Introduction

Two strings x and y are *conjugates*, if $x = uv$ and $y = vu$ for some strings u and v. A string w is said to be a *Lyndon word*, if w is lexicographically strictly smaller than all of its conjugates. The *Lyndon factorization* of a string w, denoted LF_w, is a factorization of w such that each factor is a Lyndon word and the sequence of Lyndon factors is lexicographically non-increasing [4]. Lyndon factorizations are used in a bijective variant of Burrows-Wheeler transform [10,7] and a digital geometry algorithm [3]. Given a string w of length N, LF_w can be computed on-line in $O(N)$ time [6].

When the length N of the string w is huge, even the $O(N)$-time solution may not be efficient enough. In this paper, we give two algorithms to compute LF_w, which are more efficient when w is highly compressible. Given a straight line program (SLP) $\mathcal{S}$ of size n that describes the string w, our first algorithm computes LF_w in $O(nh(n + \log N \log n))$ time and $O(n^2)$ space, where $h \leq n$ is the height of the derivation tree of $\mathcal{S}$. Since the decompressed string length $|w| = N$ can be exponentially large w.r.t. n, our solution is more efficient than the $O(N)$-time decompress-then-process solution in the worst case. In addition, our solution improves on the previous work by I et al. [9], which solves the same problem in $O(n^3(n + mh))$ time and $O(n^2)$ space, where m is the number of Lyndon factors of w. As a byproduct of our solution, we show that the number m of Lyndon factors of string w is bounded by the size n of *any* SLP representing w, i.e. $m \leq n$, which may be of independent interest.

Our second algorithm is designed for a more specific case, i.e., when the Lempel-Ziv 78 (LZ78) encoding [12] of the string w is given as input. If s is the size of the LZ78 encoding, then the algorithm computes LF_w in $O(s \log s)$ time and space.

O. Kurland, M. Lewenstein, and E. Porat (Eds.): SPIRE 2013, LNCS 8214, pp. 174–185, 2013.

For a fixed alphabet $s = O(N/\log N)$ [12]. Thus, for a fixed alphabet our $O(s \log s)$ solution is at least as efficient as Duval's $O(N)$ solution that works on uncompressed strings, and is more efficient when s is sufficiently small.

2   Preliminaries

Strings and Model of Computation. Let Σ be an ordered finite *alphabet*. An element of Σ^* is called a *string*. The length of a string w is denoted by $|w|$. The empty string ε is a string of length 0. Let Σ^+ be the set of non-empty strings, i.e., $\Sigma^+ = \Sigma^* - \{\varepsilon\}$. For a string $w = xyz$, x, y and z are called a *prefix*, *substring*, and *suffix* of w, respectively. A prefix x of w is called a *proper prefix* of w if $x \neq w$. The set of non-empty suffixes of w is denoted by $Suffix(w)$. The i-th character of a string w is denoted by $w[i]$, where $1 \leq i \leq |w|$. For a string w and two integers $1 \leq i \leq j \leq |w|$, let $w[i..j]$ denote the substring of w that begins at position i and ends at position j. For convenience, let $w[i..j] = \varepsilon$ when $i > j$. For any string w let $w^1 = w$, and for any integer $k \geq 2$ let $w^k = ww^{k-1}$, i.e., w^k is a k-time repetition of w. Let w^∞ denote an infinite repetition of w.

An integer $p \geq 1$ is said to be a *period* of a string w if $w[i] = w[i + p]$ for all $1 \leq i \leq |w| - p$. If p is a period of a string w with $p < |w|$, then $|w| - p$ is said to be a *border* of w. If w has no borders, then w is said to be *border-free*.

If character a is lexicographically smaller than another character b, then we write $a \prec b$. Let ω be the lexicographically largest character in Σ. For any strings x, y, let $lcp(x, y)$ be the length of the longest common prefix of x and y. We write $x \prec y$ iff either $x[lcp(x, y) + 1] \prec y[lcp(x, y) + 1]$ or x is a proper prefix of y. When we want to clarify that the former condition holds for some $x \prec y$, we denote $x \lhd y$. For any non-empty set $S \subseteq \Sigma^*$ of strings, let $\min_\prec S$ and $\max_\prec S$ denote the lexicographically smallest and largest strings in S, respectively.

Our model of computation is the word RAM: We assume the computer word size is at least $\lceil \log_2 |w| \rceil$, and hence, standard operations on values representing lengths and positions of string w can be manipulated in $O(1)$ time. Space complexities will be determined by the number of computer words (not bits).

Lyndon Words and Lyndon Factorization of Strings. Two strings x and y are *conjugates*, if $x = uv$ and $y = vu$ for some strings u and v. A string w is said to be a *Lyndon word*, if w is lexicographically strictly smaller than all of its conjugates. Namely, w is a Lyndon word, if for any factorization $w = uv$, it holds that $uv \prec vu$. Notice any Lyndon word is border-free.

The *Lyndon factorization* of a string w, denoted LF_w, is the factorization $\ell_1^{p_1} \cdots \ell_m^{p_m}$ of w, such that each $\ell_i \in \Sigma^+$ is a Lyndon word, $p_i \geq 1$, and $\ell_i \succ \ell_{i+1}$ for all $1 \leq i < m$. The size of LF_w is m. LF_w can be represented by the sequence $(|\ell_1|, p_1), \ldots, (|\ell_m|, p_m)$ of integer pairs, where each pair $(|\ell_i|, p_i)$ represents the i-th Lyndon factor $\ell_i^{p_i}$ of w. Note that this representation requires $O(m)$ space.

In some literature, the Lyndon factorization is defined to be a sequence of lexicographically non-increasing Lyndon words, namely, each Lyndon factor ℓ^p is decomposed into a sequence of p ℓ's. In this paper, each Lyndon word ℓ in the Lyndon factor ℓ^p is called a *decomposed Lyndon factor*.

For any string w, let $LF_w = \ell_1^{p_1} \cdots \ell_m^{p_m}$. Let $lfb_w(i)$ denote the position where the i-th Lyndon factor begins in w, i.e., $lfb_w(1) = 1$ and $lfb_w(i) = lfb_w(i-1) + |\ell_{i-1}^{p_{i-1}}|$ for any $2 \le i \le m$. For any $1 \le i \le m$, let $lfs_w(i) = \ell_i^{p_i} \ell_{i+1}^{p_{i+1}} \cdots \ell_m^{p_m}$ and $lfp_w(i) = \ell_1^{p_1} \ell_2^{p_2} \cdots \ell_i^{p_i}$. For convenience, let $lfs_w(m+1) = lfp_w(0) = \varepsilon$.

Straight Line Programs (SLPs). A *straight line program* (*SLP*) is a set of productions $\mathcal{S} = \{X_i \to expr_i\}_{i=1}^n$, where each X_i is a variable and each $expr_i$ is an expression, where $expr_i = a$ ($a \in \Sigma$), or $expr_i = X_{\ell(i)} X_{r(i)}$ ($i > \ell(i), r(i)$). Let $val(X_i)$ denote the string derived by X_i. We will sometimes associate $val(X_i)$ with X_i and denote $|val(X_i)|$ as $|X_i|$. An SLP $\mathcal{S}$ *represents* the string $w = val(X_n)$. The *size* of the program $\mathcal{S}$ is the number n of productions in $\mathcal{S}$. If N is the length of the string represented by SLP $\mathcal{S}$, then N can be as large as 2^{n-1}.

The derivation tree $T_\mathcal{S}$ of SLP $\mathcal{S}$ is a labeled ordered binary tree where each internal node is labeled with a non-terminal variable in $\{X_1, \ldots, X_n\}$, and each leaf is labeled with a terminal character in Σ. The root node has label X_n. The height of SLP $\mathcal{S}$ is the height of $T_\mathcal{S}$. We associate to each leaf of $T_\mathcal{S}$ the corresponding position in string $w = val(X_n)$.

Let $[b, e]$ be any integer interval with $1 \le b < e \le |val(X_n)| = N$. We say that node z of $T_\mathcal{S}$ *stabs* the interval $[b, e]$, if the lowest common ancestor of the leaves b and e in $T_\mathcal{S}$ is z. If X_j is the label of node z, then we also say that variable X_j stabs the interval $[b, e]$. Note that for any interval I of length at least two, there is a unique variable that stabs I.

Lempel Ziv 78 Encoding. The LZ78-factorization of a string w is a factorization $f_1 \cdots f_s$ of w, where each $f_i \in \Sigma^+$ for each $1 \le i \le s$ is defined as follows: For convenience, let $f_0 = \varepsilon$. Then, $f_i = w[p..p + |f_j|]$ where $p = |f_0 \cdots f_{i-1}| + 1$ and $f_j (0 \le j < i)$ is the longest previous factor which is a prefix of $w[p..|w|]$. The LZ78 encoding of w is a sequence $(k_1, a_1), \ldots, (k_s, a_s)$ of pairs s.t. each pair (k_i, a_i) represents the i-th LZ78 factor f_i, where k_i is the ID of the previous factor f_{k_i}, and a_i is the new character $w[|f_1 \cdots f_i|]$. The LZ78 encoding requires $O(s)$ space. Regarding this pair as a parent and edge label, the factors can also be represented as a trie of size $O(s)$.

3   Properties on Strings and Lyndon Words

In this section, we introduce some fundamental properties on strings and Lyndon words which will be used in our algorithms. Below, let $LF(w) = \ell_1^{p_1} \cdots \ell_m^{p_m}$ for string w. Proofs for Lemmas 1, 3, 6, and 7 are omitted due to lack of space.

Lemma 1. *Let $u \in \Sigma^+$ and $v \in \Sigma^*$. If $v \prec u^\infty$, $v \prec u^1 v \prec u^2 v \prec \ldots$ holds. If $v \succ u^\infty$, $v \succ u^1 v \succ u^2 v \succ \ldots$ holds.*

Lemma 2 ([6]). *For any Lyndon words u, v, uv is a Lyndon word iff $u \prec v$.*

Lemma 3. *A non-empty string w is a Lyndon word iff $w \triangleleft v$ for any non-empty proper suffix v of w.*

Lemma 4 ([6]). *For any $1 \le i < m$, $LF_w = LF_{lfp_w(i)} LF_{lfs_w(i+1)}$.*

Lemma 5 ([6]). *It holds that ℓ_1 is the longest prefix of w which is a Lyndon word and p_1 is the largest integer k such that ℓ_1^k is a prefix of w.*

The following lemmas are essentially the same as what Duval's algorithm is founded on but are tailored for explaining our algorithm.

Lemma 6. *Let $j > 1$ be any position of a string w such that $w \prec w[i..|w|]$ for any $1 < i \leq j$, and $lcp(w, w[j..|w|]) \geq 1$. Then, $w \prec w[k..|w|]$ also holds for any $j < k \leq j + lcp(w, w[j..|w|])$.*

Lemma 7. *It holds that $|\ell_1| = \hat{j} - 1$ and $p_1 = 1 + \lfloor \hat{h}/|\ell_1| \rfloor$, where $\hat{j} = \min\{j \mid w \succ w[j..|w|]\}$ and $\hat{h} = lcp(w, w[\hat{j}..|w|])$.*

Thus, computing the first Lyndon factor reduces to computing $\hat{j}$ and $\hat{h}$. From Lemmas 3 and 7, the following lemma holds.

Lemma 8. *For any $1 \leq i \leq m$ and $1 \leq j < lfb_w(i)$, $w[j..|w|] \succ lfs_w(i)$.*

4 Faster Lyndon Factorization from SLP

Here, we present a faster algorithm to compute Lyndon factorization of a string represented by an SLP. Our algorithm employs the following lemma which is used in parallel algorithms to compute Lyndon factorization of an uncompressed string. Below, let $LF_u = u_1^{p_1} \cdots u_m^{p_m}$ and $LF_v = v_1^{q_1} \cdots v_{m'}^{q_{m'}}$ for $u, v \in \Sigma^+$.

Lemma 9 ([1,5]). *$LF_{uv} = u_1^{p_1} \ldots u_c^{p_c} z^k v_{c'}^{q_{c'}} \ldots v_{m'}^{q_{m'}}$ for some $0 \leq c \leq m$, $1 < c' \leq m' + 1$ and $LF_{lfs_u(c+1)lfp_v(c'-1)} = z^k$.*

This lemma implies that we can obtain LF_{uv} from LF_u and LF_v by computing z^k since the other Lyndon factors remain unchanged in uv.

4.1 How to Compute the Medial Lyndon Factor z^k

Unfortunately, the algorithm for computing z^k given in the proof of Theorem 2.2 of [5], appears to be wrong. In the sequel, we present several combinatorial properties from which a correct and efficient algorithm to compute the medial Lyndon factor z^k follows.

Let λ_u be the minimum integer such that $lfs_u(i + 1)$ is a prefix of u_i for any $\lambda_u \leq i \leq m$.

Lemma 10. *For any $1 \leq j < \lambda_u$, $u_j \rhd lfs_u(j + 1)$.*

Proof. Since we have $lfs_u(j) \succ lfs_u(j + 1)$ from Lemma 8, we only have to show that $lcp(u_j, lfs_u(j + 1)) < \min\{|u_j|, |lfs_u(j + 1)|\}$. Note that u_j is not a prefix of $lfs_u(j + 1)$ since otherwise the jth Lyndon factor should extend to the right with at least another occurrence of u_j. It follows from the definition of λ_u that $lfs_u(\lambda_u)$ is not a prefix of u_{λ_u-1}, and hence $u_{\lambda_u-1} \rhd lfs_u(\lambda_u)$ holds. If we assume on the contrary that $lfs_u(j + 1)$ is a prefix of u_j for some $1 \leq j < \lambda_u - 1$, since $lfs_u(\lambda_u)$ appears before the $(\lambda_u - 1)$-th Lyndon factor, there exists a suffix $u[i..|u|] = lfs_u(\lambda_u)z$ of u with $i < lfb_u(\lambda_u - 1)$. It follows from $u_{\lambda_u-1} \rhd lfs_u(\lambda_u)$ that $u[i..|u|] = lfs_u(\lambda_u)z \lhd u_{\lambda_u-1} \prec lfs_u(\lambda_u - 1)$, which contradicts Lemma 8.   □

We define the set of *significant suffixes* Λ_u of u as $\Lambda_u = \{lfs_u(i) \mid \lambda_u \leq i \leq m\}$. It is clear from the definition of Lyndon factorization that for any $\lambda_u \leq i \leq m$, $u_i = lfs_u(i+1)y_i$ for some non-empty string y_i. Let x_i denote the suffix of u of length $|u_i|$. Note that $x_i = y_i lfs_u(i+1)$ and $lfs_u(i) = u_i^{p_i} lfs_u(i+1) = (lfs_u(i+1)y_i)^{p_i} lfs_u(i+1) = lfs_u(i+1)x_i^{p_i}$.

Lemma 11. *For any string u, $|\Lambda_u| = O(\log |u|)$.*

Proof. Straightforward, since $|lfs_u(i)| > 2|lfs_u(i+1)|$ for any $\lambda_u \leq i \leq m$. $\square$

Lemma 12. *For any $\lambda_u \leq i < m$, $y_i \triangleright x_{i+1}^\infty$.*

Proof. Since $u_i = lfs_u(i+1)y_i = (lfs_u(i+2)y_{i+1})^{p_{i+1}} lfs_u(i+2)y_i$, if we assume that $y_i = x_{i+1}^\infty[1..|y_i|] = (y_{i+1}lfs_u(i+2))^\infty[1..|y_i|]$ we get a contradiction that u_i has period $|lfs_u(i+2)y_{i+1}|$. Also, if $y_i \prec x_{i+1}^\infty[1..|y_i|]$, $u_{i+1}[1 + |lfs_u(i+2)y_{i+1}|..|u_{i+1}|] \prec u_{i+1}$ holds, a contradiction. $\square$

From Lemma 12, we get $x_{\lambda_u-1}^\infty \succ y_{\lambda_u} \succ x_{\lambda_u+1}^\infty \succ y_{\lambda_u+1} \succ \cdots \succ x_{m-1}^\infty \succ y_{m-1} \succ x_m^\infty = u_m^\infty$, where we assume for convenience that $x_{\lambda_u-1}^\infty = \omega^\infty$.

Lemma 13. *For any $v \in \Sigma^+$, if $sv = \min_\prec\{s'v \mid s' \in Suffix(u)\}$, then $s \in \Lambda_u$.*

Proof. We show for any $s \in (Suffix(u) - \Lambda_u)$, $sv \neq \min_\prec\{s'v \mid s' \in Suffix(u)\}$.

- If $s \neq u_i^k lfs_u(i+1)$ with $1 \leq k \leq p_i$, namely, $s = tu_i^r lfs_u(i+1)$ with $1 \leq i \leq m$ and $0 \leq r < p_i$, where t is a proper suffix of u_i. It follows from the definition of Lyndon word $t \triangleright u_i$, and hence, $sv \neq \min_\prec\{s'v \mid s' \in Suffix(u)\}$.
- If $s = u_i^k u_{i+1}'$ with $1 \leq k < p_i$. From Lemma 1, $sv \succ \min_\prec\{u_i^0 lfs_u(i+1)v, u_i^{p_i} lfs_u(i+1)v\}$. Therefore, $sv \neq \min_\prec\{s'v \mid s' \in Suffix(u)\}$. $\square$

Lemma 14. *For any string $v \in \Sigma^+$, if $x_{i-1}^\infty \succ v \succ x_i^\infty$ with $\lambda_u \leq i \leq m$, $lfs_u(i)v = \min_\prec\{sv \mid s \in Suffix(u)\}$, and $lfs_u(1)v \succ \cdots \succ lfs_u(i-1)v \succ lfs_u(i)v \prec lfs_u(i+1)v \prec \cdots \prec lfs_u(m+1)v$ holds.*

Proof. By Lemma 10, $lfs_u(1)v \triangleright \ldots \triangleright lfs_u(\lambda_w - 1)v \triangleright u_{\lambda_u}v$. By Lemma 12, $x_j^\infty \succ v$ and also $lfs_u(j+1)x_j^\infty = u_j^\infty \succ lfs_u(j+1)v$ hold for any $\lambda_u \leq j < i$, and hence, it follows from Lemma 1 that $lfs_u(j)v = u_j^{p_j} lfs_u(j+1)v \succ lfs_u(j+1)v$. Also, since $v \succ x_{j'}^\infty$ for any $i \leq j' \leq m$, $lfs_u(j')v \prec lfs_u(j'+1)v$ holds. Therefore, we get $u_1'v \succ \cdots \succ u_{i-1}'v \succ u_i'v \prec u_{i+1}'v \prec \cdots \prec u_m'v$ holds. It is clear from Lemma 13 that $u_i'v$ is the lexicographically smallest string in $\{sv \mid s \in Suffix(u)\}$. $\square$

We can compute the medial Lyndon factor as follows:

Lemma 15. *Given $LF_u = u_1^{p_1} \cdots u_m^{p_m}$ and $LF_v = v_1^{q_1} \cdots v_{m'}^{q_{m'}}$ for $u, v \in \Sigma^+$, we can compute $LF_{uv} = u_1^{p_1} \cdots u_c^{p_c} z^k v_{c'}^{q_{c'}} \cdots v_{m'}^{q_{m'}}$ by $O(\log m + \log m')$ lexicographical string comparisons.*

Proof. Clearly, it holds that $LF_{uv} = LF_u LF_v$ if $u_m \succ v_1$, and that $LF_{uv} = u_1^{p_1} \cdots u_{m-1}^{p_{m-1}} u_m^{p_m+q_1} v_2^{q_2} \cdots v_{m'}^{q_{m'}}$ if $u_m = v_1$. In what follows we consider the case when $u_m \prec v_1$. Note that $v \succ u_m^\infty$ holds in this situation.

First, we compute integer j such that $1 \leq j \leq m+1$ and $lfs_u(j)v = \min_{\prec}\{sv \mid s \in Suffix(u)\}$. From Lemma 14, for any $1 \leq i \leq m$, $j \leq i$ iff $lfs_u(i)v \prec lfs_u(i+1)v$. Hence we can find j by binary search which requires $O(\log m)$ lexicographical string comparisons. Next, we compute $j' = \min\{j'' \mid 1 \leq j'' \leq m', lfs_u(j)v \succ lfs_v(j''+1)\}$. Since $lfs_v(1) \succ lfs_v(2) \succ \ldots \succ lfs_v(m'+1)$, j' can be found by $O(\log m')$ lexicographical string comparisons with binary search.

We show $z = lfs_u(j)lfp_v(j')$ is the first decomposed Lyndon factor of $lfs_u(j)v$. By definition of j, for any position i with $lfb_u(j) < i \leq |u|$, $lfs_u(j)v \prec (uv)[i..|uv|]$. Since $lfs_u(j)v \prec lfs_v(j')$, it follows from Lemma 8 that for any $|u| < i < |u|+lfb_v(j')$, $lfs_u(j)v \prec lfs_v(j') \prec (uv)[i..|uv|]$. Next we show $v_{j'}$ is not a prefix of $lfs_u(j)v$. Assume on the contrary that $lfs_u(j)v = v_{j'}t$. The beginning position of t in uv is at most $|u|+lfb_v(j')$ since the occurrences of $v_{j'}$ cannot overlap, and hence $lfs_u(j)v \prec t$. Since $lfs_u(j)v = v_{j'}t \prec lfs_v(j')$, $lfs_u(j)v \prec t \prec v_{j'}^{q_{j'}-1}lfs_v(j'+1)$. Applying this deduction $q_{j'}$ times, we get $lfs_u(j)v \prec t \prec lfs_v(j'+1)$, a contradiction. Thus, $lfs_u(j)v \triangleleft v_{j'} \preceq (uv)[i..|uv|]$ for any $|u|+lfb_v(j') \leq i < |u|+lfb_v(j'+1)$. Since $|u| + lfb_v(j'+1)$ is the first position where the suffix becomes lexicographically smaller than $lfs_u(j)v$, the claim follows from Lemma 7.

Finally, we show $u_{j-1} \succeq z$. Assume on the contrary that $u_{j-1} \prec z$. By Lemma 2 $u_{j-1}z$ is a Lyndon word, which implies $u_{j-1}z \triangleleft z$. This contradicts $lfs_u(j)v = \min_{\prec}\{s'v \mid s' \in Suffix(u)\}$ due to $u_{j-1}lfs_u(j)v \triangleleft lfs_u(j)v$.

The above procedure correctly computes the decomposed Lyndon factorization of uv. The exponent of z can be computed by checking if $u_j = z$ and/or $u_{j'+1} = z$ and packing them together if needed. Hence the total number of lexicographical string comparisons is $O(\log m + \log m')$.   □

4.2   Computing Lyndon Factorization from SLP

Given an SLP S of size n, we process each production $X_i \rightarrow X_{\ell(i)}X_{r(i)}$ in increasing order of i, and compute LF_{X_i} from $LF_{X_{\ell(i)}}$ and $LF_{X_{r(i)}}$ using dynamic programming. We use Lemma 15 to compute the medial Lyndon factor for each variable X_i. In the final stage where $i = n$, we obtain the Lyndon factorization $LF_{X_n} = LF_w$ for the uncompressed string w. Using Lemma 15, LF_{X_i} can be computed by $O(\log m_\ell + \log m_r)$ string comparisons, where m_ℓ and m_r are respectively the number of Lyndon factors in $LF_{X_{\ell(i)}}$ and $LF_{X_{r(i)}}$. We can use the following lemma for lexicographical string comparisons on SLPs.

Lemma 16 ([8]). *We can pre-process an SLP S of size n and height h in $O(n^2h)$ time and $O(n^2)$ space, so that for any X_i and $1 \leq k_1, k_2 \leq |X_i|$, the lexicographical order and the length of the longest common prefix of $val(X_i)[k_1..|X_i|]$ and $val(X_i)[k_2..|X_i|]$ can be determined in $O(h \log N)$ time.*

What remains is to show how large the number m of Lyndon factors of a string w can be. In the sequel, we show that $m \leq n$ holds, where n is the size of *any* SLP representing the string w. The next lemma follows from Lemma 9.

Lemma 17. *Let $LF_w = \ell_1^{p_1} \cdots \ell_m^{p_m}$ for a string $w \in \Sigma^+$. Let $[b, e]$ be any interval with $1 \leq b \leq e \leq |w|$, and let $u = w[b..e]$. For any $b \leq lfb_w(i) \leq e$, there exists integer j such that $b - 1 + lfb_u(j) = lfb_w(i)$.*

Lemma 18. *Let n be the size of any SLP representing a string w. The size m of the Lyndon factorization of w is at most n.*

Proof. Let $LF_w = \ell_1^{p_1} \cdots \ell_m^{p_m}$. For any Lyndon factor $\ell_i^{p_i}$ of length at least 2, i.e. $p_i|\ell_i| \geq 2$, consider interval $I = [lfb_w(i), lfb_w(i) + p_i|\ell_i| - 1]$ which corresponds to the occurrence of $\ell_i^{p_i}$ in w. Let $X_j \rightarrow X_{\ell(j)}X_{r(j)}$ be the unique variable that stabs I. Assume on the contrary that another node of the derivation tree T_S with the same label X_j stabs a different Lyndon factor $\ell_k^{p_k}$ with $k \neq i$. Lemma 17 implies that both $\ell_i^{p_i}$ and $\ell_k^{p_k}$ are Lyndon factors of $val(X_j)$ that crosses the boundary between $X_{\ell(j)}$ and $X_{r(j)}$, a contradiction. Hence $\ell_i^{p_i}$ is a unique Lyndon factor that is stabbed by X_j in w. Moreover, by definition of Lyndon factorization, no other node of T_S with label X_j can stab the same Lyndon factor $\ell_i^{p_i}$. Therefore, if n' is the number of variables which derive a string of length at least 2, then there can be at most n' Lyndon factors of length at least 2 in w. Clearly, the number of Lyndon factors of length 1 is at most $n - n'$. Thus $m \leq n$ holds. $\square$

Theorem 1. *Given an SLP of size n and height h representing string w of length N, we can compute LF_w in $O(nh(n + \log N \log n))$ time and $O(n^2)$ space.*

Proof. By Lemmas 15, 16 and 18, for each production $X_i \rightarrow X_{\ell(i)}X_{r(i)}$ we can compute LF_{X_i} in $O(h \log N \log n)$ time, provided that $LF_{X_{\ell(i)}}$ and $LF_{X_{r(i)}}$ are already computed. Using a dynamic programming method, this takes a total of $O(nh \log N \log n)$ time. The space complexity for this dynamic programming is $O(n^2)$ since for each variable X_i we have to store at most n beginning positions of the Lyndon factors of X_i. Putting these and the pre-processing costs of Lemma 16 together, we conclude that our algorithm takes a total of $O(n^2 h + nh \log N \log n) = O(nh(n + \log N \log n))$ time and $O(n^2)$ space. $\square$

Corollary 1. *We can pre-process, in $O(nh(n + \log N \log n))$ time and $O(n^2)$ space, an SLP of size n and height h describing string w of length N so that the following query can be answered in $O(h(n + h \log N \log n))$ time: given an interval $[b, e]$ with $1 \leq b \leq e \leq N$, compute $LF_{w[b..e]}$.*

By Corollary 1, we can compute the Lyndon factorization of a query substring of w, without decompression. Corollary 1 is more efficient than applying Theorem 1 to an SLP describing substing $w[b..e]$, since it takes $O(nh(n + \log N \log n))$ time.

5   Computing Lyndon Factorization from LZ78

In this section, we show how, given an LZ78 encoding of string w, we compute the Lyndon factorization $LF(w) = \ell_1^{p_1} \cdots \ell_m^{p_m}$ of w. Our algorithm is based on Duval's algorithm [6] which computes the Lyndon factorization of a given string w of length N in $O(N)$ time by scanning w from left to right.

From Lemma 6 and Lemma 7, we can compute the first Lyndon factor by initializing $j \leftarrow 2$ and executing the following: 1) compute $h \leftarrow lcp(w, w[j..|w|])$. 2) if $w[1 + h] \prec w[j + h]$, set $j \leftarrow j + h + 1$ and go back to Step 1); otherwise, output $|\ell_1| \leftarrow j - 1$ and $p_1 \leftarrow 1 + \lfloor h/|\ell_1| \rfloor$.

Let $\hat{j}$ and $\hat{h}$ denote the last values of j and h, respectively. Duval's algorithm computes $h \leftarrow lcp(w, w[j..|w|])$ by character comparisons, and it takes a total of $O(\hat{j}+\hat{h})$ time. Note $O(\hat{j}+\hat{h}) = O(|\ell_1|p_1)$ since $\hat{j}+\hat{h} < |\ell_1|p_1 + |\ell_1|$. By Lemma 4, we can compute the second Lyndon factor by executing the above procedure with the remaining string $w[1 + |\ell_1|p_1..|w|]$. By applying this recursively, the Lyndon factorization of w can be computed in $O(\sum_{i=1}^{m} |\ell_i|p_i) = O(|w|)$ time.

In what follows, we show how to simulate, in $O(s \log s)$ time and space, the above algorithm on the LZ78 encoding of size s.

The next lemma describes one of the key ideas to achieve such complexity.

Lemma 19. *Let w be non-empty string such that $w = xvyvz$ with $v \in \Sigma^+$ and $x, y, z \in \Sigma^*$. If $|xvy| < lfb_w(k) \leq |xvyv|$ for some k, then $lfb_w(k) \in \{|xvy| + lfb_v(j) \mid \lambda_v \leq j \leq m'\}$, where $LF_v = v_1^{q_1} \cdots v_{m'}^{q'_m}$.*

Proof. By Lemma 9, $lfb_w(k) \in \{|xvy| + lfb_v(j) \mid 1 \leq j \leq m'\}$. On the contrary, assume $lfb_w(k) = |xvy| + lfb_v(j)$ with $j < \lambda_v$. By Lemma 10, $lfs_v(j) \triangleright lfs_v(\lambda_v)$ and $w[|xvy| + lfb_v(j)..|w|] \triangleright w[|x| + lfb_v(\lambda_v)..|w|]$, a contradiction due to Lemma 8. $\square$

Thanks to Lemma 19, when a string u appears multiple times without overlapping, we can utilize Λ_u to skip some suffix comparisons of Duval's algorithm. Also, we can compute Λ_u for all LZ78 factors efficiently.

Lemma 20. *Given the LZ78 encoding of size s of a string w, we can compute Λ_{f_k} for all LZ78 factors f_k, $1 \leq k \leq s$, in a total of $O(s \log s)$ time and space.*

Proof. It follows from $|f_k| \leq s$ and Lemma 11 that $|\Lambda_{f_k}| = O(\log s)$ for any $1 \leq k \leq s$. Consider any LZ78 factor $f_k = f_h a$, where $1 \leq h < k \leq s$ and $a \in \Sigma$. Let $LF_{f_h} = x_{p_1}^{p_1} \cdots x_{p_m}^{p_m}$ and $LF_{f_k} = y_{q_1}^{q_1} \cdots y_{q_{m'}}^{q_{m'}}$. We show how, given $LF_{lfs_{f_h}(\lambda_{f_h})}$, we compute $LF_{lfs_{f_k}(\lambda_{f_k})}$ in $O(\log s)$ time. We can use a simplified version of the algorithm of Lemma 15 to compute Lyndon factorization of $lfs_{f_h}(\lambda_{f_h})a$ by $O(\log \log s)$ lexicographical string comparisons. Note that $LF_{f_k} = LF_{lfp_{f_h}(\lambda_{f_h}-1)}LF_{lfs_{f_h}(\lambda_{f_h})a}$ holds from Lemmas 15 and 10. Next we compute λ_{f_k} by searching for the largest integer i such that $lfs_{f_k}(i+1)$ is not a prefix of y_i. Since $lfs_{f_k}(i+1)$ is not a prefix of y_i for any $1 \leq i < \lambda_{f_h}$ by Lemma 10, we can get λ_{f_k} without the information of $LF_{lfp_{f_h}(\lambda_{f_h}-1)}$. Hence it requires $O(\log s)$ string comparisons. Since $lfs_{f_h}(j)$ is a proper prefix of x_i for any $\lambda_{f_h} \leq i < j \leq m$, each comparison which is conducted between x_i and $lfs_{f_h}(j)a$ can be done in $O(1)$ time by using a data structure of LAQ (see Lemma 21), and hence we can compute $LF_{lfs_{f_k}(\lambda_{f_k})}$ in $O(\log s)$ time. Therefore we can compute Λ_{f_k} for all LZ78 factors f_k in a total of $O(s \log s)$ time and space. $\square$

Given the LZ78 encoding of size s corresponding to a string w, we can build the LZ78 trie T_w in $O(s)$ time. For any LZ78 factor v, let $\bar{v}$ denote the corresponding node of T_w. We use the following data structures LAQ and LCS:

Lemma 21 (Level Ancestor Query (LAQ) [2]). *We can pre-process a given rooted tree in linear time and space so that the ℓth node in the path from any node to the root can be found in $O(1)$ time for any $\ell \geq 0$, if such exists.*

The suffix tree of a reversed trie can be constructed in linear time [11]. Combined with the constant-time LCA data structure [2], we obtain the following:

Lemma 22 (Longest Common Suffix (LCS)). *We can pre-process a given trie in linear time and space so that the length of the longest common suffix of any two strings in the trie can be answered in $O(1)$ time.*

Using LAQ, given a node $\overline{v}$ of the LZ78 trie T_w, we can access any position of the corresponding LZ78 factor v in $O(1)$ time.

A string u is called an *LZ-block* w.r.t. w if u is a substring of some LZ78 factor of w. Since any node of T_w corresponds to an LZ78 factor, there exists at least one node $\overline{v}$ of the trie s.t. u is a suffix of v. Such node $\overline{v}$ is called a *handler* of u. Then u can be represented by a pair $(\overline{v}, |u|)$, in constant space. Let $\overline{p}(u)$ and $\rho(u)$ denote a handler of u and its corresponding LZ78 factor, respectively. For any LZ-block u and $1 \leq i \leq j \leq |u|$, $u[i..j]$ is also an LZ-block and its handler can be computed from $\overline{p}(u)$ in $O(1)$ time by using LAQ, i.e., when we write $u' \leftarrow u[i..j]$, it means we compute $\overline{p}(u')$ as the $(|u| - j)$th ancestor of $\overline{p}(u)$. Using LCS, we can check the equality of two given LZ-blocks u and u' of the same length in $O(1)$ time. $lcp(u, u')$ can be computed in $O(\log |u|)$ time by a binary search and finding the position where the first mismatch occurs.

Any substring of w can be represented by a sequence of LZ-blocks. Our algorithm to compute the first Lyndon factor of w maintains a sequence of LZ-blocks representing w by a dynamic linked list K, which is initially set to the LZ78 factorization of w itself but is restructured during the computation. After computing the leftmost Lyndon factor, we will also modify K to represent the remaining suffix of w in order to compute the remaining Lyndon factors. Let str_K denote the string represented by K. For any positions $i \leq j$ of str_K, let $\#_K[i, j]$ denote the number of LZ-blocks used to represent $str_K[i..j]$.

A pseudo-code of our algorithm is shown in Algorithm 1, which simulates Duval's algorithm to compute the first Lyndon factor of w on K.

The algorithm initializes $j \leftarrow 2$ and $h \leftarrow 0$, then starts with computing $lcp(w, w[j..|w|])$. Here, variables u and v are used for showing LZ-blocks which describe prefixes of $w[1 + h..|w|]$ and $w[j + h..|w|]$, respectively, where variable h shows that currently $w[1..h]$ and $w[j..j + h - 1]$ match. We can see at Lines 12-16 that the algorithm computes $lcp(w, w[j..|w|])$ by block-to-block comparisons, namely, the prefixes of length $d = \min\{|u|, |v|\}$ of u and v are cut out to LZ-blocks u' and v', and compared at Line 15. If $u' = v'$, we set $h \leftarrow h + d$ and continue matching the following LZ-blocks.

When we face LZ-blocks u' and v' that have a mismatch, we compute $h' \leftarrow lcp(u', v')$ by binary search at Line 25. At this moment $h \leftarrow h + h'$ is equal to $lcp(w, w[j..|w|])$, and w and $w[j..|w|]$ mismatch with $u'[1 + h']$ and $v'[1 + h']$. If $u'[1 + h'] \succ v'[1 + h']$, we have done the computation as Duval's algorithm does.

A major difference between our algorithm and Duval's lies in how we reset j when $u'[1 + h'] \prec v'[1 + h']$. While Duval's algorithm set $j \leftarrow j + h + 1$, our algorithm skips some positions by utilizing $\Lambda_{\rho(v[1..|v|-1])}$ in light of Lemma 19. Let $f_i = f_k v[|v|] = \rho(v)$, i.e., we are processing the i-th LZ78 factor. Since

---

Algorithm 1. Algorithm to compute the first Lyndon factor.

Input: The linked list of LZ-blocks K initialized to the sequence of the LZ78 factors of w.

Output: The first Lyndon factor $\ell_1^{p_1}$ of str_K.

// Note that variables u, u', v, v', x and y are LZ-blocks and
    manipulated via handlers.

1 $u \leftarrow K.\text{first}; \ v \leftarrow u[2..|u|];$

2 $c_u \leftarrow 0; \ c_v \leftarrow 0;$

3 $j \leftarrow 2; \ h \leftarrow 0;$

4 **while true do**

5   **if** $u = \varepsilon$ **then** $u \leftarrow \text{next}(u);$

6   **if** $v = \varepsilon$ **then** $v \leftarrow \text{next}(v);$

7   **if** v *is an LZ78 factor which is used for the first time* **then**

8           $x \leftarrow K.\text{first}; \ y \leftarrow K.\text{second};$

9           $v' \leftarrow$ the longest member in Λ_v s.t. $(xy)[1..|v'|] = v'$ if such exists, ε otherwise;

10   **if** $|x| + 1 = |v'|$ **then**

11   restructure the first LZ-block to be v', and reset u and/or v if needed;

12      $d \leftarrow \min\{|u|, |v|\};$

13      $u' \leftarrow u[1..d]; \ u \leftarrow u[d+1..|u|];$

14      $v' \leftarrow v[1..d]; \ v \leftarrow v[d+1..|v|];$

15   **if** $u' = v'$ **then**

16          $h \leftarrow h + d;$

17   **if** $u = \varepsilon \ \& \ c_u \geq 2$ **then**

18   restructure the last two LZ-blocks before u to be a single LZ-block;

19   **if** $v = \varepsilon \ \& \ c_v \geq 2$ **then**

20   restructure the last two LZ-blocks before v to be a single LZ-block;

21   **if** $u = \varepsilon \ \& \ v = \varepsilon$ **then** $c_u \leftarrow 1; \ c_v \leftarrow 1;$

22   **else if** $u = \varepsilon$ **then** $c_u \leftarrow c_u + 1; \ c_v \leftarrow 0;$

23   **else if** $v = \varepsilon$ **then** $c_v \leftarrow c_v + 1; \ c_u \leftarrow 0;$

24   **else**

25          $h' \leftarrow lcp(u', v');$

26   **if** $u'[1 + h'] \prec v'[1 + h']$ **then**

27              $c_u \leftarrow 0; \ c_v \leftarrow 0;$

28   **if** $v = \varepsilon$ **then** continue;

29              $x \leftarrow K.\text{first};$

30              $v' \leftarrow$ the longest member in $\Lambda_{\rho(v[1..|v|-1])}$ s.t. $x[1..|v'|] = v'$ if such exists, ε otherwise;

31              $j \leftarrow$ the position in str_K where the v' begins if $v' \neq \varepsilon$, the position where the v ends otherwise;

32              $h \leftarrow |v'|; \ u \leftarrow x[1 + h..|x|]; \ v \leftarrow v[|v|];$

33   **else** // $u'[1 + h'] \succ v'[1 + h']$

34              $h \leftarrow h + h';$ break;

35 **output** $|\ell_1| \leftarrow j - 1$ and $p_1 \leftarrow 1 + \lfloor h/|\ell_1| \rfloor;$

f_k is an LZ78 factor appearing before f_i, we can use Lemma 19, i.e., we only have to consider the positions where a significant suffix of f_k begins. Moreover, at Lines 8-11 we have maintained the first LZ-block x of K to be the longest member in $\bigcup_{i'=1}^{i} \Lambda_{f_{i'}}$, which is also a prefix of w. Since $x[1..|v'|] \preceq v'$ for any $v' \in \Lambda_{f_k}$, we can notice that $x[1..|v'|] \prec v'$ if $x[1..|v'|] \neq v'$, and hence we are able to skip such positions. Then we set j to be the beginning position of the longest member $v' \in \Lambda_{f_k}$ with $x[1..|v'|] = v'$ if such exists, otherwise the ending position of v, and restart suffix competition.

As for the maintenance of the first LZ-block of K at Lines 8-11, since any LZ78 factor has form $f_i = f_k a$ with $1 \leq k < i \leq s$ and $a \in \Sigma$, the length of the longest member in $\bigcup_{i'=1}^{i} \Lambda_{f_{i'}}$, which is also a prefix of w increases at most 1 when processing the new LZ78 factor. Hence the procedures at Lines 8-11 works fine as far as the first LZ-block has maintained properly.

The following is the main theorem of this section.

Theorem 2. *Given the LZ78 encoding of size s for string w, we can compute LF_w in $O(s \log s)$ time and space.*

Proof. We compute the Lyndon factorization of w from left to right using Algorithm 1 recursively. We pre-process in $O(s)$ time and space for data structures LAQ and LCS on T_w. We also compute Λ_{f_i} for all LZ78 factors f_i, from Lemma 20 it takes $O(s \log s)$ time and space.

During the whole computation, for any LZ78 factor f_i we execute Lines 8-11 just once. Since $|\Lambda_{f_i}| = O(\log s)$ it takes in total of $O(s \log s)$ time. In what follows, we consider the cost other than that comes from Lines 7-11.

Let K_1 denote the linked list of the sequence of the LZ78 factors of w. We show that Algorithm 1 computes, given K_1, the first Lyndon factor of w in $O(e_1 \log s + g_1)$ time, where $\hat{j}_1$ and $\hat{h}_1$ are respectively the last values of j and h when algorithm halts, and $e_1 = \#_{K_1}[1, \hat{j}_1]$ and $g_1 = \#_{K_1}[1, \hat{j}_1 + \hat{h}_1]$.

Firstly, let us estimate the total cost for the if-control of Line 15. Let t, t' and t'' be the numbers we execute Lines 21, 22 and 23, respectively. When we enter the if-control, any one of them must be executed. Here note that $\text{next}(v)$ is executed at most g_1. Since $\text{next}(v)$ must be executed just after either Line 21 or Line 23 is executed, $t + t'' \leq g_1$. In addition, if we execute Line 22 more than three consecutive times Line 18 reduces the number of LZ-blocks in K_1, and hence $t' \leq 3g_1$. Since the unit cost of the if-control is $O(1)$, the total cost for the if-control is $O(t + t' + t'') = O(g_1)$. Next, the else-control of Line 24 is executed $O(e_1)$ times since we either halt the computation at Line 34, or reset j to be in the last LZ-block we are processing at Line 31 and j will get over that LZ-block when Line 31 is executed next time. Since Line 25 and Line 30 take $O(\log s)$ time, the cost for the else-control is $O(e_1 \log s)$ in total. Hence the first Lyndon factor of w can be computed in $O(e_1 \log s + g_1)$ time.

After computing the first Lyndon factor, we modify K_1 to K_2 which represents the remaining suffix of w, i.e., we discard the LZ-blocks representing $w[1..|\ell_1|p_1]$. Also we maintain its first block to be the longest member in $\bigcup_{i'=1}^{i} \Lambda_{f_{i'}}$, where f_i is the last LZ78 factor we have processed. The modification takes $O(g_1)$ time.

Then we use Algorithm 1 to compute the second Lyndon factor of w, i.e., the first Lyndon factor of str_{K_1}. The computation takes $O(e_2 \log s + g_2)$ time, where $\hat{j}_2$ and $\hat{h}_2$ are respectively the last values of j and h when algorithm halts, and $e_2 = \#_{K_2}[1, \hat{j}_2]$ and $g_2 = \#_{K_2}[1, \hat{j}_2 + \hat{h}_2]$. We iterate this procedure until we get the last Lyndon factor $\ell_m^{p_m}$ of w. The sum of the cost is $O(\sum_{i=1}^m e_i \log s + \sum_{i=1}^m g_i)$. Since $lfb_w(i) + \hat{j}_i \le lfb_w(i+1)$ for any $1 \le i < m$, $\sum_{i=1}^m e_i = O(\sum_{i=1}^m \#_{K_i}[1, \hat{j}_i]) = O(\#_{K_1}[1, |w|]) = O(s)$, and hence $O(\sum_{i=1}^m e_i \log s) = O(s \log s)$.

The final concern is how we can analyze $\sum_{i=1}^m g_i = O(s)$. Since the substrings of w considered in each iteration are overlapped, e.g., $w[1..\hat{j}_1 + \hat{h}_1]$ and $w[lfb_w(2)..lfb_w(2) + \hat{j}_2 + \hat{h}_2 - 1]$ are overlapped at most $|\ell_1|$, we cannot conclude immediately that $\sum_{i=1}^m g_i = O(\#_{K_1}[1, |w|]) = O(s)$. However, we can charge the cost from the overlapped LZ-blocks to the previous LZ-blocks thanks to the restructuring at Line 20, e.g., when $w[1..\hat{j}_1 + \hat{h}_1]$ and $w[lfb_w(2)..|w|]$ are overlapped, namely $lfb_w(2) \le \hat{j}_1 + \hat{h}_1$, $\#_{K_2}[1, \hat{j}_1 + \hat{h}_1 - |\ell_1|p_1] = O(\#_{K_1}[lfb_w(2) - |\ell_1|, \hat{j}_1 + \hat{h}_1 - |\ell_1|])$. Hence, $\sum_{i=1}^m g_i = O(2 \sum_{i=1}^m \#_{K_1}[1, |\ell_i|p_i]) = O(2\#_{K_1}[1, |w|]) = O(s)$. Therefore the statement holds. $\qquad\square$

References

1. Apostolico, A., Crochemore, M.: Fast parallel Lyndon factorization with applications. Mathematical Systems Theory 28(2), 89–108 (1995)
2. Bender, M.A., Farach-Colton, M.: The level ancestor problem simplified. Theor. Comput. Sci. 321(1), 5–12 (2004)
3. Brlek, S., Lachaud, J.O., Provençal, X., Reutenauer, C.: Lyndon + Christoffel = digitally convex. Pattern Recognition 42(10), 2239–2246 (2009)
4. Chen, K.T., Fox, R.H., Lyndon, R.C.: Free differential calculus. iv. the quotient groups of the lower central series. Annals of Mathematics 68(1), 81–95 (1958)
5. Daykin, J.W., Iliopoulos, C.S., Smyth, W.F.: Parallel RAM algorithms for factorizing words. Theor. Comput. Sci. 127(1), 53–67 (1994)
6. Duval, J.P.: Factorizing words over an ordered alphabet. J. Algorithms 4(4), 363–381 (1983)
7. Gil, J.Y., Scott, D.A.: A bijective string sorting transform. CoRR abs/1201.3077 (2012)
8. I, T., Matsubara, W., Shimohira, K., Inenaga, S., Bannai, H., Takeda, M., Narisawa, K., Shinohara, A.: Detecting regularities on grammar-compressed strings. In: Chatterjee, K., Sgall, J. (eds.) MFCS 2013. LNCS, vol. 8087, pp. 571–582. Springer, Heidelberg (2013)
9. I, T., Nakashima, Y., Inenaga, S., Bannai, H., Takeda, M.: Efficient lyndon factorization of grammar compressed text. In: Fischer, J., Sanders, P. (eds.) CPM 2013. LNCS, vol. 7922, pp. 153–164. Springer, Heidelberg (2013)
10. Kufleitner, M.: On bijective variants of the Burrows-Wheeler transform. In: Proc. PSC 2009, pp. 65–79 (2009)
11. Shibuya, T.: Constructing the suffix tree of a tree with a large alphabet. IEICE Transactions on Fundamentals of Electronics, Communications and Computer Sciences E86-A(5), 1061–1066 (2003)
12. Ziv, J., Lempel, A.: Compression of individual sequences via variable-length coding. IEEE Transactions on Information Theory 24(5), 530–536 (1978)

Lossless Compression
of Rotated Maskless Lithography Images

Shmuel Tomi Klein[1], Dana Shapira[2], and Gal Shelef[1]

[1] Department of Computer Science, Bar Ilan University, Ramat Gan 52900, Israel
tomi@cs.biu.ac.il, gal.shelef@gmail.com
[2] Computer Science Department, Ashkelon Academic College, Israel
shapird@ash-college.ac.il

Abstract. A new lossless image compression algorithm is presented, aimed at maskless lithography systems with mostly right-angled regular structures. Since these images appear often in slightly rotated form, an algorithm dealing with this special case is suggested, which improves performance relative to the state of the art alternatives.

1 Introduction

The tremendous storage requirements and ever increasing resolutions of digital images, necessitate automated analysis and compression tools for information processing and extraction. Most of the images may tolerate lossy compression techniques, but there are cases in which each single pixel is significant. For example, medical X-ray images must not be changed in any way as the lost data may be critical to diagnosis. Another example, on which we wish to concentrate in this work, is microchip lithography, the process of fabricating the complex designs of a microchip onto a semiconducting substrate also known as a wafer.

Today's dominating, though quite expensive, technique for microchip fabrication is masked lithography, in which a mask is produced and then used to mass-produce wafers at a high throughput. If a change is introduced to the wafer design, then a new mask must be produced, thus incurring an even higher production cost. This inflexibility to design changes may greatly increase the costs of chip manufacturing in case of frequent design updates.

Over the last decade, a new microchip fabrication method referred to as *maskless lithography* is being developed and refined, but suffers from several limitations which stand in its way of becoming a serious competitor to masked lithography. One of these limitations is handling large-scale wafer data.

Microchip fabrication in a maskless fashion requires the microchip design data to be fed to the maskless printer which fabricates the microchip feature by feature onto the wafer. The data is essentially an image in which each pixel is a nano-scale feature. When taking into account the size of a microchip compared to the size of a feature, the resulting image measures in the hundreds of gigabits. For example, a 10×20 mm sized microchip with a 22 nm feature size would produce an image of about 385 Gb.

O. Kurland, M. Lewenstein, and E. Porat (Eds.): SPIRE 2013, LNCS 8214, pp. 186–196, 2013.

A microchip is designed using vector graphics. The vector graphics representation is very compact, but takes hours to rasterize and therefore cannot be used directly during the printing process. The image is rasterized beforehand and then fed to the maskless printer. A rasterized wafer image requires large amounts of space, which introduces a difficulty: the wafer image data is too large to be saved on-chip and therefore can only be streamed, thus creating a bottleneck in the printing process.

The throughput of the maskless writer is not competitive with masked writing, even when taking into account its flexibility advantage. One solution would be to use massively parallel writers. As alternative, simple decoding logic can be integrated on the writing chip, so the data streamed to the writers may be compressed and decoded on-chip to allow higher throughput. This problem and the data path designed to solve it are specified in detail in [7].

Generally, achieving high lossless compression for images may be impossible in some cases, but many wafer images are characterized by a simple *Manhattan-like* structure. A Manhattan structure [7] is a structure of mainly right angled polygons in which most lines are parallel to the edges of the image. This structure may be exploited in order to model the data very accurately and thus, achieve a high compression ratio. Figure 1 depicts a sample Manhattan style wafer image.

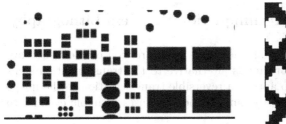

Fig. 1. Sample Manhattan style image

Fig. 2. Rotated sample image

The state of the art solutions for data throughput in maskless lithography are the algorithms Block C4 [7] and its variations, and Corner [10], both of which take a similar approach, inspired by the Burrows-Wheeler compression method. They first apply a reversible transformation on the data, which in itself does not reduce its size, but recodes it into a generally much more compressible form. Similarly here, the images are reversibly transformed so that the result, at least for this class of potential inputs, will be an extremely sparse image, i.e., one that is almost fully zeroed out. This transformed image is much more compressible than the original, and each algorithm uses some kind of entropy coder to take advantage of this fact.

However, the algorithms rely heavily on the Manhattan structure, yet not all images adhere to it. Microchips manufacturing images are sometimes deliberately slightly rotated for various hardware reasons, as, for example, in [8, 9], but these are not the only applications. We obtained rotated images from some wafer manufacturer and the reason for the choice of such a slanted layout was to allow the continuous processing of very large periodic wafers at some constant speed, without being forced to disrupt the production at regular intervals. But such rotations disrupt the Manhattan structure of the image yielding considerably worse compression savings. Even if the general layout is still dominated by the appearance of various rectangles, the sides of the polygons are not always parallel to the external edges of the image, and are rather chosen so as to form some constant, generally small, angle. A sample such rotated image appears above in Figure 2, which clearly shows how segments that must have been straight lines in some original, non-rotated image, turned into a collection of jagged saw-like forms.

The compressibility of images is obviously connected to their density (probability of a 1-bit), which is only minorly affected by rotation. However, the compressibility of the Manhattan-structured wafer image changes dramatically even after rotating by a small angle.

In the next section, we introduce a new and simple alternative to the lossless image compression techniques mentioned above. Section 3 then deals with the compression of rotated images and Section 4 brings comparative experimental results.

2   New Compression Technique for Maskless Lithography

In a first attempt to deal with the lossless image compression problem, we have designed a simple generic algorithm as alternative to the transformation steps of Block C4 and Corner. It also applies a reversible transformation to the image which zeros out large regions of the image, and then uses a simple encoder to compress the data.

Since wafer images have a tendency to contain identical or at least similar consecutive rows and columns, the idea is to exploit this similarity as in an application for the compression of correlated bitmaps [3], and to Xor each line (except the top one) with that above it, and afterwards each column (except the leftmost) with that to its left. The Xored lines and columns are expected to be much sparser than the original ones, and thus much more compressible. The transformation is of course reversible, by Xoring again, starting with the columns and then with the rows.

Using Boolean Xor has several advantages. Not only is it reversible, very fast and can exploit bit-parallelism by processing entire blocks in single operations, but its result is also a binary image, and for similar inputs, a much sparser one. Corner differentiates between several types of non-zero pixels, depending on their location (upper-left corner, etc.), and these alternatives need also to be encoded. Block C4 partitions the image into regions that can be copied from elsewhere within the image, similarly to Lempel-Ziv techniques, and other regions whose pixels are predicted by the neighboring pixels. The new bitmap produced as a result of these transformations does therefore not convey the full information necessary for the decoding, and must be supplemented by additional data. The Xoring based technique, on the other hand, yields a pure black and white image, including all the information of the original data.

Moreover, Corner works by hollowing out entire regions, transforming two-dimensional polygonal forms into their one-dimensional contours. This could be challenging if there are nested rectangular regions or non-closed geometrical forms for which the contour may be ill defined. All these are treated by Xoring in a straightforward way. On the other hand, Xoring does not deal with repeated sub-patterns, as suggested by LZ techniques, so for images containing frequent repeated regions that are not efficiently compressible on their own, a Xoring based method may be inferior.

As to the encoding, there are several alternatives, based on run length or Huffman codes, but the best results, at least on our test samples, have been obtained by adapting hierarchical bit-vector compression [4] to two dimensions. The basic idea of the one-dimensional method is the following. The data vector of size n is sectioned into blocks of size $f(n) < n$, where f is a linear function serving as parameter, such that $f(n)$ divides n. Then, a bit vector of size $n/f(n)$ is created and the value of bit i in the bit vector is set to 0 if the i-th block contains only zeroes, and to 1 otherwise. Non-zero blocks are re-partitioned into blocks of size $f(f(n))$, and again a bit vector for each block is created and concatenated to the previous vector, repeating the same process until a threshold T on the block size is reached.

If the data is mostly zeroes, then many blocks will be encoded using a single bit achieving a good compression ratio. This method may be adapted to two-dimensional data in several ways. In order to exploit the distribution of non-zero bits over the two-dimensional Xor transformed images, we adapt the bit vector compression by simply sectioning to two-dimensional blocks of size $f(w) \times f(h)$, where w is the width of the image and h is its height. The same recursive process as in the one-dimensional case is applied and a bit vector is generated which, essentially, hierarchically encodes the image by finer and finer blocks. When the block size threshold is reached, the whole block is concatenated to the bit vector in raster scan order.

We used $f(n) = n/2$ for both dimensions, and $T = 16$, i.e., the image was partitioned into 4 quarters, and each non-zero section was recursively partitioned to quarters until the section size in pixels was less or equal to 16. The formal compression algorithm for an image I, of size $w \times h$ generates a sequence of elements E forming a tree, and prints its elements in post-order, that is, printing recursively the subtrees from left to right, followed by their parent node.

Hier-compress(I, w, h)

```
1    if w · h ≤ T then output I
2    else
3        k  ←  w/f(w) · h/f(h)
4        partition I into k equal sized sections I₁, I₂, ..., Iₖ
5        define an element E of k bits
```

```
6          for j ⟵ 1 to k
7              if I_j is completely 0 then
8                  E[j] ⟵ 0
9              else
10                 E[j] ⟵ 1
11                 Hier-compress(I_j, f(w), f(h))
12         output E
```

The left side of Figure 3 is an example image with $w = h = 16$ and the right side shows the corresponding tree for $f(n) = n/2$ and $T = 4$. Note that the compressed image could either consist of the sequence of elements of this tree in pre-order, even though they were produced in post-order by the algorithm, or one might simply concatenate the elements of the tree layer by layer, top down, and in each layer, left to right. The size of each element E encoded in the layers above the lowest one is $k = wh/(f(w)f(h))$, and the size of the elements in the lowest level is T, which both are 4 bits in our example. One could of course get even better compression by applying on these blocks some entropy coding, like Huffman or arithmetic coding. Even without that additional step, the compressed size is $4 \times 15 = 60$ bits, corresponding on this example to a compression ratio 4.27. The *compression ratio* is defined as the size of the original divided by the size of the compressed image.

We shall refer to the method based on Xoring rows and columns followed by the hierarchical bit compression technique as XH.

3   Compression of Rotated Wafers

As mentioned above, wafer images are not always given in strict Manhattan style, and many of them are rotated. The following experiment shows the impact of such rotation on their compressibility. Taking as baseline the compression ratio obtained on a set of original images (rotation angle 0), the three compression methods BlockC4, Corner and XH have been applied after rotating the images by $5, 10, \ldots, 45$ degrees, and Figure 4 displays the relative compression, that is, the average compression ratio on the sample set for a given angle, divided by the corresponding compression ratio for angle 0. As can be seen, most of the values are in the 40–60% range, for all methods. This is in spite of the fact that the rotated images contain essentially the same number of 1-bits. One may thus conclude that compressibility is not only a function of the sparsity of the image, but that also the strict Manhattan structure, with lines that are parallel to the edges, has a major impact.

We suggest dealing with the problem of compressing rotated images in a way that is inspired by the *compressed matching problem* [1, 2, 6]. Given a pattern P, a text T and complementing encoding and decoding functions $\mathcal{E}$ and $\mathcal{D}$, the problem is to locate P in the compressed text $\mathcal{E}(T)$. While the obvious solution would be to decompress and then search, i.e., look for P in $\mathcal{D}(\mathcal{E}(T))$, compressed

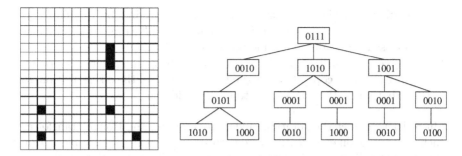

Fig. 3. Hierarchical compression of sparse binary images

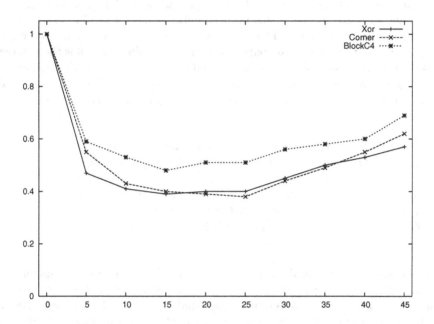

Fig. 4. Degradation of the compression ratio as function of rotation angle

matching calls for rather compressing the pattern too, and looking for $\mathcal{E}(P)$ in $\mathcal{E}(T)$, with the necessary adaptations. Similarly for our problem, we suggest that instead of investing efforts to compress a rotated — and thus less compressible — image, we first try to rectify the rotation, producing an alternative by "rotating back" into a strict Manhattan structure, and only then applying some of the known compression methods.

More specifically, the suggested algorithm consists of the following steps:

1. Detecting the rotation angle α. The angle is generally constant over the whole image, and can be detected using image processing tools.

2. Generating a "straightened" image. Rotating the original by $-\alpha$, should yield an image that is very close to being strictly Manhattan. This can then be compressed using any of the methods for such bitmaps.

3. Rectifying the (small) errors. Ideally, decompressing and rotating by α should produce an exact copy of the original, but because of the discretization, some perturbation on the edges seem to be unavoidable.

In order to detect the rotation angle, one may use the Hough transform [5], which extracts geometric features from an image. We use the simplest form to extract straight lines and then analyze the results to calculate the rotation angle. The result of applying the transform on an image is a matrix A in which each cell represents a different straight line by polar parameters. The polar representation of a line L consists of the angle θ of the perpendicular to L passing through the origin, and of r, the length of this perpendicular, as illustrated in Figure 5.

The dimensions of the matrix are thus the number of discrete steps into which the angle θ and the length r are partitioned, and the value of a cell is the number of votes that the line represented by that cell, got while applying the transform. A line gets a vote for every pixel in the original image that is on that line and is not zeroed. The formal algorithm used on an input image Img is then:

```
1    For every pixel index x from 0 to the width w of Img
2        For every pixel index y from 0 to the height h of Img
3            if Img[x, y] ≠ 0
4                For each angle θ from 0 to π in discrete steps
5                    r = ⌊x cos θ + y sin θ + ½⌋
6                    increment A[r, θ]
```

After obtaining the result matrix A, we proceed to find the local maxima in the matrix and those correspond to the lines in the image. There is, of course, some noise in the result collection of lines that is caused when a large enough group of non-zero pixels in the image are accidentally on the same straight line. To filter out this noise, only the k lines that got the most votes are taken. Each line in this group adds its votes to its angle, θ modulo 180, so parallel lines vote for the same angle. The angle that gets the most votes from the k top lines is chosen as the rotation angle. We used $k = 10$ in our experiments.

The Hough transform may be computationally too expensive to be applied to the whole image. But since we do not seek the original lines in the image, and are merely trying to calculate their angle, it is possible to scale down the image considerably while keeping the general trends, and thus apply the transformation on a significantly reduced image. This can be done, e.g., by considering a sparse enough subset of equi-spaced pixels. An alternative is partitioning the image into small squares of odd side length, which ensures an odd number of pixels in each square, and replacing each square by a single pixel set by a majority vote of the pixels in the square.

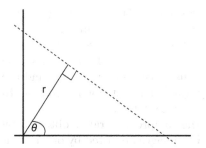

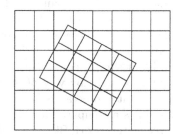

Fig. 5. Polar coordinates for Hough   **Fig. 6.** Rotated pixels

To address the problem of rectifying small perturbations, we suggest to use an *error image*, as also done, albeit differently, in BlockC4. After recognizing the rotation angle, the image is rotated back into Manhattan form, and standard compression methods designed for such images achieve then a good ratio. To uncompress, we decode and get the reverse-rotated image back, which can be rotated by the detected angle to yield an image which is almost identical to the original one, but has a few errors where interpolation has been utilized. This can be understood by realizing that each pixel of a rotated image may touch and thus influence the value of several pixels after the rotation, as shown in the schematic view of Figure 6. It is unlikely to revert to an image which is an exact copy, so one has to deal with the errors.

To repair those errors, after reverse rotating the original image, it is immediately rotated back again and Xored with the original image to get the error image E. By Xoring E with the decoded image, one essentially gets the original. Since the decoded and original images are almost identical, except for a few interpolation differences, the error image E is almost completely zeroed out. This fact enables effective compression of E similarly to what has been suggested above, since this is another instance of the same problem of compressing sparse bitmaps. The combined algorithm of detecting a rotation angle, rotating back, compressing the rectified image and adding a compressed error map, will be referred to below as rotated XH, or RXH for short.

Note that the basic idea of RXH of trying to rotate the image back into strict Manhattan form, can be applied in combination with the other compression methods as well, and not just with XH. We have therefore compared our method also to the corresponding versions of BlockC4 and Corner, which will be referred to as RBC4 and RCorner, respectively.

4   Experimental Results

Obtaining real world microchip fabrication images is not an easy task. These images are considered an industrial secret for long periods of time even after the microchips are produced, marketed and become outdated. The dataset of images used for this research is a series of 30 binary images that were kindly provided by some HighTech company. Each image takes about 150MB of storage space when uncompressed, and a fragment of one of them is depicted in Figure 1. For technical reasons, these images were split into fragments of 2048 × 10384 pixels each, and a random subset of 21 of the fragments was chosen for our tests. The chosen input sample contained images of various densities and layouts, and the results are brought by way of example only, since it might not be possible to find a set that can be agreed upon for being representative.

Figure 7 is a comparative chart, showing the compression ratios obtained by the three methods XH, BlockC4 and Corner. The images are listed by order of increasing compression ratio for XH, which goes up to about 600 (for a nearly fully blacked out image on which the hierarchical bit vector encoding compresses very efficiently). As can be seen, for certain images, the Xor based method performs up to 3 times better than BlockC4, while on others, the latter may be 10 times as effective as the former. On all the examples, XH performed better than Corner, which was also inferior to BlockC4 in most, but not all, cases. A closer look at the images on which BlockC4 performed much better than the competitors (e.g., the images indexed 1–4) revealed that they contained many repeated patterns in the form of circles, which are detected by the Lempel-Ziv copy mechanism, but deviate from the assumed Manhattan structure.

On the other hand, BlockC4 runs significantly slower when compared to Corner and our Xor based method. Table 1 shows the average obtained compression ratios, and the total encoding and decoding times, in seconds, of the three methods on the sample of 21 images. Note in particular the large encoding time used

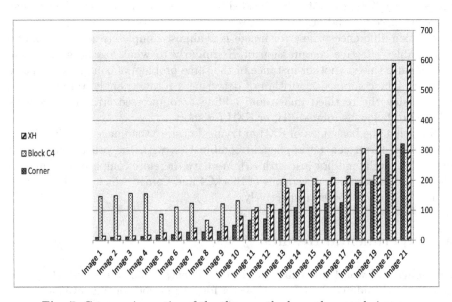

Fig. 7. Compression ratios of the three methods on the sample images

Table 1. Compression ratios and processing speed

	XH	BlockC4	Corner
compression ratio	161	160	91
encoding time	3.4	211.5	7.7
decoding time	0.7	2.2	1.2

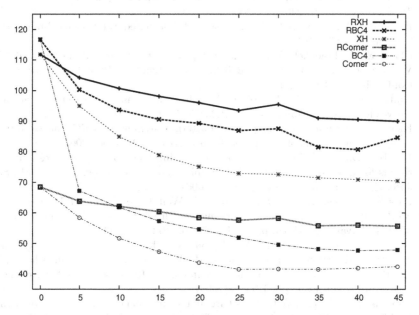

Fig. 8. Compression ratios on rotated images as function of rotation angle

by BlockC4, which is due to the search for matching sub-blocks. On all examples, XH was significantly faster.

To test the effect of RXH, Image12 was chosen as testbed, on which BlockC4 performed somewhat better than XH, and the six methods (XH, BC4, Corner and their "rotated" counterparts) were applied after having rotated the image by 5, 10, etc. up to 45 degrees. Specifically, XH, BC4 and Corner refer to the results obtained by applying the algorithms directly on the rotated image, while RXH, RBC4 and and RCorner use preprocessing by rotating back and only then attempting to compress. The results are displayed in Figure 8. As expected, there is a general loss versus the possible compression before the rotation (with angle 0). For all the rotated images, RXH gave the best results. Similar results have been obtained on other test images.

It should be noted that the authors of Corner have published improved algorithms [11, 12], which have not been compared herein, and we shall deal with it in future work.

Acknowledgements. We would like to thank Vito Dai for sharing the code of BlockC4, and Jeehong Yang for the code of Corner.

References

[1] Amir, A., Benson, G.: Efficient two-dimensional compressed matching. In: Proc. Data Compression Conference DCC 1992, Snowbird, Utah, pp. 279–288 (1992)

[2] Amir, A., Landau, G.M., Sokol, D.: In place 2D matching in compressed images. J. Algorithms 49(2), 240–261 (2003)

[3] Bookstein, A., Klein, S.T.: Compression of Correlated Bit-Vectors. Information Systems 16, 387–400 (1991)

[4] Choueka, Y., Fraenkel, A.S., Klein, S.T., Segal, E.: Improved Hierarchical Bit-Vector Compression in Document Retrieval Systems. In: Proc. 9-th ACM-SIGIR Conf., Pisa, pp. 88–97 (1986)

[5] Duda, R.O., Hart, P.E.: Use of the Hough Transformation to Detect Lines and Curves in Pictures. Comm. of the ACM 15, 11–15 (1972)

[6] Klein, S.T., Shapira, D.: Compressed Pattern Matching in JPEG Images. Intern. J. of the Foundations of Computer Science 17(6), 1297–1306 (2006)

[7] Liu, H.I., Dai, V., Zakhor, A., Nikolic, B.: Reduced complexity compression algorithms for direct-write maskless lithography systems. In: Proc. of the SPIE, San Jose, California, vol. 6151, p. 61512B (2006)

[8] Paraskevopoulos, A., Voss, S.H., Talmi, M., Walf, G.: Scalable (24–140 Gbs) optical data link, well adapted for future maskless lithography applications. In: Proc. of SPIE Advanced Lithography, vol. 7271, p. 7271–53 (2009)

[9] Petric, P., Bevis, C., Brodie, A., Carroll, A., Cheung, A., Grella, L., McCord, M., Percy, H., Standiford, K., Zywno, M.: Reflective electron-beam lithography (REBL), Alternative Lithographic Technologies. In: Proc. of SPIE Advanced Lithography, vol. 7271, pp. 7271–7 (2009)

[10] Yang, J., Savari, S.A.: A Lossless Circuit Layout Image Compression Algorithm for Maskless Lithography Systems. In: Data Compression Conference, DCC 2010, pp. 109–118 (2010)

[11] Yang, J., Savari, S.A.: Lossless circuit layout image compression algorithm for maskless direct write lithography systems. J. of Micro/Nanolithography, MEMS, and MOEMS 10(4), 043007-1–043007-13 (2011)

[12] Yang, J., Savari, S.A.: Improvements on Corner2, a lossless layout image compression algorithm for maskless lithography systems. Proc. of SPIE Advanced Lithography 8352, 83520K1–9 (2012)

Learning URL Normalization Rules Using Multiple Alignment of Sequences

Kaio Wagner Lima Rodrigues, Marco Cristo, Edleno Silva de Moura,
and Altigran Soares da Silva

Universidade Federal do Amazonas,
Department of Computer Science, Manaus, Brazil

Abstract. In this work, we present DUSTER, a new approach to detect and eliminate redundant content when crawling the web. DUSTER takes advantage of a
multi-sequence alignment strategy to learn rewriting rules able to transform URLs
to other likely to have similar content, when it is the case. We show the alignment
strategy that can lead to a reduction in the number of duplicate URLs 54% larger
than the one achieved by our best baseline.

1   Introduction

A well-know problem faced by web crawlers is the existence of large fraction of distinct URLs that correspond to pages with duplicate or near-duplicate contents. In fact,
as estimated in [7], about 29% of web pages are duplicates. Such URLs, commonly
named as DUST (Different URLs with Similar Text), represent an important problem
to search engines. Crawling DUST leads to several drawbacks: waste of resources, such
as bandwidth and disk storage; disturbance in results of link analysis algorithms; and
poor user experience due to duplicate results.

To overcome such a problem, several authors have proposed methods for detecting
and removing DUST in the past years. Whereas first efforts were focused on comparing
document content, more recent strategies inspect only the URL string without fetching
the corresponding document content to detect DUST [1, 2, 5, 9, 10]. As pointed in these
works, most of the examples of duplicate URLs can be seen as transformations on the
URL string resulting from server software. In general, a training set is provided which
contains clusters of URLs (dup-clusters), such that all the URLs of a cluster correspond
to pages with equal or similar contents. From these dup-clusters, it is possible to mine
rewrite rules that transform all URLs to a same canonical form. These rewrite rules can
be then applied to eliminate duplicates among URLs that are encountered for the first
time during the crawling.

A challenging aspect of this strategy is to derive a small set of general and precise
rewriting rules. In order to derive rules, we here present DUSTER, a new approach
which takes advantage of a multi-sequence alignment strategy to significantly improve
the quality and coverage of rewriting rules. Multiple sequence alignment is largely used
in Molecular Biology as an analyzing tool since it is easier to found patterns in aligned
sequences [3, 8]. In this work, we show that a full multi-sequence alignment of all
DUST, before rule extraction, can obtain very effective rewriting rules. By evaluating
our method in a reference collection, we observed a reduction in the number of duplicate
URLs 54% larger than the one achieved by our best baseline.

O. Kurland, M. Lewenstein, and E. Porat (Eds.): SPIRE 2013, LNCS 8214, pp. 197–205, 2013.

2 Related Work

Current research on DUST detection can be classified into content-based and URL-based. In content-based DUST detection, the similarity of two URLs is determined by comparing content signatures of the pages the URLs refer to [10, 11]. To avoid such an expensive inspection of page contents, several URL-based methods have been proposed.

The first URL-based method proposed was DustBuster [2]. In that work, the authors addressed the DUST detection problem as the problem of finding rewrite rules able to transform a given URL to other likely to have similar content. Dasgupta et. al. presented a new formalization of URL rewrite rules [5] to capture many common duplicated URL transformations ignored by DustBuster. The authors also used some heuristics to generalize the generated rules. In particular, they attempt to infer the false-positive rate of the rules to select the most precise ones. They also verify if the set of values that a certain URL component assumes is greater than a threshold value N, a heuristic which they call *fanout-N*. Their best results were obtained with $N = 10$. In this work, we refer to this method as $R_{funout-10}$. By applying the set of rules found by $R_{funout-10}$ to a number of large scale experiments on real data, the authors were able to reduce the number of duplicate URLs by 60%, whereas DustBuster achieved about 22%. We adopt this method as our first baseline because it has been compared to all other methods proposed after it. The authors in [1] extended the work in [5] to make it feasible their use at web scale. To overcome the quadratic complexity of the rule extraction, they proposed a method for deriving rules from samples of URLs. In a following paper [9], they implemented their algorithm using a distributed framework and extended the URL and rule representations to include additional patterns. They evaluated the method with 3 billion URLs showing its scalability. By comparing their method with $R_{funout-10}$, they achieved two times more reduction using 56% of the rules.

The previous methods used a bottom-up approach in which rewrite rules are learned by inducing local duplicate pairs to more general forms. Authors in [10] argue that such an approach is very sensitive to noise. Thus, they proposed a top-down approach in which a URL pattern tree (UPT) is built from clusters of duplicated URLs for a targeted website. They evaluated their approach in a collection with 70 million URLs and showed that their method was able to outperform $R_{funout-10}$ achieving about twice the reduction using 46% of the rules and consuming half of the learning time. As far as we know, this is the best method in literature, which led us to adopt it as our second baseline. In this work, we refer to it as R_{tree}.

As $R_{funout-10}$ [5], we use a bottom-up approach. Differently from other bottom-up methods, we do not derive candidate rules from URLs pairs. We perform a *full* alignment of all URLs within the dup-cluster *before* the rule generation. We do not address large dup-clusters since they are very rare. For such a cases, we can use efficient heuristics based on the ones proposed in [9].

3 Sequence Alignment

A sequence alignment can be seen as a way of arranging n sequences in order to identify similar regions between them. Given the sequences X and Y with m and n characters respectively, the alignment process can be described by using a matrix S of size $(m + 1) \times (n + 1)$, so that S cells are filled as follows:

$$S_{i,j} = \begin{cases} 0 & if\ i=0\ or\ j=0 \\ max \begin{pmatrix} S_{i-1,j-1} + sf(X_i,Y_j), \\ S_{i-1,j}, \\ S_{i,j-1} \end{pmatrix} & otherwise \end{cases} \quad (1)$$

where $sf(X_i, Y_j)$ is a scoring function that defines a similarity between the pairs of symbols (X_i, Y_j). This function gives points for matching tokens and penalties for any gap.

Given $k > 2$ sequences $\mathscr{S} = \{S_1, S_2, ..., S_k\}$, a Multiple Sequence Alignment of $\mathscr{S}$ can be considered a natural generalization of the pairwise alignment problem.

Definition 1. *(Multiple Sequence Alignment) Let $\{S_1, S_2, S_3, ..., S_k\}$ be sequences of characters, and let $|S_i|$ represent the size of S_i. The Multiple Sequence Alignment among S_1 to S_k is a mapping of $\{S_1, S_2, ..., S_k\}$ to other sequences $\{S'_1, S'_2, ..., S'_k\}$ such that S'_i has the same characters of S_i in the same order with possibly the adition of spaces (also known as gaps) and $|S'_1| = |S'_2| = ... = |S'_k|$.*

As the Multiple Sequence Alignment problem is known to be NP-hard, we use a heuristic solution known as *Progressive Alignment* [6]. In general terms, the method first performs the alignment between two previously selected sequences. Then a new sequence is chosen and aligned with the first alignment obtained or other pair of sequences is selected and aligned. This process repeats until all sequences have been aligned, giving rise to the final multiple alignment.

4   Duster

In this section, we present DUSTER, our method to generate rewrite rules from dup-clusters provide as input. The method (1) obtains a *consensus sequence* (or CS, for short) for each dup-cluster in training set; (2) merges similar consensus sequences and extract rules from them; (3) generalizes rules; and (4) removes redundant rules. In the following, each of these steps is described in turn.

4.1   Finding CSs

We perform this task by aligning the URLs in each cluster and then generating the CS as a result of this alignment. Following, we first show how we align two URLs and then continue showing how to generate a CS involving all URLs within a dup-cluster.

Pair-wise URL Alignment. The output of our alignment process is a sequence of sets which we refer to as *CS*. Unlike previous work that treated URLs as strings generated according to W3C grammar[1], we adopt a simpler representation: each URL to be aligned is initially decomposed into a sequence of t URL tokens set (referred to as *tokenization*), whose types are either *alphabetic, digit* or *punctuation*. For example, URL $u = $ http://www.example.com/01.html is represented by the following sequence of 14 tokens set:

⟨{*http*}, {:}, {/}, {/}, {*www*}, {.}, {*example*}, {.}, {*com*}, {/}, {0}, {1}, {.}, {*html*}⟩

Definition 2. *(Consensus Sequence) Let $\{u_1, u_2, ..., u_n\}$ be a set of n tokenized and aligned URLs, such that $|u_1| = |u_2| = ... = |u_n| = k$. Let t_{ui} represents the token of URL u at position i. A CS is a sequence of k token sets $\langle T_1, ..., T_k \rangle$ such that $T_i = \cup_{\forall u} t_{ui}$.*

---

[1] http://www.w3.org/Addressing/URL/5_BNF.html

CS of n sequences is composed by the union of the tokens in the corresponding positions of the n aligned sequences. To obtain the CS of two URLs u_1 and u_2, we first tokenize them in two sequences X and Y, associated with u_1 and u_2 respectively, with m and n tokens. Sequences X and Y are then aligned by inserting gaps, either into or at the ends of them. To determine where gaps should be inserted, matrix S in Equation 1 has to be calculated. To accomplish this, it is necessary to define a score function sf to measure the distance between the URL tokens set. The scoring function we adopt is given by Equation 2, which prioritizes the alignment of tokens located at the same positions.

$$sf(X_i, Y_j) = \begin{cases} 7 & \text{if } X_i \cap Y_j \neq \emptyset \wedge i = j \\ 5 & \text{if } X_i \cap Y_j \neq \emptyset \wedge i \neq j \\ 1 & \text{if } \exists (x_i, y_j) \in X_i \times Y_j | \tau(x_i) = \tau(y_j) \\ -1 & \text{otherwise} \end{cases} \tag{2}$$

where $\tau : \mathscr{T} \rightarrow \{a, d, p\}$ is a function which maps a token to its token type, $\mathscr{T}$ is the token space and $\{a, d, p\}$ are the token types (a for alphabetic, d for digit, and p for punctuation).

The basic idea of the scoring function is the following: (i) if token sets (X_i, Y_j) contain at least one token in common[2] at the same position, the score should be higher (we adopted 7 in our experiments); (ii) if token sets (X_i, Y_j) contain tokens in common (at least one) only at *different* positions, the score should be high, but smaller than in the first case (we adopted 5 in our experiments); (iii) if the token sets have no tokens in common but have tokens of same type, the score value is small (we adopted 1 in our experiments). Finally, (iv) the score value indicates a penalty (we adopted -1 as a penalty in our experiments) in any other case.

For instance, given URLs $u_1 = \text{http://www.ex/}$ and $u_2 = \text{http://www.un/}$ home, X and Y are given by: $X = \langle \{http\}, \{:\}, \{/\}, \{/\}, \{www\}, \{.\}, \{ex\}, \{/\} \rangle$ and $Y = \langle \{http\}, \{:\}, \{/\}, \{/\}, \{www\}, \{.\}, \{un\}, \{/\}, \{home\} \rangle$. Thus, given Definition 2, the CS is given by: $CS = \langle \{http\}, \{:\}, \{/\}, \{/\}, \{www\}, \{.\}, \{ex, un\}, \{/\}, \{\lambda, home\} \rangle$, where λ indicates a gap. To avoid a cumbersome notation, from now on, we will omit curly brackets and commas to denote token sets with just one token. Thus, CS will be written as: $CS = \langle http : //www.\{ex, un\}/\{\lambda, home\} \rangle$

Multiple URL Alignment. To solve the problem of align more than two URLs, we take advantage of the progressive alignment strategy [6]. At each iteration of the progressive alignment, we align two sequences and obtain a new sequence from this alignment. For the initial alignment when the input is composed by two URLs, we proceed as described above. This process is a variation of the technique described in [6] and is described in Algorithm 1.

The goal of URL normalization is to distinguish the URL tokens that impact on page content from the ones that do not or are irrelevant. In this way, given a CS = $\langle T_1, T_2, ..., T_k \rangle$, inferred from a set of URLs U, token set T_i is classified as follows:

(a) **Irrelevant**: T_i is *irrelevant* if $\lambda \in T_i$, that is, some token of T_i was aligned with a gap during the multiple alignment process. Theses tokens are considered irrelevant, i.e., the page content is the same independently of their presence in the URL.

(b) **Invariant**: T_i is *invariant* if $|T_i - \{\lambda\}| = 1$ (inside U) and it is present in all URLs of U. Such invariant tokens in a CP can be considered as identifiers of the replicated page, e.g., the token *disclaimer* from the urls $*\text{disclaimer.html}$;

---

[2] The comparison here is case-insensitive.

Algorithm 1. GenerateConsensSequence	**Algorithm 2.** HAC
Input: C: list of n URLs $\{u_1,...,u_n\}$ in a dup-cluster **Output:** Consensus Pattern p of C 1: **for** $i \leftarrow 0$ to n **do** {Distance matrix S is created} 2: **for** $j \leftarrow i+1$ to n **do** 3: add PairwiseSeqAlignment(u_i, u_j) to S 4: **end for** 5: **end for** 6: **while** $size(C) > 1$ **do** 7: Let $S_{i,j}$ be the highest score in matrix S 8: $p \leftarrow$ InferConsensusPattern(u_i, u_j) {pair is aligned and a CP p is inferred} 9: add p to C 10: remove u_i and u_j from C 11: eliminate alignments for u_i and u_j from S 12: $\forall_{u_j \in C \mid u_j \neq p}$ (add PairwiseSeqAlignment(p, u_j) to S) {Distance between this new sequence and all the remaining ones} 13: **end while** 14: **return** p	**Input:** $\mathscr{C}$: list of n CSs with same signature **Output:** $\mathscr{R}$: set of rules to be extracted from $\mathscr{C}$ 1: **if** $\lvert \mathscr{C} \rvert > 1$ **then** 2: $\forall_{i,j} D_{i,j} \leftarrow d(\mathscr{C}_i, \mathscr{C}_j)$ 3: **while** $D \neq \emptyset$ **do** 4: $(a,b) \leftarrow \arg\max_{i,j} D_{i,j}$ 5: **if** merge $\mathscr{C}_a$ and $\mathscr{C}_b$ is *safe* **then** 6: $\mathscr{C}_m \leftarrow$ MergePatterns(C_a, C_b) 7: eliminate $\mathscr{C}_a$ and $\mathscr{C}_b$ from $\mathscr{C}$ 8: eliminate row a and column b from D 9: add $\mathscr{C}_m$ to $\mathscr{C}$ 10: $\forall_j D_{m,j} \leftarrow d(\mathscr{C}_m, \mathscr{C}_j)$ 11: **else** 12: eliminate $D_{a,b}$ from D 13: **end if** 14: **end while** 15: **end if** 16: $\mathscr{R} \leftarrow \cup_{\forall c \in \mathscr{C}} Rule(c)$ 17: **return** $\mathscr{R}$

(c) **Converted**: T_i is *converted* if $\lvert T_i - \{\lambda\} \rvert = 1$ (inside U) just after their tokens being converted to lowercase. Depending of how a web server manages names of directories and files, we can find tokens in URLs as *disclaimer*, *DISCLAIMER* or *Disclaimer*;

(d) **Variant**: T_i is *variant* if $\lvert T_i - \{\lambda\} \rvert > 1$ even after their tokens being converted to lowercase. Differently from irrelevant tokens, we can not remove them from the URLs, being necessary to choose one of them. However, the content of the page will not change depending on the chosen alternative. As examples of theses tokens, we cite tokens denoting directories where files were copied redundantly, multiple domain names from the same website or session id lists used to identify users;

4.2   Phase 2: Generating Rules

After inferring a CS for each dup-cluster, our method generates the rewrite rules to be used to normalize URLs. Here, we precisely define rewrite rules and show how to generate them from a set of similar CSs.

Rewrite Rules. A rewrite rule is a description of the conditions and operations necessary to translate a URL into a canonical form. More formally, a rewrite rule is a pair $r = (c,t)$, where c and t are regular expressions (regexes) named *context* and *transformation*, respectively. The context c of a rule is a regex that represents the set of URLs to which the rule can be applied. The transformation t is a regex applied to transform a URL into a general, or canonical, form.

As an example, consider a the rewrite rule $r = (c,t)$, where $c = $ ^http://www(1| 2|3)?.([A-Za-z]+).edu/(a|b)/([A-Za-z]+) is the context regex and $t = $ http://www(1|2|3)?.\%2.edu/(a|b)/\$4 is the transformation regex. Thus, URL $u = $ http://www1.EX.edu/a/d will match context c. Note that %2 will store the string matched in the second group after converted to lowercase and $4 will store the string matched in the fourth group. After applying transformation t, u is transformed in http://www(1|2|3)?.ex.edu/(a|b)/d.

Rule Generation. Before generate rewrite rules, we first need to merge similar CSs. To accomplish this, we apply a strategy based on a Hierarchical Agglomerative Clustering (HAC) technique. The application of a HAC process over the n patterns inferred in the previous phase can be prohibitive at web scale. To cope with this problem, the method generates a mask (signature) to represent each CS and then creates lists of CS with the same masks. After that, the HAC process is applied in each list to derive rules. These rules are then added to the final pool of rules. Function Signature that generates such masks is explained as follows: Given the consensus sequence $CS = \langle T_1, T_2, ..., T_k \rangle$, the signature of CS is the string $s_1 s_2 ... s_k$, where s_i is given by Equation 3.

$$s_i = \left\{ \begin{array}{ll} t & \text{if } |T_i| = 1 \wedge \tau(t) = p \\ \tau(t) & \text{otherwise} \end{array} \right\} \tag{3}$$

where $t \in T_i$. For instance, given the CP $P = \langle http://www.example.com.br/\{a,b,c\}/A - 210.html \rangle$, the signature of P is given by $a://a.a.a.a/a/a-ddd.a$.

The HAC algorithm adopts a bottom-up strategy in which CSs are gradually unified until all the similar patterns have been merged. A stopping criterion is used to avoid merging CSs from different contexts. Algorithm 2 describes such processes in details.

We start by evaluating all pairwise distances between the CSs, building distance matrix D (Line 2). The metric we use to evaluate the distances between CSs is $d(C_a, C_b) = \sum_{i=1}^{k} \frac{|X_i \cap Y_i|}{min(X_i, Y_i)}$, where X_i and Y_i represent the ith token sets from consensus sequences C_a and C_b respectively. During the execution of HAC, similar CSs should be merged. The basic idea behind this process is simple. Given two patterns $C_a = \langle X_1, ..., X_k \rangle$ and $C_b = \langle Y_1, ..., Y_k \rangle$, their merge corresponds to the union of their token sets, that is, $C_a \cup C_b = \langle X_1 \cup Y_1, ..., X_k \cup Y_k \rangle$. However, for this merging be considered safe, and to avoid merging CSs that do not capture the same structured transformations[3], *the union involving a variant token set is allowed only if one of the sets is a subset of the another.* For example, if $\{m, n, o\}$ is a variant token set and $\{n\}$ is an invariant one, their merge is safe since $\{n\} \subseteq \{m, n, o\}$.

Note that, after the merging of two CSs $C_a = \langle X_1, ..., X_k \rangle$ and $C_b = \langle Y_1, ..., Y_k \rangle$, the classification of any token set Z_i in the resulting consensus pattern $C_m = \langle Z_1, ..., Z_k \rangle$ is defined by the first of the following conditions to hold: (i) the same of X_i and Y_i if they share the same classification; (ii) optional if X_i or Y_i is optional; (iii) obligatory if X_i or Y_i is obligatory; (iv) converted if X_i or Y_i is converted.

Conversion of CSs to Rewrite Rules. CSs can be straightforwardly converted to rewrite rules. To this, each token set T_i in the consensus sequence CS has to be converted, as described as follows: (a) If T_i is invariant and $|T_i| = 1$, the token $t \in T_i$ is added to the rule context and transformation. If $|T_i| > 1$ (that is, P was derived from the union of CSs), all tokens from T_i are grouped between parentheses, separated by $|$ in the rule context. In rule transformation, a *backreference* $N is included in order to allow the reuse of the corresponding URL to be matched; (b) If T_i is irrelevant, all tokens from T_i are grouped between parentheses, separated by $|$, with a question mark ? placed after the closing bracket. The regular expression is included in both rule context and transformation; (c) If T_i is converted and $|T_i| > 1$, all tokens from T_i are grouped between parentheses, separated by $|$ in the rule context. In rule transformation, a *backreference* %N is included in order to allow the reuse of the corresponding URL to be matched, converted to lowercase. (d) If T_i is variant, all tokens from T_i are grouped inside parentheses separated by $|$ in the rule context and transformation. (e) Finally, anchors are included in the final expression.

For example, given the $CP = \langle http://www\{1,2,3,\lambda\}.\{ex, EX\}.edu/\{a,b\}/\{c,d\} \rangle$ the corresponding rule is $r = (c, t)$, where $c = $ ^http://www(1|2|3)?.(ex|EX)

---

[3] Transformations normally resulting from some underlying process being carried out by for instance a web server or a web designer.

`.edu/(a|b)/(c|d)$` is the context and $t =$ `http://www(1|2|3)?.%2.edu/`
`(a|b)/$4` is the transformation pattern.

Generalizing Rules. To generalize rule r, we proceed as follows. We first find a group of alternative tokens with more than a certain number of tokens (threshold value) and generalize such list converting it to a regular expression according to the type of its tokens: (a) Alphabetic: if the token group has at least one token with more than one letter, it is substituted by the regular expression $([a-zA-Z]+)$. Otherwise, it is substituted by the regular expression $([a-zA-Z])$; (b) Digit: the group is substituted by $([0-9])$; (c) Punctuation: no normalization is done.

Different threshold values are used according to the classification of the group of tokens. We use a larger value for variant and irrelevant tokens (we adopted 2 in our experiments) than for invariant and converted tokens (we adopted 5 in our experiments).

Refining Rules. As a final step of our method, we refine our rule set by eliminating redundant and very specific rules. To detect such cases, we divide the example dataset into training and validation sets. We infer the rules from the training set and apply the inferred rules to the validation set. Whenever a rule α affects a subset of another more general rule in the validation set, we eliminate α from our set of rules. Finally, we also remove specific rules, which are detected by removing rules which affect no URL at the validation set.

5 Experimental Evaluation

Experimental Setup. The list of duplicate URLs we use in this work was derived from the GOV2 TREC Dataset [4]. This dataset consists of about 25 million pages from the US government domains. This dataset contains about 3.42 million duplicate URLs divided into about 1.43 million dup-clusters. To evaluate the effectiveness of our method and the baselines, we adopted the same metrics used in [9]:

Precision: let f be the number of URLs for which the application of a rule has failed, that is, two URLs from different clusters are converted to the same canonical form. Now, let r_{cov} be the number of URLs that match the context of some rule. Given f and r_{cov}, precision is defined as $1 - (f/r_{cov})$;

Compression: this metric measures the reduction ratio of the number of URLs to be really crawled, after the removal of duplicates. It is defined as $\frac{|U_{orig}| - |U_{norm}|}{|U_{orig}|}$, where U_{orig} is the original URL set and U_{norm} is the normalized URL set;

In all the experiments, we calculated these metrics as the average obtained for ten sub-sets of URLs according to a 10-fold cross-validation strategy [12], as follows. We randomly divide the duplicate clusters of GOV2 into ten approximately equal-size sets. From each of these ten subsets, 50% of the URLs were retained as training set, 40% as validation set, and the remaining 10%, as test set. We then performed 10 runs, rounding the sets such that a same URL was never used as test example in different runs. We used the training set to generate the rules, the validation set to refine them, and the test set to evaluate them. We adopted this strategy for our method and the baselines. The first baseline is the work by Dasgupta et al [5], which we implemented using the *fanout-10* heuristic. Second baseline is the method proposed in [10], which we here refer to as R_{tree}.

Results. The elimination of redundant rules resulted in a reduction from about 11% to about 15% in the total number of rules when varying the estimated precision from 100% to 80%. Such result shows the importance of performing the filtering of redundant rules, thus avoiding the generation of a large number of rules which would not be useful for the DUST detection.

Table 1 shows a comparison between DUSTER and the baseline methods regarding the task of DUST detection. This table shows, for each precision level p and method, the number of rules generated that achieved this precision level (#Rules column), along with its respective proportion over all valid rules (% of Rules). Column Compression shows the compression ratio, i.e., the reduction in the amount of URLs crawled, obtained with these rules.

As for precision, the performance of DUSTER was fair superior when compared to the baselines at all precision levels experimented. We consider the 100% level as the most important one, since it includes rules that did not fail in any of the test URLs in the validation set. At this level, DUSTER was able to reduce the amount of URLs crawled in 26.54%, while the best baseline ($R_{fanout}10$) achieved only 17.22%. Furthermore, these rules correspond to about 77% of our valid rules against about 55% of $R_{fanout}10$. This means that, besides achieving a higher compression rate, the rules generated by DUSTER are more accurate than the ones generated by $R_{fanout}10$. The necessity of splitting the URLs according to each target site can partially explain the poor performance of R_{tree}. This prevents the generation of rules involving multiple domains. Unlike R_{tree}, our method and $R_{fanout}10$ are able to generate such rules.

These results indicate that DUSTER is a quite effective and viable alternative for solving the DUST detection problem. When considering other precision levels experimented, again DUSTER was able to outperform the baselines. The gains in compression ratio of DUSTER over the best baseline, $R_{fanout}10$, were about 54% for 100%-precise rules, about 59% for over-95%-precise rules, and about 75% for rules with precision between 80% and 95%.

Table 1. Compression obtained at each precision level by $R_{fanout-10}$, R_{tree} and DUSTER

Precision	R_{tree}			$R_{fanout-10}$			DUSTER		
Level	#Rules	% of Rules	Compression	#Rules	% of Rules	Compression	#Rules	% of Rules	Compression
1.0	1531.7	5.79	6.83	14601.9	55.21	17.22	20034.1	76.74	**26.54**
>= 0.95	1632.1	5.46	9.45	15912.4	60.16	20.18	20668.1	79.16	**32.08**
>= 0.90	1677.1	5.61	10.45	16972.2	64.17	21.03	20828.9	79.78	**36.88**
>= 0.80	1726.3	5.77	10.46	18475	69.85	21.97	21036.9	80.58	**38.48**

6   Conclusions and Future Work

DUSTER generates rewrite rules that are very precise in converting distinct URLs which refer to a same content to a common canonical form. By analyzing the alignments obtained, high-quality rewrite rules can be generated, as demonstrated in our experiments. DUSTER also shown a clearly superior performance in the coverage of the rules generated. As future work, we intend to improve the scalability and precision of our method, as well as to evaluated it using other datasets.

References

1. Agarwal, A., Koppula, H.S., Leela, K.P., Chitrapura, K.P., Garg, S., GM, P.K., Haty, C., Roy, A., Sasturkar, A.: Url normalization for de-duplication of web pages. In: CIKM 2009, pp. 1987–1990. ACM, New York (2009)
2. Bar-Yossef, Z., Keidar, I., Schonfeld, U.: Do not crawl in the dust: Different urls with similar text. ACM Trans. Web 3(1), 3:1–3:31 (2009)

3. Blackshields, G., Sievers, F., Shi, W., Wilm, A., Higgins, D.G.: Sequence embedding for fast construction of guide trees for multiple sequence alignment. Algorithms Mol. Biol. 5, 21 (2010)
4. Clarke, C.L.A., Craswell, N., Soboroff, I.: Overview of the trec 2004 terabyte track. In: Voorhees, E.M., Buckland, L.P. (eds.) TREC, Volume Special Publication 500-261. NIST (2004)
5. Dasgupta, A., Kumar, R., Sasturkar, A.: De-duping urls via rewrite rules. In: KDD 2008, pp. 186–194. ACM, New York (2008)
6. Feng, D.F., Doolittle, R.F.: Progressive sequence alignment as a prerequisite to correct phylogenetic trees. Journal of molecular evolution 25(4), 351–360 (1987)
7. Fetterly, D., Manasse, M., Najork, M.: On the evolution of clusters of near-duplicate web pages. In: LA-WEB 2003, pp. 37–45. IEEE Computer Society, Washington, DC (2003)
8. Katoh, K., Misawa, K., Kuma, K., Miyata, T.: MAFFT: a novel method for rapid multiple sequence alignment based on fast Fourier transform. Nucleic Acids Research 30(14), 3059–3066 (2002)
9. Koppula, H.S., Leela, K.P., Agarwal, A., Chitrapura, K.P., Garg, S., Sasturkar, A.: Learning url patterns for webpage de-duplication. In: WSDM 2010, pp. 381–390. ACM, New York (2010)
10. Lei, T., Cai, R., Yang, J.-M., Ke, Y., Fan, X., Zhang, L.: A pattern tree-based approach to learning url normalization rules. In: WWW 2010, pp. 611–620. ACM Press, New York (2010)
11. Mao, X., Liu, X., Di, N., Li, X., Yan, H.: SizeSpotSigs: An effective deduplicate algorithm considering the size of page content. In: Huang, J.Z., Cao, L., Srivastava, J. (eds.) PAKDD 2011, Part I. LNCS, vol. 6634, pp. 537–548. Springer, Heidelberg (2011)
12. Mitchell, T.M.: Machine Learning. McGraw-Hill, Inc., New York (1997)

On Two-Dimensional Lyndon Words

Shoshana Marcus[1] and Dina Sokol[2],*

[1] Simons Center for Quantitative Biology,
Cold Spring Harbor Laboratory, Cold Spring Harbor, NY, 11724
smarcus@cshl.edu
[2] Department of Computer and Information Science,
Brooklyn College of the City University of New York, Brooklyn, NY, 11210
sokol@sci.brooklyn.cuny.edu

Abstract. A Lyndon word is a primitive string which is lexicographically small-est among cyclic permutations of its characters. Lyndon words are used for con-structing bases in free algebras, constructing de Bruijn sequences, finding the lexicographically smallest or largest substring in a string, and succinct suffix-prefix matching of highly periodic strings. In this paper, we extend the concept of the Lyndon word to two dimensions. We introduce the 2D Lyndon word and use it to capture 2D horizontal periodicity of a matrix in which each row is highly periodic, and to efficiently solve 2D horizontal suffix-prefix matching among a set of patterns. This yields a succinct and efficient algorithm for 2D dictionary matching. We present several algorithms that compute the 2D Lyndon word that represents a matrix. The final algorithm achieves linear time complexity even when the least common multiple of the periods of the rows is exponential in the matrix width.

1 Introduction

Two strings are *conjugate* if they differ only by a cyclic permutation of their charac-ters. A string is said to be *primitive* if it cannot be expressed as u^k for any integer $k > 1$. A *Lyndon word* is a primitive string which is the smallest of its conjugates for the alphabetic ordering [13]. For example, abba and aabb are conjugate; aabb is a Lyndon word, while abba is not. Lyndon words are useful for constructing bases in free Lie algebras [14], constructing de Bruijn sequences [8], musicology [5], computing the lexicographically smallest or largest substring in a string [4], computing the shortest superstring for a set of strings [15], searching for tandem approximate repeats [6], and succinct suffix-prefix matching of highly periodic strings [16].

A string S is *periodic* if it can be expressed as $u^j u'$ where u' is a proper prefix of u, and $j \geq 2$. When u is primitive we call it "the period" of S. Depending on the context, we use the term *period* to refer to either u or $|u|$.

Lyndon word naming classifies highly periodic strings by the conjugacy of their pe-riods and uses the Lyndon word as the class representative. Once Lyndon word naming has been performed, a string can be represented by the name of its period's class and its *LWpos*, the position at which the Lyndon word first occurs in the string. For example,

* This work was supported in part by PSC-CUNY research award 65112-0043.

O. Kurland, M. Lewenstein, and E. Porat (Eds.): SPIRE 2013, LNCS 8214, pp. 206–217, 2013.
© Springer International Publishing Switzerland 2013

Table 1. A summary of the algorithms presented in this paper to compute the 2D Lyndon word that represents an $m \times m$ matrix

	Best Time Complexity	Worst Time Complexity	Described In
Naive Algorithm	$O(m \cdot LCM_m)$	$O(m \cdot LCM_m)$	Section 3
Algorithm 1	$O(m \log^2 m + LCM_m)$	$O(m \log^2 m + (LCM_m + m)\frac{m}{\log m})$	Section 3
Algorithm 2	$O(m \log^2 m)$	$O(m \log^2 m + \frac{m^2}{\log m})$	Section 4

the strings T_1 = abbaabbaabbaabbaab and T_2 = aabbaabbaabbaabbaa are in the same class and the class representative is aabb. *LWpos* of T_1 is 3 while *LWpos* of T_2 is 0 since it begins with the Lyndon word that represents its period [16].

In this paper, we extend the concept of the Lyndon word to two dimensions. We first define the *2D Lyndon word*. We then introduce a new classification scheme which captures the horizontal suffix-prefix matches among a set of matrices for matrices whose rows are highly periodic. We focus on matrices whose rows are highly periodic since a non-periodic pattern row would allow us to quickly narrow down the possible pattern occurrences in a text yielding much fewer possibilities of overlap.

We present several algorithms for computing the 2D Lyndon word that represents a matrix; see Table 1 for a summary of time complexities. The input to these algorithms is the output of Lyndon word naming[1] on the rows, i.e., the period size of each row, and the offset of the first Lyndon word occurrence in each row. LCM_m denotes the least common multiple of the periods of all rows. Since LCM_m may be exponential in m, the straightforward algorithms take exponential time for certain inputs. Thus, we use modular arithmetic in Algorithm 2 to develop an extremely efficient algorithm that does not need to process the actual matrix and is independent of LCM_m.

The classification technique that we introduce is a new way of capturing horizontal 2D periodicity. Amir and Benson introduced the concept of 2D periodicity [1, 2] and it serves as the basis for an efficient 2D pattern matching algorithm [3]. However, their approach to 2D periodicity is not suitable for multiple pattern matching, which requires suffix-prefix matching between pairs of different patterns. The *all-pairs suffix-prefix matching problem* is the problem of finding, for any pair of strings in a given set, the longest suffix of one which is a prefix of the other. Lyndon word naming is an efficient tool to identify suffix-prefix matches between highly periodic strings. 2D Lyndon word naming is equally meaningful for horizontal suffix-prefix matching in matrices, resulting in efficient dictionary matching. Both one and two dimensional Lyndon word naming have the additional benefit over other algorithms, e.g. [9, 10, 17], of being on-line and of using very little working space. While this paper focuses on square matrices for ease of exposition, the new concepts and techniques apply to rectangles that are of uniform size in at least one dimension.

The remainder of this paper is organized as follows. We begin by defining the 2D Lyndon word in Section 2. In Section 3, we present an algorithm that calculates the 2D Lyndon word directly from the actual matrix. In Section 4, we present a more efficient algorithm that calculates the 2D Lyndon word using modular arithmetic. In Section 5, we apply this technique and show how it is useful for several applications.

---

[1] Lyndon word naming of the matrix rows takes linear $O(m^2)$ time [16].

a	b	a	b	a	b	a	b	LWpos
a	b	a	b	a	b	a	b	0
b	c	a	b	c	a	b	c	2
a	a	a	a	a	a	a	a	0
c	a	b	c	a	b	c	a	1
c	b	c	c	b	c	c	b	7
b	a	b	a	b	a	b	a	1
c	c	a	c	c	a	c	c	2
c	b	c	b	c	b	c	b	1

b	a	b	a	b	a	b	a	LWpos
b	a	b	a	b	a	b	a	1
c	b	c	a	b	c	a	b	3
a	a	a	a	a	a	a	a	0
a	c	a	b	c	a	b	c	2
b	c	b	c	c	b	c	c	0
a	b	a	b	a	b	a	b	0
c	c	c	a	c	c	a	c	3
b	c	b	c	b	c	b	c	0

a	b	a	b	a	b	a	b	LWpos
a	b	a	b	a	b	a	b	0
a	b	c	a	b	c	b	c	0
a	a	a	a	a	a	a	a	0
b	c	a	b	c	a	c	a	7
c	c	b	c	c	b	c	b	5
b	a	b	a	b	a	b	a	1
a	c	c	a	c	c	c	c	0
c	b	c	b	c	b	c	b	1

Fig. 1. Matrices that are horizontal 2D conjugate, along with each *LWpos* array. In its first conjugate, the matrix is shifted right by one column and in its second conjugate it is shifted left by two columns. The matrix on the right is a 2D Lyndon word.

2 Main Idea

2.1 Definition of 2D Lyndon Word

We say that a matrix M has a *horizontal prefix* (resp. suffix) U if U is an initial (resp. ending) sequence of contiguous columns in M.

Definition 1. *Two matrices, M_1 and M_2, are* horizontal 2D conjugate *if $M_1 = UV$, $M_2 = VU$ for some horizontal prefix U and horizontal suffix V of M_1.*

In other words, we say that two matrices are *horizontal 2D conjugate* if they differ only by a cyclic permutation of entire columns. When it is clear from the context, we simply use the word conjugate to refer to horizontal 2D conjugate.

Lemma 1. *Horizontal 2D conjugacy is an equivalence relation among matrices.*

The proof is omitted due to lack of space and will appear in the journal version.

We represent each horizontal 2D conjugate as a sequence $c_1, c_2, \ldots, c_m$. If row i is non-primitive, i.e. u^k, then c_i represents the minimum number of characters that need to be cyclically permuted in the primitive root u to obtain a 1D Lyndon word. If row i is primitive, c_i is the minimum number of characters that need to be cyclically permuted in row i to obtain a 1D Lyndon word. For example, if row i is the string $T = uv$, u is a prefix of T, v is a suffix of T, and vu is a Lyndon word, $c_i = |u|$. We refer to the sequence of a conjugate as the *LWpos* array since it is essentially an array of Lyndon word positions in the matrix.

We order horizontal 2D conjugates by comparing their *LWpos* arrays. Three matrices that are horizontal 2D conjugate are depicted in Fig. 1, along with their *LWpos* arrays. The order of the matrices in ascending order is: the matrix on the right, the matrix on the left, and then the matrix in the center.

Definition 2. *A matrix M is* horizontally primitive, *or* h-primitive, *if it cannot be expressed as U^k, where U is a horizontal prefix of M and k is an integer, $k > 1$.*

Definition 3. *A* 2D Lyndon word *is an h-primitive matrix that is the smallest of its horizontal 2D conjugates for the numerical ordering of* LWpos *arrays.*

The 2D Lyndon word is defined for all h-primitive matrices. It is a compact representation of the Lyndon word that is conjugate to each row, combined with the relative alignments of the Lyndon words among the matrix rows. In one dimension, Lyndon word naming provides a classification of highly periodic strings based upon the Lyndon word of the *period* of the string, which is by definition primitive. Analogous to this in two dimensions, we compute the representative 2D Lyndon word of the 2D horizontal period, or more specifically, the h-primitive *LCM-matrix*, as described in the next subsection.

2.2 Classification Scheme

In this section, we present a new classification scheme for matrices whose rows are all highly periodic (i.e. for a matrix of width m, all rows have period $\leq m/4$). In a matrix whose rows are all highly periodic, the columns may also repeat at regular intervals. This matrix-wide repetition is at a distance of the lowest common multiple (LCM) of the periods of the rows, as we prove in Lemma 2. If the columns repeat, we focus on a submatrix that spans the first LCM_m columns of the original matrix. If LCM_m spans up to $m/2$ columns, then LCM_m is in fact a horizontal period of the matrix. If the width of the matrix is smaller than LCM_m, then we can enlarge the matrix to LCM_m columns by extending the period in each row. We refer to the (possibly enlarged) matrix of width LCM_m as the *LCM-matrix* of the original matrix. Fig. 2 shows a matrix with its *LCM-matrix* highlighted. We use a matrix with a small LCM_m in the example to illustrate the definitions, however, all the definitions and algorithms apply to a matrix with a large LCM_m as well.

Note that we cannot simply view the columns as metacharacters, name them, and use 1D Lyndon word naming to classify matrices. This approach[2] would work only for matrices whose horizontal 1D pattern of metacharacters is highly periodic. Our techniques exploit the periodicity of the rows to classify 2D patterns irrespective of whether there exists a horizontal period.

Lemma 2. *In a matrix with m rows, each of which is a periodic string, the columns repeat every LCM_m columns, where LCM_m denotes the least common multiple of the periods of all rows.*

Proof. Every row repeats in columns that are multiples of its period. LCM_m is a multiple of every row's period. Since every row repeats at LCM_m columns, the entire matrix repeats at LCM_m columns.   □

It follows from Lemma 2 that each of the LCM_m conjugates of an *LCM-matrix* has a distinct *LWpos* array. The key property of an *LCM-matrix* is that horizontal 2D conjugacy preserves row periodicity. We prove this in Lemma 3 by showing that a cyclic permutation of the columns in an *LCM-matrix* results in a cyclic permutation of each row's period.

Lemma 3. *Two* LCM-matrices *that are horizontal 2D conjugate have periods in corresponding rows that are 1D conjugate.*

---

[2] Processing columns even for this type of pattern would have several additional drawbacks, including working only for matrices that are uniform in both dimensions.

Fig. 2. A matrix with its *LCM-matrix* highlighted. The periods of the rows are of length 1, 2 and 3. $LCM_m = 6$, yielding an *LCM-matrix* that is 6 columns wide.

Fig. 3. Matrices whose *LCM-matrices* are horizontal 2D conjugate, along with each *LCM-matrix*'s *LWpos* array. These matrices are not horizontal 2D conjugate. In each of these matrices, $LCM_m = 6$. The first occurrence of the Lyndon word in each row of the *LCM-matrices* is highlighted. The first matrix appears in Fig. 2, where the focus is on its *LCM-matrix*. In its first conjugate, the *LCM-matrix* is shifted left by one column and in its second conjugate it is shifted left by two columns.

The proof is omitted due to lack of space and will appear in the journal version.

In our new classification scheme, each equivalence class consists of matrices whose *LCM-matrices* are horizontal 2D conjugate. We use the 2D Lyndon word in each class as the class representative, in a similar manner to the 1D equivalence relation that uses the *Lyndon word* as the class representative. Three matrices whose *LCM-matrices* are horizontal 2D conjugate are shown in Fig. 3, along with their *LWpos* arrays. To classify matrix M as belonging to exactly one horizontal 2D conjugacy class, we compute the conjugate of M's *LCM-matrix* that is a 2D Lyndon word. This classification allows us to represent a matrix compactly, with a constant number of 1D arrays. At the same time, this representation allows us to quickly answer horizontal suffix-prefix queries on a set of classified matrices. In the following two sections, we present several algorithms for computing the 2D Lyndon word that represents a given matrix.

3   Simple Algorithm for Computing 2D Lyndon Word

In this section we develop an intuitive algorithm that efficiently computes the 2D Lyndon word to represent an $m \times m$ matrix whose rows are highly periodic. We present algorithms that are run after Lyndon word naming has been performed on each row of the matrix. That is, the input to each of these algorithms is an $m \times m$ matrix represented by 3 arrays of size m, the 1D Lyndon word names for each row, the period size

0	1	0	1	0	1
2	1	0	2	1	0
0	0	0	0	0	0
1	0	2	1	0	2
1	0	2	1	0	2
1	0	1	0	1	0
2	1	0	2	1	0
1	0	1	0	1	0

Fig. 4. The set of *LWpos* arrays for the conjugates of an *LCM-matrix*. Each column in this table contains the *LWpos* array of the conjugate that begins with that column. The column that begins the 2D Lyndon word is highlighted. This *LCM-matrix* corresponds to the matrix in Fig. 2 and the leftmost matrix in Fig. 3.

of each row, and an *LWpos* array. Lyndon word naming of the matrix rows takes linear $O(m^2)$ time [16]. 1D Lyndon word naming was designed for highly periodic strings, with periods $\leq m/4$. As in [16], these ideas can be extended from squares to rectangles of uniform size in at least one dimension.

We have already seen that each conjugate of an *LCM-matrix* can be obtained by a cyclic permutation of columns in the *LCM-matrix*. As a result, computing the 2D Lyndon word that represents a matrix is a search for the cyclic permutation of its *LCM-matrix* at which the *LWpos* array is smallest.

We can naively compute the 2D Lyndon word that represents a matrix by computing the *LWpos* array for each conjugate of its *LCM-matrix* and then finding the minimum sequence. We show in Lemma 4 that the conjugates can be obtained by shifting the matrix rows. Thus, we can generate each conjugate's *LWpos* array from the matrix's *LWpos* array combined with the periods of the rows and then select the minimum *LWpos* array in this set. Since the *LCM-matrix* has LCM_m conjugates to consider (by Lemma 2), the naive algorithm runs in time proportional to the size of the *LCM-matrix*, $O(m \cdot LCM_m)$ time. Fig. 4 shows the set of *LWpos* arrays for the conjugates of the *LCM-matrix* depicted in Fig. 2. Each column represents the *LWpos* array of the conjugate that begins with that column. The columns of this table are compared from top-down and the minimum is selected as the 2D Lyndon word. In this example, the conjugate that begins with the third column is the 2D Lyndon word that represents the matrix depicted in Fig. 2.

Lemma 4. *Two matrices have* LCM-matrices *that are* horizontal 2D conjugate *iff the* LWpos *entries for each row are shifted by* C (mod *period*$[i]$), *where* C *is an integer and period*$[i]$ *is the period size of row* i.

The proof is omitted due to lack of space and will appear in the journal version.

We improve on the naive algorithm and present an $O(m+LCM_m)$ time algorithm for calculating the 2D Lyndon word that represents an $m \times m$ matrix. This procedure is delineated in Algorithm 1 and described in the following paragraphs.

We can systematically compute the numerically smallest *LWpos* array among the conjugates of the *LCM-matrix* without actually generating the complete *LWpos* arrays. The computation is incremental and considers one row at a time. Initially, before we examine the first row, all columns of the *LCM-matrix* are potentially the beginning of the 2D Lyndon word. As we proceed through the rows, we discard columns that cannot be the beginning of the 2D Lyndon word. Once a column is discarded, it is never considered again.

Algorithm 1. Computing a 2D Lyndon Word

Input: $LWpos[1...m], period[1...m]$ for matrix M.
Output: 2D Lyndon word $2D\_LW[1...m]$, shift z (i.e. column number in LCM-matrix of M).

$2D\_LW[1] \leftarrow 0$
$z \leftarrow LWpos[1]$
$\{LWpos[1]$ is first column of shift $0\}$
$\{$columns $z, z + period[1], z + 2 * period[1], \ldots$ can be 2D Lyndon word$\}$
$LCM[1] \leftarrow period[1]$
for $i \leftarrow 2$ to m **do**
   $GCD \leftarrow \gcd(LCM[i-1], period[i])$
   $LCM[i] \leftarrow LCM[i-1] * period[i]/GCD$
   if $LCM[i-1] \equiv 0 \pmod{period[i]}$ **then**
      $\{$if period of row i is a factor of cumulative LCM$\}$
      $2D\_LW[i] \leftarrow (LWpos[i] - z) \pmod{period[i]}$
   else
      $\{LCM[i] > LCM[i-1]\}$
      $firstShift \leftarrow (LWpos[i] - z) \pmod{period[i]}$
      $\{$shift $LWpos[i]$ to column $z\}$
      $2D\_LW[i] \leftarrow \min((firstShift - x * LCM[i-1]) \pmod{period[i]})$
      $\{$minimize over $x \geq 0$ such that $z + x * LCM[i-1] \leq LCM[m]\}$
      $z+ = x * LCM[i-1]$
      $\{$adjust z by x that minimizes shift in previous equation$\}$
   end if
end for

In general, we begin by eliminating all but the columns at which the Lyndon word of the first row begins. Suppose the first $LWpos$ entry is z and the period of the first row is u. Columns $z, z + u, z + 2u, \ldots$ are the only columns at which the 2D Lyndon word can begin; the other columns are immediately eliminated.

Subsequently, for each row i there are two possibilities. The first possibility is that the period of row i is a factor of the least common multiple of the periods of the first $i - 1$ rows, which we denote by $LCM[i - 1]$. In this case, the Lyndon word offset is identical in all remaining columns. We calculate the $LWpos$ entry without eliminating any columns. The other possibility is that the period of row i is not a factor of $LCM[i-1]$, i.e., $LCM[i]$ is larger than $LCM[i-1]$. In this case, we calculate the $LWpos$ value in each remaining column, select the minimum, and update z to be the first column that attains this minimum value. Then, columns $z, z + u, z + 2u, \ldots$, where $u = LCM[i]$, are the only columns at which the 2D Lyndon word can begin, since the columns of the first i rows in the table of $LWpos$ arrays recur every $LCM[i]$ columns, by Lemma 2.

This process continues until the last row is reached and only one column remains, since the columns in an *LCM-matrix* are distinct.

Lemma 5. *Let M be an $m \times m$ matrix and let α denote the time complexity of a single arithmetic operation on LCM_m of the matrix and a second operand that is $\leq m$. Algorithm 1 computes the 2D Lyndon word that represents M in $O(m \log^2 m + (LCM_m + m)\alpha)$ time and uses $O(m \log m)$ bits of working space.*

Proof. The greatest common divisor of $LCM[i-1]$ and $period[i]$ can be computed in $O(\log^2 m)$ time since the Euclidean algorithm takes $O(\log^2 m)$ time to compute the greatest common divisor of two integers when the smaller operand is stored in $\log m$ bits [11], after the first modulus step that requires $O(\alpha)$ time. In this case, $period[i] \leq m/4$ is stored in $\log m$ bits and $LCM[i-1]$ may be larger. Subsequently, the least common multiple of $LCM[i-1]$ and $period[i]$ is computed from their greatest common divisor in $O(\alpha)$ time. Over all rows, the total time spent on LCM computations is $O(m \log^2 m + m\alpha)$.

The *LCM-matrix* has LCM_m distinct columns, by Lemma 2. Thus, Algorithm 1 begins with a set of LCM_m columns at which the 2D Lyndon word can begin. As row i is examined, the if statement in Algorithm 1 has two possibilities:
(i) Its period is a factor of $LCM[i-1]$: computation completes in $O(\alpha)$ time.
(ii) $LCM[i] > LCM[i-1]$: we examine the *LWpos* arrays for the conjugates beginning in several columns. The values are compared and all but the columns of minimal value are discarded. Since $LCM(x, y) > 2x$ where $x > y$, and y is not a factor of x, at least half the possibilities are discarded, and we can charge the computation of *LWpos* values in row i to the discarded columns. Over all rows, at most LCM_m columns can be discarded. The computation of an *LWpos* value takes $O(\alpha)$ time.

Thus, the overall time complexity, aside from the LCM computation, is $O((LCM_m + m)\alpha)$. In terms of space, the representative 2D Lyndon word is stored in $O(m \log m)$ bits since it is an array of m integers, each of which is between 0 and $m/4$. Along the way, the only extra information we store are the column number, z, and *LWpos* values for the active column and for the minimum in the preceding columns of the row. $\square$

In the best case, LCM_m is linear or polynomial in m, thus it can be stored in $O(\log m)$ bits and fits in one word of memory. Then, $\alpha = O(1)$, and the algorithm runs in $O(m \log^2 m + LCM_m)$ time. In the worst case, LCM_m can grow exponentially, yet an upper bound of 3^m has been proven for the LCM of the numbers 1 through m [7]. Thus, the least common multiples can always be stored in $O(m)$ bits and α is at most $O(m/\log m)$. Hence, the worst case running time of Algorithm 1 is $O(m \log^2 m + (LCM_m + m)\frac{m}{\log m})$.

Since Algorithm 1 requires exponential time with respect to the input size in the worst case, in the next section we present a different algorithm whose time complexity is dependent on the number of bits needed to store LCM_m, yielding a worst case linear time algorithm for computing a 2D Lyndon word. We compare the time complexities of these algorithms in Table 1.

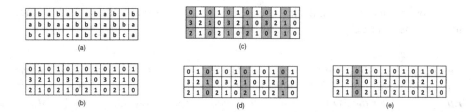

Fig. 5. (a) An *LCM-matrix*. (b) Its table of *LWpos* arrays in which each column contains the *LWpos* array of the conjugate that begins with that column. (c)-(e) The computation of the 2D Lyndon word that represents the matrix. The columns that remain after each iteration of the algorithm are highlighted. (c) After examining the first row, the columns beginning with 0 remain. (d) After examining the first two rows, the columns beginning with 01 remain. (e) The only remaining column is the 2D Lyndon word.

4   Computation of 2D Lyndon Word by Modular Arithmetic

In this section we derive a more efficient algorithm to compute the 2D Lyndon word that represents a matrix. The naive algorithm generates the *LWpos* array for each conjugate of the *LCM-matrix*, as shown in Fig. 5(b), and selects the minimum. Algorithm 1 only partially computes the *LWpos* arrays and narrows in on the column at which the 2D Lyndon word begins, as computation proceeds through the rows of the matrix. This is depicted in Fig. 5(c),(d), and (e). In this section, we present Algorithm 2, which uses modular arithmetic to directly compute each *LWpos* entry of the 2D Lyndon word, avoiding the comparison of any *LWpos* entries of the *LCM-matrix*'s conjugates. We show a reduction of the computation of a representative 2D Lyndon word to an algebraic problem that is solved with modular arithmetic.

In Algorithm 1, we obtain the representative 2D Lyndon word by computing *LWpos* values for some of the conjugates of the *LCM-matrix* and then selecting the minimum value. We transform this to a sequence for each row i, $2 \leq i \leq m$,

$$S_i = \{f - \ell x \pmod{p}\}_{x=0}^{p-1}$$

where ℓ is $LCM[i-1]$, p is $period[i]$, and f is the first column for which we consider an *LWpos* entry for row i. The objective of Algorithm 1 is to iteratively find the minimum value in S_i and the value of x at its first occurrence. We use properties of modular arithmetic to solve this problem.

Definition 4. *[12] The* modular inverse *of an integer ℓ (mod p) is an integer ℓ^{-1} such that $\ell(\ell^{-1}) \equiv 1 \pmod{p}$. More simply, we refer to ℓ^{-1} as an inverse.*

The modular inverse of ℓ (mod p) exists iff $\gcd(\ell, p) = 1$. In other words, ℓ (mod p) has an inverse when ℓ and p are *relatively prime* [12].

When ℓ and p are relatively prime, 0 is the minimum value in the sequence and ℓ^{-1} (mod p) is the first position x for which $S_i(x) = 1$. Multiplying ℓ^{-1} by the first value in the sequence, $\ell^{-1} * f$ (mod p), locates the first position x such that $S_i(x) = 0$, the first minimum in the sequence.

Algorithm 2. Computing a 2D Lyndon Word More Efficiently

Input: $LWpos[1...m], period[1...m]$ for matrix M.
Output: 2D Lyndon word $2D\_LW[1...m]$, shift z (i.e. column number in LCM-matrix of M).

$2D\_LW[1] \leftarrow 0$
$z \leftarrow LWpos[1]$
$LCM[1] \leftarrow period[1]$
for $i \leftarrow 2$ to m **do**
    $GCD \leftarrow \gcd(LCM[i-1], period[i])$
    $\ell \leftarrow LCM[i-1]/GCD$
    $p \leftarrow period[i]/GCD$
    $\ell Inv \leftarrow$ inverse of $\ell \pmod p$
    $LCM[i] \leftarrow \ell * period[i]$
    $firstShift \leftarrow (LWpos[i]\text{-}z) \pmod{period[i]}$
    $\{$shift $LWpos[i]$ to $z\}$
    $divFirstShift \leftarrow \lfloor firstShift / GCD \rfloor$
    $x \leftarrow (\ell Inv * divFirstShift) \pmod p$
    $2D\_LW[i] \leftarrow (firstShift - x*LCM[i-1]) \pmod{period[i]}$
    $z{+}{=} x*LCM[i-1]$
end for

When ℓ and p are not relatively prime, the minimum value in S_i may not be 0. We can convert S_i to a sequence with a minimum of 0 by dividing both ℓ and p by their greatest common divisor. Then 0 is surely in S_i and we can use ℓ^{-1} to locate its first occurrence, as before. The process of computing the representative 2D Lyndon word by modular arithmetic is delineated in Algorithm 2.

We illustrate Algorithm 2 using the example in Fig. 5. In the first row, the location of the first 0 is $LWpos[1]$, thus, no computation is necessary for S_1. $S_2 = \{1 - 2x \pmod 2\} = \{3, 1\}$, which are the highlighted values in the second row of Fig. 5(c). S_2 does not include the values 0 and 2 since $period[2] = 4$ and $LCM[1] = 2$ have a GCD of 2. Thus, we convert S_2 to a sequence in which the minimum value is 0 and find that the minimum occurs in the third column. $S_3 = \{3 - 4x \pmod 4\} = \{0, 2, 1\}$, which are the highlighted values in the third row of Fig. 5(d). S_3 includes all the original values since $period[3] = 3$ and $LCM[2] = 4$ are relatively prime. Thus, for the third row, $\ell^{-1} * f \pmod p = 1 * 3 \pmod 3 = 0$ so the first 0 occurs in the first remaining column and the third column is the 2D Lyndon word that represents this matrix.

Lemma 6. *Let M be an $m \times m$ matrix and let α denote the time complexity of a single arithmetic operation on LCM_m of the matrix and a second operand that is $\leq m$. Algorithm 2 computes the 2D Lyndon word that represents M in $O(m \log^2 m + m\alpha)$ time and uses $O(m \log m)$ bits of working space.*

The proof is omitted due to lack of space and will appear in the journal version.

The best case is where LCM_m is polynomial in m, so $\alpha = O(1)$ and Algorithm 2 runs in sublinear $O(m \log^2 m)$ time. The worst case is where $LCM_m = O(3^m)$, resulting in $\alpha = O(m/\log m)$, yielding worst case time complexity of $O(m \log^2 m + \frac{m^2}{\log m})$.

5   Applications

2D Periodicity

The classification technique that we introduce in this paper is a new perspective on horizontal 2D periodicity. When Amir and Benson introduced the concept of 2D periodicity [1, 2], they presented matrix periodicity as self-overlap that covers the center of the matrix. Our classification scheme is based on horizontal periodicity in a matrix. Just as Amir and Benson's 2D periodicity is the basis for an efficient 2D pattern matching algorithm [3], so is horizontal periodicity. Our new techniques have the benefit of being compact since we do not need to store a 2D witness table for each pattern. Furthermore, our techniques generalize nicely to multiple patterns, as we show in the next two sections.

Suffix-Prefix Matching

The 2D Lyndon word naming technique contributes the first efficient tool for horizontal suffix-prefix matching in a set of matrices whose rows are all highly periodic. Two $m \times m$ matrices whose rows are all periodic, M_1 and M_2, can have a horizontal suffix-prefix match of $\geq m/2$ columns if the *LCM-matrices* of M_1 and M_2 are horizontal 2D conjugate. When two matrices are in the same class, the difference between the number of columns that are cyclically permuted in each *LCM-matrix* determines whether there is a horizontal suffix-prefix match, and if so, by how many columns. After linear time preprocessing classifies each matrix in a set, horizontal suffix-prefix queries between two matrices are answered in constant time. This algorithm is succinct since it uses only $O(km)$ extra space for input of size $O(km^2)$. It is online since matrices can be classified as they arrive and horizontal suffix-prefix matches can be announced at any time.

Succinct Dictionary Matching with the 2D Lyndon Word

The *dictionary matching* problem is to search a text for all occurrences of patterns in a given set. Our new classification technique improves the succinct 2D dictionary matching algorithm of [13], which has a strict implicit assumption that the period of the first row of each pattern matches the horizontal period of the pattern. For succinct dictionary matching on 2D data whose rows are highly periodic, we need to compare segments of the larger text to many patterns simultaneously. We can work with $3m/2 \times 3m/2$ blocks of the text to conserve working space. It takes too much time to classify every $m \times m$ submatrix of the text. However, we can partially compute the 2D Lyndon word that represents each submatrix of the text in linear time even when LCM_m is exponential.

Let r be the minimum value for which $LCM[r] > m$. We calculate the 2D Lyndon word that represents the first r rows of the text block using Algorithm 2 in $O(1)$ time per row. We compare $2D\_LW[1...r]$ of the text block to $2D\_LW[1...r]$ of patterns with the same 1D name simultaneously by traversing a compressed trie. Then we compute the *LWpos* array in the first two columns of the text block's *LCM-matrix* at which $2D\_LW[1...r]$ occurs. It suffices to compare these *LWpos* arrays to the *LWpos* array at the first occurrence of $2D\_LW[1...r]$ in each of the patterns to determine if a pattern can occur in a text block since a difference in the *LWpos* shifts results in patterns that cannot overlap. This process completes in time proportional to the size of the text block.

Acknowledgements. The authors would like to thank Binyomin Balsam for his helpful discussions and his insight into the modular arithmetic solution.

References

[1] Amir, A., Benson, G.: Two-dimensional periodicity and its applications. In: ACM-SIAM Symposium on Discrete Algorithms (SODA), pp. 440–452 (1992)

[2] Amir, A., Benson, G.: Two-dimensional periodicity in rectangular arrays. SIAM Journal on Computing 27(1), 90–106 (1998)

[3] Amir, A., Benson, G., Farach, M.: An alphabet independent approach to two-dimensional pattern matching. SIAM Journal on Computing 23(2), 313–323 (1994)

[4] Apostolico, A., Crochemore, M.: Fast parallel Lyndon factorization with applications. Mathematical Systems Theory 28(2), 89–108 (1995)

[5] Chemillier, M.: Periodic musical sequences and Lyndon words. Soft Computing 8(9), 611–616 (2004)

[6] Delgrange, O., Rivals, E.: Star: an algorithm to search for tandem approximate repeats. Bioinformatics 20(16), 2812–2820 (2004)

[7] Farhi, B.: Nontrivial lower bounds for the least common multiple of some finite sequences of integers. Journal of Number Theory 125(2), 393–411 (2007)

[8] Fredricksen, H., Maiorana, J.: Necklaces of beads in k colors and k-ary de Bruijn sequences. Discrete Mathematics 23(3), 207–210 (1978)

[9] Gusfield, D., Landau, G.M., Schieber, B.: An efficient algorithm for the all pairs suffix-prefix problem. Information Processing Letters 41(4), 181–185 (1992)

[10] Kedem, Z.M., Landau, G.M., Palem, K.V.: Parallel suffix-prefix-matching algorithm and applications. SIAM Journal on Computing 25(5), 998–1023 (1996)

[11] Knuth, D.E.: The Art of Computer Programming, vol. 2. Addison Wesley (1998)

[12] Koshy, T.: Elementary Number Theory with Applications, 2nd edn. Academic Press (2001)

[13] Lothaire, M.: Applied Combinatorics on Words (Encyclopedia of Mathematics and its Applications). Cambridge University Press, New York (2005)

[14] Lyndon, R.C.: On burnside's problem. Transactions of the American Mathematical Society 77(2), 212–215 (1954)

[15] Mucha, M.: Lyndon words and short superstrings. In: ACM-SIAM Symposium on Discrete Algorithms (SODA), pp. 958–972 (2013)

[16] Neuburger, S., Sokol, D.: Succinct 2D dictionary matching. Algorithmica 65(3), 662–684 (2013)

[17] Ohlebusch, E., Gog, S.: Efficient algorithms for the all-pairs suffix-prefix problem and the all-pairs substring-prefix problem. Information Processing Letters 110(3), 123–128 (2010)

Fully-Online Grammar Compression*

Shirou Maruyama[1], Yasuo Tabei[2], Hiroshi Sakamoto[3], and Kunihiko Sadakane[4]

[1] Preferred Infrastructure, Inc.
maruyama@preferred.jp
[2] ERATO Minato Project, JST
yasuo.tabei@gmail.com
[3] Kyushu Institute of Technology
hiroshi@donald.ai.kyutech.ac.jp
[4] National Institute of Informatics
sada@nii.ac.jp

Abstract. We present a fully-online algorithm for constructing straight-line programs (SLPs). A naive array representation of an SLP with n variables on an alphabet of size σ requires $2n \lg(n + \sigma)$ bits. As already shown in [Tabei et al., CPM'13], in offline setting, this size can be reduced to $n \lg(n + \sigma) + 2n + o(n)$, which is asymptotically equal to the information-theoretic lower bound. Our algorithm achieves the same size in online setting, i.e., characters of an input string are given one by one to update the current SLP. With an auxiliary position array of size $n \lg(N/n) + 3n + o(n)$ bits, our representation supports substring extractions in $O((m + h)t)$ time where N is the length of the input string, m is the length of a substring extracted, $h = O(\lg N)$ is the height of the SLP, $t = O(1)$ in offline case, and $t = O(\lg n/ \lg \lg n)$ in online case. The working space is bounded by $(1 + \alpha)n \lg(n + \sigma) + n(3 + \lg(\alpha n))$ bits depending on a constant $\alpha \in (0, 1]$, which is a load factor of hash tables. We compared our algorithm to LZend in experiments using real world repetitive texts.

1 Introduction

Recently, large-scale and highly repetitive text collections have become ubiquitous. Examples are next-generation sequence data of individual human genomes, version controlled documents, source codes in repositories. In addition, such text data is increasingly generated in one after another, i.e., stream data. There is therefore a strong demand to process such text data in an online fashion.

Grammar compression builds a small context free grammar (CFG) that generates only an input text, and represents the obtained CFG as compactly as possible. The method reveals high compressive and processing abilities for highly repetitive texts in e.g. pattern matching [18,19], edit-distance computation [4], q-gram mining [2] and mining characteristic substrings [5,10] and so on. Basically, existing methods first build a complete CFG from an input text, and then

---

* This work was supported by JSPS KAKENHI(24700140,23680016,23240002).

O. Kurland, M. Lewenstein, and E. Porat (Eds.): SPIRE 2013, LNCS 8214, pp. 218–229, 2013.

Table 1. Comparison with known algorithms. Here N is the length of the input string, σ is the alphabet size, n is the number of generated rules, and α is a parameter between 0 and 1 (the load factor of hash tables). The size of the auxiliary index for efficient substring decoding is excluded. The expected time complexities are due to the use of a hash function.

compression time	working space (bits)	Ref.
$O(N/\alpha)$ expected	$(3+\alpha)n\lg(n+\sigma)$	[9]
$O(N/\alpha)$ expected	$(\frac{11}{4}+\alpha)n\lg(n+\sigma)$	[17]
$O(N\lg n)$	$2n\lg n(1+o(1))+2n\lg\rho\ (\rho\leq 2\sqrt{n})$	[16]
$O(\frac{N\lg n}{\alpha\lg\lg n})$ expected	$(1+\alpha)n\lg(n+\sigma)+n(3+\lg(\alpha n))$	Theorem 1
$O(N\lg n)$	$2n\lg n(1+o(1))+2n$	Theorem 2

encode it into a compact representation. A crucial drawback of those methods is to require a large working space consumed for building a CFG and its encoding. Even worse, they can not deal with stream data because of their static property.

Maruyama et al. [9] solved the inefficiency problem of large working space and the static property by introducing an online grammar compression called *online LCA* (OLCA). OLCA assumes *straight line program* (SLP), a canonical form of CFG, and builds production rules in an SLP by gradually reading an input text. Although OLCA achieves a good worst-case approximation ratio of $O(\lg^2 N)$ to the smallest CFG for an input string of length N, it has a serious issue of large working space and its inability of direct encoding of an SLP into a succinct representation, resulting in limited scalability. Later, Takabatake et al. [17] presented an online encoding scheme of an SLP of n variables built from OLCA into a succinct representation achieving $\frac{7}{4}n\lg n + 4n + o(n)$ bits, which was still larger than an information-theoretic lower bound of $n\lg n+n+o(n)$ bits recently presented in [16]. Moreover, they did not present a space-efficient *reverse dictionary*, a crucial data structure for checking whether or not a production rule in an SLP already exists in execution, which has been implemented using a chaining hash table having a load factor α. The space is $\alpha n\lg n + \alpha$ bits for the hash table and $n\lg n + \sigma$ bits for the lists, resulting in the total working space of $(\frac{11}{4}+\alpha)n\lg n + \alpha$ bits. Though this scheme is fully-online, its working space is larger than an information-theoretic lower bound. Since available data of highly repetitive texts is ever increasing, developing a fully-online grammar compression for building an SLP using the minimum space remains a challenge.

We present a fully-online grammar compression building an SLP and directly encoding it into a succinct representation in an online manner. Our online algorithm called *fully-online LCA* (FOLCA) is a modification of OLCA [9] that builds a *post order SLP* (POSLP), a special form of an SLP having post-order internal nodes in the partial parse tree. A major advantage of a POSLP is enabling a direct encoding into a succinct representation, while keeping the approximation ratio $O(\lg^2 N)$ of OLCA. The memory of our representation is at most $n\lg n + 2n + o(n)$ bits for an SLP, which is asymptotically equal to the information-theoretic lower bound of $n\lg n + n + o(n)$ bits, while supporting random access in $O(\lg n)$ time. To keep working space small, we also present a novel representation for reverse dictionary by leveraging a nice property of each list as a strictly increasing sequence of integers in the hash table. Thus, FOLCA also computes POSLP in an online manner using small space (see Theorems 1 and 2). Table 1 summarizes results of ours and existing algorithms.

Experiments were performed on extracting substrings from real-word repetitive texts. The performance comparison with LZend [8], a state-of-the-art compression and substring extraction algorithm for repetitive texts, demonstrates significant reduction of working space to build SLPs and high ability of substring extractions in FOLCA.

2 Preliminaries

2.1 Basic Notations

We assume a finite alphabet Σ for the symbols forming input texts throughout this paper. Σ has total order relation. The set of all strings over Σ is denoted by Σ^*, and Σ^i denotes the set of all strings of length i. The length of $w \in \Sigma^*$ is denoted by $|w|$, and the cardinality of a set C is similarly denoted by $|C|$. $\mathcal{X}$ is a recursively enumerable set of variables with $\Sigma \cap \mathcal{X} = \emptyset$. A sequence of symbols from $\Sigma \cup \mathcal{X}$ is also called a string, and a pair of symbols from $\Sigma \cup \mathcal{X}$ is called a digram. Strings x and z are said to be a prefix and suffix of the string $w = xyz$, respectively. Also, x, y, z are called substrings of w. The i-th symbol of w is denoted by $w[i]$ $(1 \leq i \leq |w|)$. For integers i, j with $1 \leq i \leq j \leq |w|$, the substring of w from $w[i]$ to $w[j]$ is denoted by $w[i, j]$. $\lg n$ stands for $\log_2 n$.

2.2 Grammar-Based Compression

A context-free grammar (CFG) is a quadruple $G = (\Sigma, V, D, X_s)$ where V is a finite subset of $\mathcal{X}$, D is a finite subset of $V \times (V \cup \Sigma)^*$ of production rules, and the start symbol $X_s \in V$. Variables in V are called nonterminals. Let $val(X_i)$ represent the string derived from $X_i \in V$, and let $|val(X_i)|$ be the length of $val(X_i)$. We assume a total order over $\Sigma \cup N$. The set of strings in Σ^* derived from X_s by G is denoted by $L(G)$. A CFG G is called *admissible* if exactly one $X \to \gamma \in D$ exists and $|L(G)| = 1$. An admissible G deriving a text S is called a grammar compression of S. The size of G is the total length of strings on the right hand sides of all production rules, and is denoted by $|G|$. The problem of grammar compression is formalized as follows:

Definition 1 (Grammar Compression). *Given a string $w \in \Sigma^*$, compute the smallest and admissible G that derives the only w.*

In the following, we assume the case $|\gamma| = 2$ for any production rule $X \to \gamma$. This assumption is reasonable because any grammar compression with n variables can be transformed into such restricted CFG with at most $2n$ variables.

The parse tree of G is represented as a rooted ordered binary tree such that internal nodes are labeled by variables in V and the *yields*, i.e., the sequence of labels of leaves is equal to S. In a parse tree, any internal node $Z \in V$ corresponds

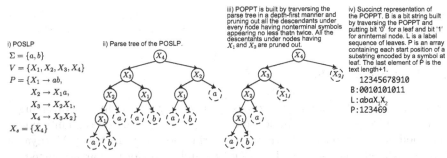

Fig. 1. Example of post order SLP (POSLP), parse tree, post order partial parse tree (POPPT), and succinct representation of POPPT

to the production rule $Z \to XY$, and has a left child labeled X and a right child labeled Y. Let $height(X_i)$ be the height of the subtree having the root X_i in the parse tree. We assume a *straight line program* (SLP) for a CFG as follows.

Definition 2. *(Karpinski-Rytter-Shinohara [7]) An SLP is a grammar compression over $\Sigma \cup V$ whose production rules are formed by $X_k \to X_i X_j$, where $X_i, X_j \in \Sigma \cup V$ and $1 \le i, j < k \le |V| + |\Sigma|$.*

Note that although our definition of an SLP is different from the original definition [7] in that our production rules derive only digrams, they are equivalent. In this paper we use our definition for notational convenience.

2.3 Phrase/Reverse Dictionaries

For a set P of production rules, a *phrase dictionary* D is a data structure for directly accessing the phrase $X_i X_j$ for any $X_k \in V$ if $X_k \to X_i X_j \in P$. A plain representation of D by an ordinal array requires $2n \log n$ bits of space to store n production rules. A reverse dictionary D^{-1} is a data structure for directly accessing the variable X_k given $X_i X_j$ for a production rule $X_k \to X_i X_j \in P$. Thus, $D^{-1}(X_i X_j)$ returns X_k if $X_k \to X_i X_j \in P$.

2.4 Rank/Select Dictionary

Our method represents an SLP using a rank/select dictionary, a succinct data structure for a bit string B [6] supporting the following queries: $\mathrm{rank}_c(B, i)$ returns the number of occurrences of $c \in \{0, 1\}$ in $B[0, i]$ and $\mathrm{select}_c(B, i)$ returns the position of the i-th occurrence of $c \in \{0, 1\}$ in B. Although naive approaches require the $O(|B|)$ time to compute a rank, data structures with only the $|B| + o(|B|)$ bit storage to achieve $O(1)$ time rank and select queries [14] have been presented. Practical implementations have been also presented [11,13].

3   Post-order SLPs and Succinct Encoding

Our algorithm builds *post-order SLP* (POPPT) as a special form of an SLP that generates a post-order partial parse tree. A POSLP is directly encoded into its succinct representation, which enables online substring extractions. In this section, we present the definition of POSLP and its succinct representation.

3.1   Post-order SLPs

Rytter [15] defined a partial parse tree as a binary tree built by traversing a parse tree in a depth-first manner and pruning out all the descendants under every node of a nonterminal symbol appearing no less than twice. Maruyama et al. [9] defined a post-order SLP (POSLP) and a post-order partial parse tree (POPPT) as follows.

Definition 3 (Post-order SLP (POSLP) and post-order partial parse tree (POPPT)). *A post-order partial parse tree is a partial parse tree whose internal nodes have post-order variables. A post-order SLP is an SLP whose partial parse tree is a post-order partial parse tree.*

For a POSLP of n variables, the numbers of nodes in the POPPT is $2n + 1$, because the numbers of internal nodes and leaves are n and $n + 1$, respectively. Figure 1-(i)(iii) shows an example of POSLP and POPPT, respectively. In this example, all the descendants under every node having the second X_1 and X_2 in the parse tree (ii) are pruned out. The resulting POPPT (iii) has internal nodes consisting of post-order variables.

Maruyama et al. [9] proposed OLCA, which outputs the POPPT of a given string, with a decoding algorithm of the input string from the POPPT.

Lemma 1 ([9]). *A POSLP of n variables can be represented in at most $n \lg(n + \sigma) + 2n + o(n)$ bits of space. The string represented by the POSLP is decoded in time proportional to the string length.*

However, their encoding algorithm is not fully-online. That is, their algorithm first constructs the phrase dictionary that occupies $2n \lg(n + \sigma)$ bits of space, then converts it into the POSLP. Furthermore, in addition to the space for the phrase dictionary, their algorithm also uses at least $n(1 + \alpha) \lg(n + \sigma)$ bits of space for the reverse dictionary where $0 < \alpha < 1$ is a fixed parameter called a *load factor* of a hash table.

In this paper, we propose *fully-online LCA* (FOLCA), which improves the OLCA in the following sense:

- A string is given in an online manner; that is, characters of the string is given one by one from left to right. At anytime, a POSLP of n variables representing the current string and its phrase dictionary can be encoded in at most $n \lg(n + \sigma) + 2n + o(n)$ bits of space.
- The reverse dictionary uses less space than that of the OLCA.
- At anytime during compression, we can decode any substring efficiently.

The details are described in the following subsections.

3.2   Succinct Representation of POSLP and the Phrase Dictionary

A major advantage of POSLP is that we can encode the corresponding POPPT and the phrase dictionary into a succinct representation consisting of a succinct tree B and a label sequence L. We build a succinct tree B as a bit string by traversing POPPT in post-order, and putting $'0'$ if a node is a leaf and $'1'$ otherwise. The last bit $'1'$ in B represents a virtual node. Our succinct tree uses the following four operations: $left\_child(B, i)$ returns the left child of a node i; $right\_child(B, i)$ returns the right child of a node i; $leftmost\_leaf(B, i)$ returns the leftmost leaf of a node i; $rightmost\_leaf(B, i)$ returns the rightmost leaf of a node i. They are computed in $O(1)$ time for a static case [14] and $O(\lg n/\lg \lg n)$ time for a dynamic case [12] for an n-node tree. Let $T_{\text{tree}}(n)$ denote those time complexities. The space for our succinct tree is at most $2n + o(n)$ bits for both offline and online cases, because a POPPT of a POSLP of n variables consists of n internal nodes, $n + 1$ leaves and a virtual node.

Label sequence L keeps symbols in leaves of a POPPT. The length of L is $n + 1$ and the space is $(n + 1)\lceil \lg(n + \sigma) \rceil$ bits. Accessing to $L[i]$ for $i \in [1, n+1]$ is performed by several rank/select operations on B in $O(T_{\text{tree}}(n))$ time. Given a variable X_i, the computation of its right-hand side is as follows: We first compute the positions $p = select_0(B, i)$, and then compute the left child $q_l = left\_child(B, p)$ and the right child $q_r = right\_child(B, p)$. If q_l (respectively q_r) is an internal node, i.e., $B[q_l] = 1$ (respectively $B[q_r] = 1$), we compute the corresponding symbol by $rank_1(B, q_l)$ (respectively $rank_1(B, q_r)$). Otherwise, we compute $L[select_0(B, q_l)]$ (respectively $L[select_0(B, q_r)]$).

Thus, the following lemma holds.

Lemma 2. *A POSLP of n variables and its phrase dictionary can be represented in at most $n \lg(n + \sigma) + 2n + o(n)$ bits of space while supporting access to the right-hand side of any rule in $O(T_{tree}(n))$ time.*

3.3   Smaller Reverse Dictionary

The reverse dictionary of the OLCA is a chaining hash table. Let α be a constant called a load factor. The hash table has αn entries and each entry stores a list of integers i representing the left-hand side of a rule $X_i \to X_j X_k$. For the rule $X_i \to X_j X_k$, the hash value is computed from j and k. Then the list corresponding to the hash value is scanned to search for i. For each value i' in the list, we compute the right-hand side using the phrase dictionary, and if it is $X_j X_k$, the value i' is what we search for. The size of the data structure is $\alpha n \lg(n + \sigma)$ bits for the hash table and $n \lg(n + \sigma)$ bits for the lists. Therefore the total size is $n(1 + \alpha) \lg(n + \sigma)$ bits. The access time is expected $O(1/\alpha)$ time.

We reduce the space for the lists by using gap-encoding. Because we use post-order for rules, newer rules have larger post-orders. When we insert a new value i into a list in the hash table, we append it at the end of the list. Then each list in the hash table consists of a strictly increasing sequence of integers. Instead of encoding i as it is, we encode the difference between i and the preceding one, say i', by the delta code. The difference $i - i'$ is encoded in $1 + \lfloor \lg(i - i') \rfloor + $

$2\lfloor \lg \lfloor 1 + \lg(i - i') \rfloor \rfloor$ bits. For all n rules, the space for the lists is upper-bounded by $n(1 + \lg(\alpha n) + 2 \lg \lg(\alpha n))$. Note that this is the worst-case bound and does not depend on the skewness of hash values. In total, the reverse dictionary uses $\alpha n \lg(n + \sigma) + n(1 + \lg(\alpha n) + 2 \lg \lg(\alpha n))$ bits. We can rewrite the formula to $\alpha n \lg(n + \sigma) + n(1 + \lg(\alpha n))$ by multiplying the original α by a constant. The access time is expected $O(1/\alpha)$ time, the same as OLCA.

By adding the space for the phrase and reverse dictionaries, we obtain the main result:

Theorem 1. *For a string of length N, a POSLP of n variables and its phrase and reverse dictionaries can be constructed in $O(\frac{N \lg n}{\alpha \lg \lg n})$ expected time using $(1 + \alpha)n \lg(n + \sigma) + n(3 + \lg(\alpha n))$ bit working space.*

Tabei et al. [16] proposed another representation of the phrase and the reverse dictionaries based on the wavelet tree. The dictionaries can be stored in $2n \lg n(1 + o(1))$ bits and a query to either the phrase or the reverse dictionary is done in worst-case $O(\lg n)$ time. Our data structure is also smaller than it, though our query time is an expected time. Furthermore, by using our POSLP, we can reduce the working space of Tabei et al. [16]. We use the bit vector B to encode the tree shape of the grammar in $2n + o(n)$ bits. Instead of using the label sequence L and the hash table for the reverse dictionary, we can use their data structure for the phrase and the reverse dictionaries using $2n \lg n(1 + o(1))$ bits. Thus we obtain the following:

Theorem 2. *For a string of length N, a POSLP of n variables and its phrase and reverse dictionaries can be constructed in $O(N \lg n)$ time using $2n \lg n(1 + o(1)) + 2n$ bit working space.*

3.4   Substring Extraction

We need to compute the length $|val(X_i)|$ of the string $val(X_i)$ derived from any X_i efficiently for substring extractions. Here, we use position array P of length $n + 2$ to store all $|val(X_i)|$, defined as follows: $P[i] = 1$ if $i = 1$ and $P[i] = \sum_{k=1}^{i-1} |val(X_k)|$ otherwise. A naive representation of P is a standard array which requires $n \lceil \lg N \rceil$ bits of space for the length N of an input text. A crucial observation is that P is a monotonic increasing sequence where $P[n+2] = N+1$. For such a sequence, we can apply a compressed integer representation [3] whose bits of space is at most $n(\lceil \lg N \rceil - \lfloor \lg n \rfloor + 2) + o(n) \le n \lg \frac{N}{n} + 3n + o(n)$, while supporting constant time access to $P[i]$ for any position i.

We compute the length $|var(X_i)|$ of the substring encoded by a variable X_i as follows: We access the position $k = select_0(B, i)$, and then compute its leftmost and rightmost leaves by $l = leftmost\_leaf(B, k)$ and $r = rightmost\_leaf(B, k)$, respectively. Thus, $|var(X_i)| = P[r+1] - P[l]$. The computation time is $T_{tree}(n)$ and thus the following lemma holds.

Lemma 3. *A POSLP of n variables and the length N of an input text is represented by $n \lg \frac{N}{n} + 3n + o(n)$ bits of space, while $|var(X_i)|$ for any variable X_i can be computed in $O(T_{tree}(n))$-time.*

Theorem 3. *The POSLP, which derives w, of n variables and height h, and its position array can be represented in at most $n \lg N + n \lg(1 + \frac{\sigma}{n}) + 5n + o(n)$ bits of space, while enabling extractions of substring $w[l, r]$ in $O((r - l + h)T_{tree}(n))$ time.*

Proof. The time complexity is clear, because the access to right-hand side and the length for any variable are computed in $T_{tree}(n)$-time. The required bits of space consists of at most $n \lg(n + \sigma) + 2n + o(n)$ bits for a POSLP and $n \lg \frac{N}{n} + 3n + o(n)$ for its position array. Thus, the total bits of space is $n \lg \frac{N}{n} + n \lg(n + \sigma) + 5n + o(n) = n \lg N + n \lg(1 + \frac{\sigma}{n}) + 5n + o(n)$.

To extract a substring $val(X_i)[l, r]$, we apply a well-known method for SLPs by simulating descending in the parse tree having the root X_i [1]. Let $X_i \rightarrow X_j X_k$. If $|val(X_i)| > l$, we descend to X_l, otherwise, we descend to X_r and $l := l - |val(X_j)|$. This process is recursively repeated until we reach the first character $val(X_i)[l]$. Note that the time to access $val(X_i)[l]$ is $O(height(X_i)T_{tree}(n))$, if the right-hand side of any rules and the length of any variables are computed in $T_{tree}(n)$ time. To extract the substring $val(X_i)[l, r]$, we first access to position l, and then extract its right-side characters while backtracking the descending path. The extraction time for any substring $w[l, r] = val(X_i)[l, r]$ is $O((r - l + h)T_{tree}(n))$, because the length of the descending path in the tree is $O(h)$ from the root to $w[l]$ and the number of nodes covering $w[l, r]$ is bounded by $O(r - l)$.

In practice, we first find the index $i = max\{j|P[j] \leq l, 1 \leq j \leq n + 2\}$ which is computed by a binary search on $P[1, n + 2]$ in $O(\lg n)$ time. The substring of $val(L[i])$ includes a prefix of $w[l, r]$, We then extract only the prefix $w[l, p]$ from $val(L[i])$ and decode $w[p + 1, r]$ from the right-side leaves of $L[i]$. This procedure enables practically fast substring extractions, because the first descending process from the root X_n is omitted in the most case.

Note the height h can be as large as $\Omega(n)$, this would cause inefficient substring extraction in the worst case. In the next section, we present a direct construction algorithm of POPPTs that height is bounded by $O(\lg N)$.

4   Fully-Online Grammar Compression

Maruyama *et al.* [9] proposed an online grammar-transform algorithm, called *online LCA (OLCA)*, which generates SLPs with height $h = O(\lg N)$ and number of variables $n = O(n_* \lg^2 N)$, where n_* is the optimal grammar size, N is the input string length. In this section, we present *fully-online LCA (FOLCA)* as a modification of OLCA. In OLCA, a function $\ell : [m] \times \{\Sigma \cup V\}^m \rightarrow \{0, 1\}$ is defined to check whether or not the i-th character $u[i]$ in the string $u \in \{\Sigma \cup V\}^m$ has a *landmark* [1]. We say that the character $u[i]$ has a landmark if $\ell(i, u) = 1$. The appearances of landmark are almost synchronized in long and common substrings to minimize the number of different variables generated in common substrings. $\ell(i, u)$ can be determined by a constant range in one symbol to the

---

[1] Precisely, the function ℓ is a trivial modification to detect *special pairs* in the original.

Algorithm 1. FOLCA: Fully-Online LCA. D: phrase dictionary, D^{-1}: reverse dictionary, q_k: queue at level k.

1: $D := \emptyset$, $D^{-1} := \emptyset$, initialize queues;
2: **while** Read a new character c and c is not the end of the file **do**
3:     process_symbol(q_1, c)
4: **end while**

Algorithm 2. process_symbol(q_k, x): a queue q_k and a symbol $X \in \Sigma \cup V$.

1: $q_k.enque(X)$;
2: **if** $q_k.size() = 4$ **then**
3:     **if** $\ell(2, q_k) = 0$ **then**
4:         $Y := D^{-1}(q_i[3], q_i[4])$; $D := D \cup \{Y \to q_i[3]q_i[4]\}$;
5:         process_symbol(q_{k+1}, Y);
6:         $q_k.deque()$; $q_k.deque()$;
7:     **end if**
8: **else if** $q_k.size() = 5$ **then**
9:     $Y := D^{-1}(q_k[4], q_k[5])$; $D := D \cup \{Y \to q_i[4]q_i[5]\}$;
10:     $Z := D^{-1}(q_k[3], Y)$; $D := D \cup \{Z \to q_i[3]Y\}$;
11:     process_symbol(q_{k+1}, Z);
12:     $q_k.deque()$; $q_k.deque()$; $q_k.deque()$;
13: **end if**

---

left and two symbols to the right of $u[i]$, and for any adjacent landmark positions i and j in u, we have $2 \leq |j - i| \leq 2 \lg |\Sigma \cup V|$. However, we omit the detail since the basic idea is the same as the OLCA (See [9]). For a sufficiently large N, we assume a sequence of queues $q_1, \cdots, q_k$ such that $k \leq \lg N$ and the length of any q_k $(1 \leq i \leq k)$ is fixed by a constant.

For an input string $w \in \Sigma^*$, when the prefix $w[1, N]$ has been processed by OLCA, the next character $w[N + 1]$ is enqueued into q_1. In the k-th queue q_k, the enqueued symbols is recursively processed by following operations: (i) decide a replaced digram XY in q_k by computing landmarks. (ii) generate a rule $Z \to XY \in D$ for $q_k[i, i+1] = XY$ and a new variable Z if $Z \notin V$, (iii) dequeue XY from q_k, and (iv) enqueue Z into the upper queue q_{k+1}. The final dictionary D is returned as an SLP.

We modify OLCA so that D represents a post-order SLP, that is, the algorithm gradually constructs a parse tree represented as D by simulating post-order traversal. The new algorithm is described in Algorithms 1 and 2. In FOLCA, each queue q is initialized by two *dummy symbol* $d \notin \{\Sigma \cup V\}$ so that $q[1, 2] = dd$ and $q.size() = 2$. The length of each queue is at most five such that $q[1, 2]$ is used for deciding a new landmark by the function ℓ, and depending on the result, one of digrams in $q[3, 5]$ is selected and it is replaced by an appropriate variable. In particular, when $q[4, 5]$ is replaced to a variable X, the resulting digram $q[3]X$ is immediately replaced to a variable Y. In this case, the variable Y is enqueued to the upper queue. This process is described in Algorithm 2.

Theorem 4. *Let P be the set of production rules generated by FOLCA, and let T be the partial parse tree corresponding to P. Then, T is a POPPT.*

Proof. Without loss of generality, we can assume T is identical to the parsing tree itself, that is, T has no variables appearing no less than twice. Let V_k is the set of variables generated with the queue q_k. Then, we show that T_X for any $X \in V_k$ is a POPPT by induction on k where T_X is the subtree of T whose root is X. The base step is clear because any X is a leaf. Suppose the hypothesis on some $k \geq 1$ and let $q_k[3,5] = ABC$ for some $A, B, C \in V_k$. For the trigram ABC, there are two cases for parsing: (1) $X \to AB \in P$ and $X \in q_{k+1}$, and (2) $X \to BC, Y \to AX \in P$ and $X, Y \in q_{k+1}$. By the induction hypothesis, $X > A, B$ holds in Case (1) and similarly, $X > B, C$ and $Y > A, X$ hold in Case (2), that is, T_X and T_Y are POPPTs. Hence, T_X is a POPPT for any $X \in V_{k+1}$. This concludes that the whole tree T is a POPPT.

By Theorem 4, Algorithm 1 computes a phrase dictionary equivalent to a POPPT T deriving $S \in \Sigma^*$. Given a succinct representation of T, we can update it for Sa ($a \in \Sigma$) in $O(\lg n / \lg \lg n)$ time, using the results in the previous section. Thus, FOLCA can be regarded as a fully-online algorithm for succinct grammar compression. We note the theoretical performance of FOLCA is the same as OLCA, however, we omit the details because it is most of the same.

5 Experiments

We tested our method of substring extractions in comparison with LZend [8]. We used highly repetitive texts Ecoli and kernel downloadable from http://pizzachili.dcc.uchile.cl/repcorpus/real/ as a test dataset. The sizes of the Ecoli and kernel texts were 108MB and 247MB, and their alphabet sizes were 15 and 160, respectively. We also used compression/extraction times and working space as evaluation measures. FOLCA and LZend were implemented using C++, and all experiments were performed on one core of an aeight core Intel Xeon(R) CPU E5640 (2.67GHz) machine with 100GB memory.

Table 5 and table 5 show compression time and working memory on the Ecoli and kernel texts, respectively. We chose a load factor from $\{0.01, 0.05, 0.1, 0.3, 0.5\}$, and evaluated the space-time trade-off of FOLCA. Although LZend is known as a state-of-the-art substring extraction method based on LZ77-encoding, it consumes a large amount of working memory of about 2.5GB and 4.6GB for constructing the index, respectively. This is because LZend needs suffix array, inverse suffix array, and FM-index for compression. Such a large working space prevents practical usage of LZend. The working space of FOLCA were 20-202 times smaller than that of LZend, while FOLCA was 2-5 times faster than LZend. The most of the space was consumed by the hash table for each load factor. The space of hash table varied from 23MB to 90MB according to the load factor on the Ecoli and kernel texts, while there was a large reduction of compression time from $1,328$ seconds to 408 seconds.

Table 2. Experimental results on the Ecoli text. The table details compression time in seconds and working space of hash table (H), dictionary (D) and position array (P) in mega bytes. **LZend** does not have the parameter load factor.

Method					LZend	
load factor	time (sec)	H (MB)	$H + D$ (MB)	$H + D + P$ (MB)	time (sec)	space (MB)
0.01	1,328	23	45	50		
0.05	728	37	59	64		
0.1	553	48	70	75	2,217	2,410
0.3	416	65	87	92		
0.5	408	90	112	117		

Table 3. Experimental results on the kernel text

Method					LZend	
load factor	time (sec)	H (MB)	$H + D$ (MB)	$H + D + P$ (MB)	time (sec)	space(MB)
0.01	2,891	11	21	23		
0.05	2,071	13	23	25		
0.1	1,472	16	26	28	4,547	4,653
0.3	951	30	40	42		
0.5	882	42	52	54		

Table 4. Time for substring extraction on the Ecoli and the kernel texts

	Ecoli		kernel	
length	FOLCA(sec)	LZend(sec)	FOLCA(sec)	LZend(sec)
10^1	0.00007	0.00002	0.00010	0.00003
10^2	0.00026	0.00011	0.00029	0.00012
10^3	0.00224	0.00100	0.00224	0.00098
10^4	0.02176	0.00954	0.02182	0.00901
10^5	0.21328	0.09215	0.21622	0.09418

Table 5. Working memory for substring extraction in megabyte on the Ecoli and the kernel texts

	Ecoli	kernel
FOLCA	27	12
LZend	23	14

Table 5 and table 5 show time and working space for substring extractions. FOLCA was about three times slower than LZend, while the working space of FOLCA was slightly larger than that of LZend. These results were reasonable when considering the large reduction of construction space of FOLCA.

6   Conclusion

We first proposed a special representation of SLP, called *POSLP*, which supports basic functionalities in addition to the space being asymptotically close to an information-theoretic lower bound. Any grammar-compressed string can be converted into POSLP for *algorithms on SLP-compressed strings* because, once loaded a POSLP on main memory, they can be immediately run without transforming into plain SLPs. Therefore, our representation would be suitable for any grammar compression as a standard format.

We improved an online grammar-transform algorithm of Maruyama *et al.* [9] into a fully-online version named *FOLCA*, that can directly construct a POSLP while keeping the theoretical performance of the original. The FOLCA can efficiently compress highly-redundant data in the space of compressed string, not the space of input string. In the real world, representative examples of compressible data consist of genome collection of individual species, versioned documents, web texts and so on. Such data would be generated day after day, one right after the other, and thus our study would be the foundation of storage systems for highly-redundant data collections in online setting.

References

1. Claude, F., Navarro, G.: Self-indexed grammar-based compression. Fundamenta Informaticae 111, 313–337 (2010)
2. Goto, K., Bannai, H., Inenaga, S., Takeda, M.: Fast q-gram mining on slp compressed strings. J. Discrete Algorithms 18, 89–99 (2013)
3. Grossi, R., Gupta, A., Vitter, J.S.: High-order entropy-compressed text indexes. In: SODA, pp. 636–645 (2003)
4. Hermelin, D., Landau, G.M., Landau, S., Weimann, O.: A unified algorithm for accelerating edit-distance computation via text-compression. In: STACS, pp. 26–28 (2009)
5. Inenaga, S., Bannai, H.: Finding characteristic substrings from compressed texts. In: PSC, pp. 40–54 (2009)
6. Jacobson, G.: Space-efficient static trees and graphs. In: FOCS, pp. 549–554 (1989)
7. Karpinski, M., Rytter, W., Shinohara, A.: An efficient pattern-matching algorithm for strings with short descriptions. Nordic J. Comp. 4(2), 172–186 (1997)
8. Kreft, S., Navarro, G.: On compressing and indexing repetitive sequences. Theoretical Computer Science 483, 115–133 (2013)
9. Maruyama, S., Sakamoto, H., Takeda, M.: An online algorithm for lightweight grammar-based compression. Algorithms 5(2), 213–235 (2012)
10. Matsubara, W., Inenaga, S., Ishino, A., Shinohara, A., Nakamura, T., Hashimoto, K.: Efficient algorithms to compute compressed longest common substrings and compressed palindromes. Theoretical Computer Science 410(8-10), 900–913 (2009)
11. Navarro, G., Providel, E.: Fast, small, simple rank/select on bitmaps. In: Proc. SEA, pp. 295–306 (2012)
12. Navarro, G., Sadakane, K.: Fully-functional static and dynamic succinct trees. ACM Transactions on Algorithms (2010), Accepted A preliminary version appeared in SODA 2010
13. Okanohara, D., Sadakane, K.: Practical entropy-compressed rank/select dictionary. In: Workshop on Algorithm Engineering & Experiments (2007)
14. Raman, R., Rao, S.S., Raman, V.: Succinct indexable dictionaries with applications to encoding k-ary trees, prefix sums and multisets. ACM Transactions on Algorithms 3 (2007)
15. Rytter, W.: Application of Lempel-Ziv factorization to the approximation of grammar-based compression. Theor. Comput. Sci. 302(1-3), 211–222 (2003)
16. Tabei, Y., Takabatake, Y., Sakamoto, H.: A succinct grammar compression. In: Fischer, J., Sanders, P. (eds.) CPM 2013. LNCS, vol. 7922, pp. 235–246. Springer, Heidelberg (2013)
17. Takabatake, Y., Tabei, Y., Sakamoto, H.: Variable-length codes for space-efficient grammar-based compression. In: SPIRE, pp. 398–410 (2012)
18. Tiskin, A.: Towards approximate matching in compressed strings: Local subsequence recognition. In: Kulikov, A., Vereshchagin, N. (eds.) CSR 2011. LNCS, vol. 6651, pp. 401–414. Springer, Heidelberg (2011)
19. Yamamoto, T., Bannai, H., Inenaga, S., Takeda, M.: Faster subsequence and don't-care pattern matching on compressed texts. In: Giancarlo, R., Manzini, G. (eds.) CPM 2011. LNCS, vol. 6661, pp. 309–322. Springer, Heidelberg (2011)

Solving Graph Isomorphism
Using Parameterized Matching

Juan Mendivelso[1], Sunghwan Kim[2], Sameh Elnikety[3], Yuxiong He[3],
Seung-won Hwang[2], and Yoan Pinzón[1]

[1] Universidad Nacional de Colombia, Colombia
[2] POSTECH, Republic of Korea
[3] Microsoft Research, Redmond, WA, USA

Abstract. We propose a new approach to solve graph isomorphism using parameterized matching. To find isomorphism between two graphs,
one graph is linearized, *i.e.*, represented as a graph walk that covers all
nodes and edges such that each element is represented by a parameter.
Next, we match the graph linearization on the second graph, searching
for a bijective function that maps each element of the first graph to an
element of the second graph. We develop an efficient linearization algorithm that generates short linearization with an approximation guarantee, and develop a graph matching algorithm. We evaluate our approach
experimentally on graphs of different types and sizes, and compare to
the performance of VF2, which is a prominent algorithm for graph isomorphism. Our empirical measurements show that graph linearization
finds a matching graph faster than VF2 in many cases because of better
pruning of the search space.

1 Introduction and Related Work

Graphs are widely used in many application domains, and graph isomorphism
is a fundamental problem that appears in graph processing techniques of many
applications including pattern analysis, pattern recognition and computer vision
as discussed in a recent survey [8]. Graph isomorphism is a challenging problem:
Given two graphs, we search for a bijective mapping from each element of the
first graph to an element of the second graph such that both data and structural
properties match. Data properties include node and edge attributes and types,
and structural properties maintain the adjacency relations.

A naive solution could search for all possible mappings, facing an exponential
search space. Surprisingly, the exact complexity of graph isomorphism is not
determined yet [9], but likely to be NP-Complete. Notice however, graph sub-
isomorphism, which is a closely related but a different problem, is NP-Complete
[9]. Existing algorithms for graph isomorphism include Nauty Algorithm [19],
Ullmann Algorithm [22] and VF2, a more recent algorithm [9]. All these algorithms have exponential worst case performance (since isomorphism is a hard
problem). Except for some easy cases, solving isomorphism generally takes much
longer time if there is no match, since all possible mappings are progressively

O. Kurland, M. Lewenstein, and E. Porat (Eds.): SPIRE 2013, LNCS 8214, pp. 230–242, 2013.

searched until shown not to lead to an isomorphism. Several heuristics, however, are employed to find likely mappings quickly. A good algorithm for graph isomorphism should find isomorphic graphs quickly in many cases.

In this paper we apply parameterized matching to solve graph isomorphism. Parameterized matching [4] was introduced to efficiently track down duplicate code in large software systems. It determines if two strings have the same structure. Specifically, two equal-length strings parameterized-match if there exists a bijective function f for which every text symbol in one string is equal to the image under f of the corresponding symbol in the other string. Brenda Baker [4] introduced this problem in 1993, and research work [1–3, 5–7, 10, 12–17, 20, 21] extends parameterized matching. A survey on parameterized matching is presented in [18].

Our approach to solve graph isomorphism has two main steps, linearization, and matching. First, in the linearization step, one of the graphs is represented as a graph walk that visits each node and edge, such that each element is represented as a parameter. This linear sequence is used in the second step for matching, which parameterized-matches the graph linearization against the other graph, to search for mapping.

This approach allows us to incorporate optimizations for both linearization and parameterized matching steps. Although we focus on presenting and evaluating the fundamental approach, we point out several attractive features of this approach. For example, this approach supports general graph models, such as attributed multi-graphs (in which nodes and edges may have arbitrary attributes, and several edges may connect two nodes). The graph statistics, such as node degree distribution and histograms of attribute values can be easily integrated in the linearization step to provide better linearization. The matching algorithm is embarrassingly parallel, enabling efficient implementation on multi-core machine and distributed frameworks.

We present the algorithms, correctness and complexity analysis of these two steps and implement them for experimental evaluation using graph of several types and sizes. We also compare to an optimized implementation of VF2, which is one of most widely used algorithms for graph isomorphism. Our empirical results show that in many cases, the graph linearization approach provides shorter response times, and the improvements increase with the graph size.

Our contributions are the following: (1) We propose a new approach to graph isomorphism using parameterized matching (Section 3). (2) We develop an efficient linearization algorithm to represent a graph as a parameterized walk, and we establish a bound on the linearization length (Section 4). (3) We introduce an algorithm to parameterized-match the linearization on graph (Section 5). (4) We evaluate our approach experimentally (Section 6).

2   Preliminaries

This section defines the graph isomorphism problem and points out its similarity to parameterized matching in strings. In this paper, we consider multigraphs.

A multigraph $G(V, E)$ is comprised of a set of vertices V, $n = |V|$, and a set of undirected edges $E \subseteq V \times V$, $m = |E|$, where multiple edges between two distinct vertices and self loops are permitted. We distinguish the edges that have the same end vertices by the notation of the edge; for example, $e = (u, v)$ and $e' = (u, v)$. Let $\mathcal{E}_G = V \cup E$ denote the set of *graph elements* of G, i.e. the set of vertices and edges in G. Also, let $u.degree$ denote the number of adjacent edges that vertex $u \in V$ has. In this paper, we consider undirected multigraphs; however our algorithms can be easily extended to support directed multigraphs. Next we define the *Graph Isomorphism* problem.

Problem 1 (Graph Isomorphism). Let $G_1(V_1, E_1)$ and $G_2(V_2, E_2)$ be two multigraphs such that $n = |V_1| = |V_2|$ and $m = |E_1| = |E_2|$. The graph isomorphism problem determines whether there exists a bijective mapping function $f : \mathcal{E}_{G_1} \to \mathcal{E}_{G_2}$, such that $\forall_{u,v \in V_1}, e = (u, v) \in E_1 \iff f(u), f(v) \in V_2 \wedge f(e) = (f(u), f(v)) \in E_2$.

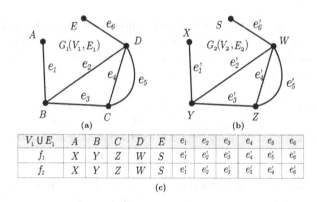

Fig. 1. Isomorphism example: the multigraphs presented in (a) and (b) are isomorphic; the functions that define the isomorphism are presented in (c). The difference between f_1 and f_2 is that $f_1(e_4) = e_4'$ and $f_1(e_5) = e_5'$ while $f_2(e_4) = e_5'$ and $f_2(e_5) = e_4'$.

For example, the graphs in Figure 1(a,b) are isomorphic; furthermore there are two possible mapping functions that define the isomorphism (see Figure 1(c)). Notice that the graph isomorphism determines whether the topological structures of two multigraphs are the same. It is very similar to what parameterized matching does with strings: checking whether two strings have the same structure. Next we define parameterized matching on strings:

Definition 1. *Let $X = X_{1...\ell}$ and $Y = Y_{1...\ell}$ be two equal-length strings defined over alphabet Σ. Each symbol in the alphabet is called a parameter. Strings X and Y are said to parameterized-match iff there exists a bijective function $f : \Sigma \to \Sigma$ such that $f(X_i) = Y_i$, for all $1 \leq i \leq \ell$.*

For example, let $X = abacab$ and $Y = bcbabc$ be two strings defined over $\Sigma = \{a, b, c\}$. They parameterized-match as X is equal to Y by means of f : $(a, b, c) \rightarrow (b, c, a)$. In Section 3.1, we define parameterized matching for walks to solve the graph isomorphism problem.

3   Graph Linearization

Our approach for solving graph isomorphism consists of two main steps: (i) linearizing G_1 into a walk $p_{1...\ell}$; and (ii) exploring all the walks in G_2 to determine whether there is one that parameterized matches $p_{1...\ell}$. In this section, we define graph linearization and parameterized matching on graph walks (Section 3.1). Then, we discuss characteristics and algorithms for linearization (Section 3.2).

3.1   Definition of Graph Linearization

Definition 2 (Linearization). *Let $G(V, E)$ be a connected undirected multigraph. A walk $p = p_{1...\ell}$ of vertices and edges is a linearization of G iff:*

1. *p_i is a vertex $v \in V$ if i is odd, $1 \leq i \leq \ell$.*
2. *p_i is an edge $e \in E$ if i is even, $1 \leq i \leq \ell$, such that $e = (p_{i-1}, p_{i+1})$.*
3. *Each vertex $v \in V$ and each edge $e \in E$ appears at least once in p.*

Our motivation for defining graph linearization is to represent the topology of a multigraph through a walk. Specifically, the linearization p of G is a walk that represents all its adjacency relation, which we use to solve the graph isomorphism problem by comparing walks instead of multigraphs. For this purpose, we define parameterized matching on walks as follows:

Definition 3 (Parameterized Matching on Graph Walks). *Let $G_1(V_1, E_1)$ and $G_2(V_2, E_2)$ be two connected undirected multigraphs. Also, let $V_1' \subseteq V_1$ and $E_1' \subseteq E_1$ be subsets of vertices and edges in G_1; similarly, $V_2' \subseteq V_2$ and $E_2' \subseteq E_2$ are subsets of vertices and edges in G_2. Consider the walk $p_{1...k}$ in G_1 and the walk $q_{1...k}$ in G_2. The walk $p_{1...k}$ is said to parameterized-match the walk $q_{1...k}$ if and only if there exists a bijective function $f : \mathcal{E}_{G_1} \rightarrow \mathcal{E}_{G_2}$ such that $q_i = f(p_i)$ for $1 \leq i \leq k$.*

The core idea of using parameterized matching to solve the graph isomorphism problem is as follows. Let p be a linearization of G_1. Recall that, p represents the topology of G_1. Thus, if a walk q in G_2 parameterized-matches p, then p and q have the same topology. Furthermore, as q represents G_2, we conclude that G_1 and G_2 are isomorphic. This is formally presented in the next theorem:

Theorem 1. *Let $G_1(V_1, E_1)$ and $G_2(V_2, E_2)$ be two connected undirected multigraphs such that $n = |V_1| = |V_2|$ and $m = |E_1| = |E_2|$. Also, let $p_{1...\ell}$ be the linearization of G_1. Then, G_1 and G_2 are isomorphic if and only if there exists a walk $q_{1...\ell}$ in G_2 such that $p_{1...\ell}$ parameterized-matches $q_{1...\ell}$.*

3.2   Characteristics and Algorithms for Graph Linearization

There may be many linearizations that represent the same graph. However, a compact representation is preferable. For solving graph isomorphism, the length of the linearization is an important measure on the matching time. This is because a shorter linearization often leads to a smaller cost at the matching stage. Next, we define *length-optimal linearization*.

Definition 4 (Length-Optimal Linearization). *The linearization $p = p_{1...\ell}$ of a connected undirected multigraph is length-optimal if the length of p, i.e. ℓ, is minimum.*

The *Graph Linearization* problem is very similar to the *Chinese Postman Problem* (CPP). CPP finds a walk that visits all the edges (and all the vertices) in the multigraph at least once; the only difference is that Graph Linearization does not require the starting vertex to be the same final vertex. In [11], an $O(n^3 + m^2)$ algorithm for the CPP was proposed. We can adapt this algorithm to calculate a length-optimal linearization. However, for large multigraphs, it is desirable to have algorithms with lower time complexity even if they do not produce length-optimal linearizations. As an attractive trade-off between length-optimality and efficiency, we propose a greedy approximation algorithm with an approximation guarantee.

4   Graph Linearization Algorithm - GLA

This section presents the GLA or *Graph Linearization Algorithm*. First, we describe the key ideas of the algorithm in Section 4.1; then we go through the details in Section 4.2. In Section 4.3 we present an upper bound for the length of GLA linearizations. Finally, in Section 4.4, we present the complexity analysis.

4.1   Key Ideas

One of the challenges of linearization algorithms is visiting all the edges with short linearization length. To address the challenge, we develop three heuristics: (1) the traversal starts from the vertex with the lowest degree; (2) the unexplored edges that lead to already explored vertices are visited before the ones that lead to unexplored vertices; and (3) the edges that lead to unexplored vertices are considered sorted, in ascending order, on the number of unexplored edges they have. Heuristics (1) and (3) aim to put the vertices that are close to be covered in the top levels of the DFS tree. Furthermore, heuristic (2) aims to cover the vertices in the highest levels of the DFS tree at an early stage. The three heuristics make the traversal explore one region of the multigraph before visiting another one; then, the produced linearization is shorter.

The proposed linearization approach also allows us to incorporate optimizations for both linearization and parameterized matching steps. For instance, the matching time will not only depend on the length of the linearization, but also on the order of comparisons. Specifically, the graph statistics of the multigraphs can be used to produce a linearization that prunes the search space during the matching phase. For example, if the frequency of some vertices of a certain degree (or a certain attribute in attributed graphs) is low, it would be appropriate to start the linearization from such vertices. However, for clarity, in this paper, we focus on the fundamental approach only.

4.2   Algorithm

The pseudocode of the Graph Linearization Algorithm (GLA) is listed in Figures 2 and 3. The linearization produced by GLA for the graph presented in Figure 1(a) is $Ae_1Be_3Ce_4De_5Ce_5De_2\ Be_2De_6E$; its length is 17.

Algorithm 1: GLA Algorithm

Input: $G(V, E)$ **Output:** p

1.   **for every** $e \in E$ **do** $e.Explored \leftarrow false$
2.   **for every** $v \in V$ **do**
3. $v.Explored \leftarrow false$
4. $S \leftarrow \{(u, v) \mid v \in V \wedge (u, v) \in E\}$
5. $v.NumUnexploredEdges \leftarrow |S|$
6.   **choose** $u \in V_P$ **with** $min(u.NumUnexploredEdges)$
7. $p \leftarrow \langle \rangle,\ unexplGE \leftarrow |V| + |E|$
8. $TraverseGraph(G, u, p, unexplGE)$
9.   **return** p

Fig. 2. GLA Algorithm

4.3   Length of GLA Linearization

Theorem 2 shows that given the multigraph $G = (V, E)$, the length of the walk generated by GLA is at most 2 times the length of an optimal linearization. Therefore, the length produced by GLA is asympotically optimal.

Theorem 2. GLA is 2-approximate with respect to the length of the length-optimal linearization.

This theorem is based on the fact that each edge in the multigraph G appears at most twice in the linearization $p = p_{1...\ell}$ generated by GLA. Then, ℓ is compared to a lower bound that visits each edge only once to show worst-case approximation ratio. However, even an optimal linearization may not achieve the lower bound for many graph structures. Thus, for average cases in practice, GLA linearization is much closer to the optimal.

Algorithm 2: TRAVERSEGRAPH **Procedure**

Input: $G(V, E), u, p, unexplGE$

1. $p.Add(u), u.Explored \leftarrow true, unexplGE$--
2.   **for every** $e \in E$ such that $e = (u, v)$ **do**
3.   **if** $!e.Explored \wedge v.Explored$ **then**
4. $p.Add(e), e.Explored \leftarrow true, unexplGE$--$, p.Add(v)$
5. $u.NumUnexplEdges$--$, v.NumUnexplEdges$--
6.   **if** $unexplGE > 0$ **do**
7. $p.Add(e), p.Add(u)$
8.   **while** there are unexplored edges $e = (u, v)$
9.   **choose** e **with** $min(v.NumUnexploredEdges)$
10. $p.Add(e), e.Explored \leftarrow true, unexplGE$--
11. $u.NumUnexplEdges$--$, v.NumUnexplEdges$--
12. $TraverseGraph(G, v, p, unexplGE)$
13.   **if** $unexplGE = 0$ **then break**
14. $p.Add(e), p.Add(u)$

Fig. 3. TRAVERSEGRAPH Procedure

4.4 Complexity Analysis

The complexity of GLA is dominated by the walk traversed (line 8, Figure 2) which corresponds to the linearization. Notice that p has at most $2m$ edges and $2m + 1$ vertices. Each insertion takes constant time as it is always done at the end of p. But when a vertex is inserted for the first time, it is necessary to consider the unexplored adjacent edges e that lead to unexplored vertices v sorted on $v.NumUnexplEdges$ (lines $8 - 9$, Figure 3). This sorting operation takes $O(d \lg d)$, where d is the maximum degree of the vertices in G_1; specifically $d = \max_{v \in V_1} v.degree$. Thus, the time complexity of GLA is $O(2m + (2m + 1)(d \lg d)) = O(dm \lg d)$.

5   Matching a Linearized Graph

The *Parameterized Matching on multi-Graphs* (PMG) algorithm uses a linearization of $G_1(V_1, E_1)$, denoted as $p = p_{1...\ell}$, and matches it against $G_2(V_2, E_2)$ to determine whether G_1 and G_2 are isomorphic by using Theorem 1.

5.1   Key Ideas

PMG considers all the possible injective functions $f : \mathcal{E}_{G_1} \rightarrow \mathcal{E}_{G_2}$ to determine whether there is mapping with two properties: (i) f is bijective; and (ii) there exists a walk $q_{1...\ell}$ in G_2 for which $q_i = f(p_i)$ (*i.e.* q parameterized-matches p). These possible injective functions are explored by traversing p and G_2 simultaneously; specifically, a graph element p_i is compared to a graph element ge

in G_2 to determine whether an injective mapping is possible. We progressively extend a successful mapping by considering p_{i+1} and an adjacent graph element of ge. The graph elements of G_2 are traversed in a depth-first manner while p is traversed from left to right. Let us consider the DFS tree that represents the traversal of G_2. Then, the idea of this traversal of G_2 is considering the possible injective mappings by attempting to set $f(p_i) = ge$ where $ge \in \mathcal{E}_{G_2}$ is a graph element at level i of the DFS tree. Notice that the walk from the root to a leaf in the DFS tree parameterized-matches $p_{1\ldots\ell}$ under f; hence G_1 and G_2 are isomorphic.

Next, we show our heuristics to prune the search space. At each step of the process, a vertex $u \in V_2$ and a vertex in p_i are compared. Let us say that we set $f(p_i) = u$. In order to extend the match, we use vertex degrees and previous assignments in f to prune the search space. Specifically, we consider two cases:

Case 1: Vertex p_{i+2} is unassigned. We consider all the possible assignments $f(p_{i+1}) = e$ and $f(p_{i+2}) = v$ for edges $e = (u, v) \in E_2$ such that: (i) both e and v are unassigned; and (ii) $v.degree = p_{i+2}.degree$. Condition (i) is to guarantee that f is injective; condition (ii) is a pruning criterion based on that fact that, if G_1 and G_2 are isomorphic, then analogous vertices must have the same degree. Notice that if p_{i+2} is unassigned, p_{i+1} is unassigned as well; this is because the assignment of an edge in p is done at the same time (or after) the assignment of its end vertices. The process continues by considering p_{i+2} and each v.

Case 2: Vertex p_{i+2} is assigned to $v \in V_2$. There are two sub-cases. (a) Edge p_{i+1} is already assigned: it is not necessary to check adjacency as this was done when the mapping was set. We continue by considering p_{i+2} and v. (b) Edge p_{i+1} is unassigned: the algorithm considers all the possible assignments $f(p_{i+1}) = e$ for the unassigned edges $e = (u, v)$. The process continues at p_{i+2} and v.

If the algorithm reaches a successful assignment for p_ℓ, then the algorithm reports that the multigraphs are isomorphic.

5.2   Pseudocode

Figure 4 lists the pseudocode of PMG. The mapping function is represented as the array f. On the other hand, boolean array g indicates if each graph element in $\mathcal{E}_{G_2}$ is already assigned to a graph element in $\mathcal{E}_{G_1}$ (through function f). When we run PMG for G_2 and the linearization $p = Ae_1Be_3Ce_4De_5Ce_5De_2Be_2De_6E$ of G_1, the match is returned when any of the following walks are traversed: $q_1 = Xe_1'Ye_3'Ze_4'We_5'Ze_5'We_2'Ye_2'We_6'S$ or $q_2 = Xe_1'Ye_3'Ze_5'We_4'Ze_4'We_2'Ye_2'We_6'S$. Notice that both q_1 and q_2 parameterized-match p. The mapping functions of these matches correspond to the functions f_1 and f_2 presented in Figure 1(c).

5.3   Complexity Analysis

The time complexity of PMG is given by the number of executions of the recursive procedure EXTENDMATCH; each execution requires constant time. This number is equal to the number of vertices and edges in the DFS search trees.

Algorithm 3: PMG Algorithm

Input: $G_1(V_1, E_1), G_2(V_2, E_2)$ **Output:** $true/false$

1. $p = GLA(G_1)$
2.   **for every** $ge \in (V_1 \cup E_1)$ **do** $f[ge] \leftarrow undef$
3.   **for every** $ge \in (V_2 \cup E_2)$ **do** $g[ge] \leftarrow false$
4.   **for every** $u \in V_2$ **do**
5.   **if** $u.degree = p_1.degree$
6. $f' \leftarrow copyOf(f),\ \ f'[p_1] \leftarrow u$
7. $g' \leftarrow copyOf(g),\ \ g'[u] \leftarrow true$
8.   **if** $ExtendMatch(u, p, 1, f', g', G_2) = true$
9.   **return** $true$
10.   **return** $false$

Fig. 4. PMG Algorithm

Algorithm 4: EXTENDMATCH Algorithm

Input: $u, p = p_{1...\ell}, i, f, g, G_2(V_2, E_2)$ **Output:** $true/false$

1.   **if** $i = \ell$ **then return** $true$
2.   **if** $f[p_{i+2}] = undef$
3.   **for every** $e = (u, v) \in E_2$ **do**
4.   **if** $g[v] = false$ **and** $g[e] = false$ **and** $v.degree = p_{i+2}.degree$
5. $f' \leftarrow copyOf(f),\ \ f'[p_{i+1}] \leftarrow e,\ \ f'[p_{i+2}] \leftarrow v$
6. $g' \leftarrow copyOf(g),\ \ g'[e] \leftarrow true,\ \ g'[v] \leftarrow true$
7.   **if** $ExtendMatch(v, p, i + 2, f', g', G_2) = true$
8.   **return** $true$
9.   **else**
10. $v = f[p_{i+2}]$
11.   **if** $p_{i+1} = undef$
12.   **for every** $e = (u, v) \in E_2$ **such that** $g[e] = false$
13. $f' \leftarrow copyOf(f),\ \ f'[p_{i+1}] \leftarrow e$
14. $g' \leftarrow copyOf(g),\ \ g'[e] \leftarrow true$
15.   **if** $ExtendMatch(v, p, i + 2, f', g', G_2) = true$
16.   **return** $true$
17.   **else**
18.   **if** $ExtendMatch(v, p, i + 2, f, g, G_2) = true$
19.   **return** $true$
20.   **return** $false$

Fig. 5. EXTENDMATCH Algorithm

As the number of edges in a DFS tree is equivalent to the number of vertices — each vertex, except the root, is associated to an edge that leads to its parent, the asymptotic behavior of PMG depends on the number of vertices in the DFS trees. Next theorem gives an upper bound for this number.

Theorem 3. *Let $p_{1...\ell}$ be a linearization of G_1. Also, let d be the maximum degree of the vertices in G_2; specifically $d = \max_{v \in V_1} v.degree$. The DFS tree that represents the traversal of G_2 done by PMG has at most $O(d^{\lfloor \ell/2 \rfloor})$ vertices.*

This theorem is based on the following facts: (i) there are $O(\lfloor \ell/2 \rfloor)$ branching vertices in the DFS search tree associated to the vertices in the linearization $p = p_{1...\ell}$; and (ii) the branching factor for each of such vertices in the search tree is $O(d)$. As a DFS tree starts at each vertex in G_2, the total number of vertices visited, and hence the time complexity of PMG, is $O(nd^{\lfloor \ell/2 \rfloor})$. Note that if G_2 is complete, *i.e.*, $d = n - 1$, the time complexity is $O(n(n-1)^{\lfloor \ell/2 \rfloor}) = O(n^{\lceil \ell/2 \rceil})$.

However, it is important to remark that Theorem 3 gives an upper bound for the worst-case complexity. It assumes that, at every level of vertices, all the possible neighbors are explored. The average-case situations in practice are often not that "bad" because (i) when a vertex p_i has already been assigned, only such assigned vertex is considered; and (ii) when the multigraph has varied vertex degrees, the pruning criterion highly reduces the number of adjacent vertex to be visited.

6   Experimental Evalution

We assess the performance of our proposed approach experimentally. We implement the linearization and the matching algorithms in C#. We employ a set of synthetic graphs generated for benchmarking. We compare our approach to VF2, using an optimized implementation from the networkX library [1]. All evaluations are performed on a server running under a Windows platform on a 3.40GHz CPU with 16GB memory.

For graph generation, we deliberately avoid the "trivial cases". For example, consider a graph where vertex v_i is connected to $v_1, \ldots, v_{i-1}$. As the degree of each node is unique, testing isomorphism can be done trivially by a simple heuristic like sorting nodes by degree. In contrast, we consider cases where no such simple heuristic wins. Graphs where every node has the identical degree would much more challenging in that sense.

Meanwhile, we also avoid topologies that are always isomorphic, such as a grid or a complete graph. For this reason, we generate random graph pairs of identical-degree nodes. As the complexity of isomorphism testing algorithm is reported to vary significantly over degree, from $O(n^2)$ to $O(nn!)$ [9], we consider both low- and high-degree cases to evaluate algorithms in a wide spectrum of settings. The lower end of this spectrum is observed when the matching graphs are found early in a sparse graph, while the opposite case of dense graphs often leads to long running times. More specifically, we generate sparse and dense identical-degree graphs as follow: **1-Sparse:** We generate a random graph G where every node has degree three, with $3N$ total edges for N nodes. We first build a random binary tree with $N - 1$ edges. Then, the nodes with the degree

---

[1] http://networkx.github.io

less than three connect to another such node chosen at random. **2-Dense:** We generate graph G' by subtracting G from a complete graph. Every node of G has the same degree of $N - 4$.

In each setting, we vary the number of nodes from 16 to 256, and evaluate the response time of GLA (our proposed approach) and VF2 (baseline). For each point in the figures, we randomly generate 45 graphs and report the median response time. We choose median response time as our performance metric because the running time on different graphs significantly varies over graph complexity (as discussed above) while the optimization margin is narrow for easy cases and hard extremes. Our target problems are thus neither of these, and using the average or min/max as the main performance metric would bias the results to represent either extreme. In contrast, median would filter out extreme results.

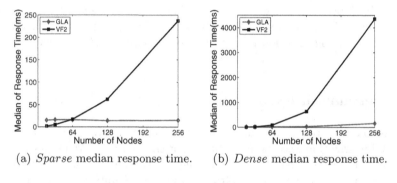

(a) *Sparse* median response time.   (b) *Dense* median response time.

Fig. 6. Response time of GLA and VF2 on sparse and dense graphs

Figure 6(a) and (b) show the results for *Sparse* and *Dense* respectively. The X-axis is the number of nodes (in log scale) and Y-axis is the median response time in milliseconds. Note the two graphs have different scales, and number of edges is linear with the number of nodes for sparse graphs and quadratic for dense graphs. In Figure 6(a), the median running time of GLA remains more or less constant to 10 milliseconds, despite the increase in graph size. As a result, when $v = 256$, GLA outperforms VF2 by an order of magnitude. In Figure 6(b), we observe a consistent trend, except that the performance gap is larger. In particular, for $N = 256$, GLA is faster by two orders of magnitude. These figures show that GLA has low response time, shorter than VF2, by effectively pruning the notoriously large search space, guided by linearization rules leveraging node degree and exploration history.

7   Conclusions

This paper presents a novel approach to solve graph isomorphism. The key idea is to linearize one graph into a parameterized sequence — a walk that covers every node and edge — and parameterized-match the linearization on the second graph. We develop a fast linearization algorithm that produces a short linearization, and a parameterized matching algorithm. We implement the algorithms and evaluate them experimentally against VF2, and observe lower response times for sparse and dense graphs with varying sizes.

References

1. Amir, A., Aumann, Y., Cole, R., Lewenstein, M., Porat, E.: Function matching: Algorithms, applications, and a lower bound. In: Proc. 30th International Colloquium on Automata, Languages and Programming (2003)
2. Amir, A., Farach, M., Muthukrishnan, S.: Alphabet dependence in parameterized matching. Information Processing Letters 49(3), 111–115 (1994)
3. Apostolico, A., Giancarlo, R.: Periodicity and repetitions in parameterized strings. Discrete Applied Mathematics 156(9), 1389–1398 (2008)
4. Baker, B.: A theory of parameterized pattern matching: algorithms and applications. In: Proc. 25th Annual Symposium on Theory of Computing (1993)
5. Baker, B.: Parameterized pattern matching by Boyer-Moore-type algorithms. In: Proceedings of the Sixth Annual ACM-SIAM Symposium on Discrete Algorithms, p. 550. Society for Industrial and Applied Mathematics (1995)
6. Baker, B.: Parameterized pattern matching: algorithms and applications. J. Comput. Syst. Sci. 52(1), 28–42 (1996)
7. Baker, B.: Parameterized duplication in strings: algorithms and an application to softwaremaintenance. SIAM Journal on Computing 26(5), 1343–1362 (1997)
8. Conte, D., Foggia, P., Sansone, C., Vento, M.: Thirty years of graph matching in pattern recognition. International Journal of Pattern Recognition and Artificial Intelligence 18(03), 265–298 (2004)
9. Cordella, L.P., Foggia, P., Sansone, C., Vento, M.: A (sub) graph isomorphism algorithm for matching large graphs. IEEE Transactions on Pattern Analysis and Machine Intelligence 26(10), 1367–1372 (2004)
10. Du Mouza, C., Rigaux, P., Scholl, M.: Parameterized pattern queries. Data & Knowledge Engineering 63(2), 433–456 (2007)
11. Edmonds, J., Johnson, E.L.: Matching, euler tours and the chinese postman. Mathematical Programming 5(1), 88–124 (1973)
12. Fredriksson, K., Mozgovoy, M.: Efficient parameterized string matching. Information Processing Letters 100(3), 91–96 (2006)
13. Hazay, C.: Parameterized matching. Master's thesis, Bar-Ilan University (2004)
14. Hazay, C., Lewenstein, M., Sokol, D.: Approximate parameterized matching. ACM Transactions on Algorithms 3(3), 29 (2007)
15. Hazay, C., Lewenstein, M., Tsur, D.: Two dimensional parameterized matching. In: CPM, pp. 266–279 (2005)
16. Kosaraju, S.: Faster algorithms for the construction of parameterized suffix trees. In: Proceedings of the 36th Annual Symposium on Foundations of Computer Science. IEEE Computer Society Press, Washington, DC (1995)
17. Lee, I., Mendivelso, J., Pinzón, Y.J.: $\delta\gamma$ – parameterized matching. In: Amir, A., Turpin, A., Moffat, A. (eds.) SPIRE 2008. LNCS, vol. 5280, pp. 236–248. Springer, Heidelberg (2008)

18. Lewenstein, M.: Parameterized matching. In: Encyclopedia of Algorithms. Springer (2008)
19. McKay, B.D.: Practical graph isomorphism. Congressus Numerantium 30, 45 (1981)
20. Mendivelso, J., Lee, I., Pinzón, Y.J.: Approximate function matching under δ- and γ- distances. In: Calderón-Benavides, L., González-Caro, C., Chávez, E., Ziviani, N. (eds.) SPIRE 2012. LNCS, vol. 7608, pp. 348–359. Springer, Heidelberg (2012)
21. Salmela, L., Tarhio, J.: Sublinear algorithms for parameterized matching. In: Proc. 17th Annual Symposium on Combinatorial Pattern Matching (2006)
22. Ullmann, J.R.: An algorithm for subgraph isomorphism. Journal of the ACM (JACM) 23(1), 31–42 (1976)

Suffix Array of Alignment: A Practical Index for Similar Data

Joong Chae Na[1], Heejin Park[2], Sunho Lee[3], Minsung Hong[3],
Thierry Lecroq[4], Laurent Mouchard[4], and Kunsoo Park[3,*]

[1] Department of Computer Science and Engineering, Sejong University, Korea
jcna@sejong.ac.kr
[2] College of Information and Communications, Hanyang University, Korea
hjpark@hanyang.ac.kr
[3] School of Computer Science and Engineering, Seoul National University, Korea
{slee,mshong,kpark}@theory.snu.ac.kr
[4] Department of Computer Science, University of Rouen, France
{Thierry.Lecroq,Laurent.Mouchard}@univ-rouen.fr

Abstract. The *suffix tree of alignment* is an index data structure for similar strings. Given an alignment of similar strings, it stores all suffixes of the alignment, called *alignment-suffixes*. An alignment-suffix represents one suffix of a string or suffixes of multiple strings starting at the same position in the alignment. The suffix tree of alignment makes good use of similarity in strings theoretically. However, suffix trees are not widely used in biological applications because of their huge space requirements, and instead suffix arrays are used in practice.

In this paper we propose a space-economical version of the suffix tree of alignment, named the *suffix array of alignment (SAA)*. Given an alignment ρ of similar strings, the SAA for ρ is a lexicographically sorted list of all the alignment-suffixes of ρ. The SAA supports pattern search as efficiently as the *generalized suffix array*. Our experiments show that our index uses only 14% of the space used by the generalized suffix array to index 11 human genome sequences. The space efficiency of our index increases as the number of the genome sequences increases. We also present an efficient algorithm for constructing the SAA.

Keywords: Indexes for similar data, suffix arrays, alignments.

1 Introduction

The 1000 Genomes project [4] is aiming at building a database of 1092 individual human genome sequences using a cheap and fast sequencing, called Next Generation Sequencing (NGS). To sequence an individual genome using the NGS, the individual genome is divided into short segments (called reads) and they are aligned to the Human reference Genome. This is possible because an individual genome is more than 99% identical to the Human reference Genome. The similarity also enables us to store individual genomes efficiently. Instead of storing

---

* Corresponding author.

O. Kurland, M. Lewenstein, and E. Porat (Eds.): SPIRE 2013, LNCS 8214, pp. 243–254, 2013.

1000 whole individual sequences, only 1% different regions of each individual genome can be stored.

Not only efficient storing techniques but also efficient indexing techniques for similar strings have been developed. The first such index was proposed by Mäkinen et al. [16,17]. Their index uses run-length encoding, a suffix array, and BWT [3]. Huang et al. [9] indexed similar strings by building separate data structures for common regions and non-common regions. In addition, indexes based on Lemple-Ziv compression schemes [15,21] have been developed [5,14]. Some of these indexes are surveyed in [20]. The space reductions of these indexes are achieved mostly by using classical compressed indexes. However, the indexes do not support efficient pattern search or require auxiliary data structures to improve the pattern search time.

Recently, a suffix tree for similar strings, called a *suffix tree of alignment* [19], have been proposed without sacrificing the pattern search time, i.e., the suffix tree of alignment supports linear-time pattern search. Given an alignment of similar strings, the suffix tree of alignment stores suffixes of an alignment, called *alignment-suffixes* (for short a-suffixes) rather than suffixes of a string. An a-suffix may represent suffixes of multiple strings starting at the same position in an alignment. The suffix tree of alignment makes good use of similarity in strings theoretically. Although suffix trees support many functionalities [2,8], however, they are not widely used in biological applications because of the huge space requirement. Instead, suffix arrays [18] (including their compressed forms [6,7]) are widely used in practice.

In this paper we propose the *suffix array of alignment (SAA)*, an array version of the suffix tree of alignment. Given an alignment ρ, the SAA for ρ is a lexicographically sorted list of all the a-suffixes of ρ. We show that the sorted order of the a-suffixes is well defined and the longest common prefix (lcp) of two a-suffixes is also well defined. Assume that given strings consist of common regions and non-common regions alternatively, e.g., three strings A, B, and C can be represented as $A = \alpha_1\beta_1\ldots\alpha_k\beta_k\alpha_{k+1}$, $B = \alpha_1\delta_1\ldots\alpha_k\delta_k\alpha_{k+1}$, and $C = \alpha_1\vartheta_1\ldots\alpha_k\vartheta_k\alpha_{k+1}$, where α_i's are common regions and β_i's, δ_i's and ϑ_i's are non-common regions. Then, the SAA requires $O(|A|+\sum_{i=1}^{k}(2|\alpha_i^*|+|\delta_i|+|\vartheta_i|))$ space, where α_i^* is the longest suffix of α_i appearing at least twice in A, in B or in C. (For simplicity, three strings are considered but our results work well for more than three strings.) The space requirement of the SAA is asymptotically the same as that of the suffix tree of alignment, but the SAA is more space-efficient practically. Furthermore, our suffix array supports pattern search as efficiently as the generalized suffix array (GSA).

Moreover, we show by experiments that our index is space-efficient for similar data in practice by analyzing and comparing the space requirements of the SAA and the GSA, which support the same efficiency of pattern search. The space requirement of our index is influenced by the lengths of α_i^* and non-common regions. Our experiments show that these lengths are short in practice and thus our index consumes very small space. We used 11 human genome sequences, one reference sequence and 10 individual sequences from the 1000 Genomes project website[1]. In the genome sequences, non-common regions are only 0.3% of the entire positions, i.e., these sequences are very similar. Moreover, the α_i^*'s, which is a main factor for the space requirement of the SAA, occupy 5% of the entire positions and the length of α_i^* is 16.64 on average. Conclusively, the SAA requires only 14% of the space required by the GSA for indexing the 11 sequences. It should be noted that the space efficiency of our index increases as the number of the genome sequences increases.

We also present an efficient algorithm for constructing the SAA. One might think the SAA can be simply constructed by simulating the algorithm for constructing the suffix tree of alignment in [19]. However, it is not easy because the algorithm heavily uses the dynamic property of the suffix tree and makes use of suffix links. The core of the tree construction algorithm is how to compute α_i^* efficiently, which is solved using a property satisfied in a partial suffix tree containing suffixes derived from several strings. Thus, we developed a new algorithm to compute α_i^* using only suffix arrays. For this, we generalize the property dedicated to the suffix tree so that the property is satisfied in substrings of input strings. Conclusively, we can compute α_i^* and thus construct the SAA as efficiently as the algorithm in [19].

2   Suffix Array of Alignment (SAA)

In this section we define the suffix array of alignment (SAA) and present how to construct the SAA. For simplicity, we consider only alignments of three strings but our definitions and algorithms can be easily extended to more than three strings. We first consider alignments with one non-common chunk and then general alignments with more than one non-common chunk.

2.1   Definition of SAA

Let A, B, and C be similar strings such that $A = \alpha\beta\gamma$, $B = \alpha\delta\gamma$, and $C = \alpha\vartheta\gamma$, where α and γ are common regions in all strings, and β, δ, and ϑ are non-common regions. Then, these regions represent an alignment of the strings and each string can be transformed to another string by replacing non-common regions. We denote this alignment of the three strings by $\rho = \alpha(\beta/\delta/\vartheta)\gamma$. For simplicity, we assume that all strings end with a special symbol $\# \in \Sigma$ occurring nowhere else in the strings.

The suffixes of the alignment ρ, called *alignment-suffixes* (for short *a-suffixes*), are defined as in [19]. Let α^a, α^b, and α^c be the longest suffixes of α occurring at least twice in the strings A, B, and C, respectively. Let α^* be the longest of α^a, α^b, and α^c, i.e., α^* is the longest suffix of α occurring at least twice in A, in B, or in C. Then, these are a-suffixes of ρ, which are classified into 5 types.

1. a suffix of γ,
2. $\omega^a\gamma$, where ω^a is a (non-empty) suffix of $\alpha^*\beta$.

---

[1] http://www.1000genomes.org/

idx	POS	LCP	a-suffixes (type)
1	$(1, 9)$	-	# (1)
2	$(2, 6)$	0	a̲ab# (3)
3	$(1, 4)$	3	a̲a̲b̲ab# (2)
4	$(3, 4)$	2	a̲a̲c̲ab# (4)
5	$(1, 7)$	1	a̲b# (1)
6	$(2, 4)$	2	a̲b̲aab# (3)
7	$(1, 5)$	3	a̲b̲ab# (2)
8	$(1, 1)$	2	a̲b̲ca(ab/ba/ac)ab# (5)
9	$(3, 5)$	1	a̲c̲ab# (4)
⋮	⋮	⋮	⋮

Fig. 1. The SAA of abca(a̲b/ba/ac)ab#. A pair (a, b) in POS represents the string number a and the starting position b of an a-suffix. $LCP[i]$ is the length of lcp between two a-suffixes of $POS[i-1]$ and $POS[i]$.

3. $\omega^b\gamma$, where ω^b is a (non-empty) suffix of $\alpha^*\delta$.
4. $\omega^c\gamma$, where ω^c is a (non-empty) suffix of $\alpha^*\vartheta$.
5. $\alpha'(\beta/\delta/\vartheta)\gamma$, where α' is a suffix of α longer than α^*.

For example, assume that an alignment abca(a̲b/ba/ac)ab# is given. Then, $\alpha^a = \alpha^b = $ a and $\alpha^c = $ ca. Since α^* is ca, caa̲b̲ab# is an a-suffix of type 2 and bca(a̲b/ba/ac)ab# is an a-suffix of type 5.

The *suffix array of alignment (SAA)* for ρ is a lexicographically sorted list of all the a-suffixes of ρ. It is clear what the sorted order for a-suffixes of types 1-4 is since an a-suffix of types 1-4 represents one string. On the other hand, an a-suffix of type 5, e.g., $\omega = \alpha'(\beta/\delta/\vartheta)\gamma$ where $|\alpha'| > |\alpha^*|$ represents three strings $\alpha'\beta\gamma$, $\alpha'\delta\gamma$, and $\alpha'\vartheta\gamma$ derived from A, B, and C, respectively. However, it does not cause trouble when determining the order of ω between the a-suffixes of ρ. Since α' occurs only once in each string, i.e, as prefix of $\alpha'\beta\gamma$, $\alpha'\delta\gamma$, and $\alpha'\vartheta\gamma$, the order of ω is determined by α'. Thus, the lexicographically sorted order between the a-suffixes is well defined and the longest common prefix (lcp), an additional information often used together with the suffix arrays, between a-suffixes of ρ is also well defined. See Figure 1 for an example.

The space requirement of the SAA is linear to the number of a-suffixes. There are $|\gamma|$ a-suffixes of type 1. The number of a-suffixes of types 2, 3, and 4 is $|\alpha^*\beta| + |\alpha^*\delta| + |\alpha^*\vartheta|)$ and the number of a-suffixes of type 5 is $|\alpha| - |\alpha^*|$. Since $|A| = |\alpha| + |\beta| + |\gamma|$, the SAA of ρ requires $O(|A| + 2|\alpha^*| + |\delta| + |\vartheta|)$ space.

2.2 Construction of SAA

One method for constructing the SAA of ρ is using the suffix tree of the alignment ρ [19] as an intermediate index. However, this method does not make full use of the space-efficiency of suffix arrays because suffix trees require much more space than suffix arrays. Another method, without constructing suffix trees, is constructing first the generalized suffix array (GSA) for the three strings as an intermediate index and then deleting suffixes that are not a-suffixes in the GSA. However, this method also is not efficient in working space as well as in construction time because the time and space requirement of the GSA is proportional to the total length of the strings regardless of similarity among the strings. The more number of strings are in the alignment and the more similar the strings are, the more is the inefficiency.

We present how to construct the SAA efficiently in time and space. Our algorithm for constructing the SAA of ρ consists of three steps. Let γ^a, γ^b, and γ^c be the longest prefixes of γ occurring at least twice in the strings A, B, and C, respectively. Let γ^* be the longest of γ^a, γ^b, and γ^c. (Note that these definitions are symmetrical with those of α^a, α^b, α^c, and α^*, and are different from the definition of $\hat{\gamma}$ used in [19].) Then, the outline of our algorithm is as follows:

1. Compute $|\alpha^*|$ and $|\gamma^*|$.
2. Construct the GSA for three strings A, $\alpha^*\delta\gamma^*d$, and $\alpha^*\vartheta\gamma^*d$, where d is the symbol following γ^* in γ.
3. Delete suffixes of γ^*d derived from $\alpha^*\delta\gamma^*d$ and $\alpha^*\vartheta\gamma^*d$.

Step 1 is the core step of our algorithm. We mainly focus on the problem of computing $|\alpha^*|$ since $|\gamma^*|$ can be computed symmetrically. For a string S, let S^R be the reversed string of S. We can compute $|\alpha^a|$ by searching for α^R in the suffix array of A^R. Thus, one method to compute $|\alpha^*|$ is constructing the suffix array of each reversed string and computing $|\alpha^a|$, $|\alpha^b|$, and $|\alpha^c|$. However, this method requires the time proportional to the total length of the three strings due to constructing the three suffix arrays.

To compute $|\alpha^*|$ more efficiently, we make use of the similarity in the strings. Consider the strings A and B. The following lemma says that, given $|\alpha^a|$, a substring including δ is sufficient for computing $\max(|\alpha^a|, |\alpha^b|)$ instead of the entire of B. (Note that we do not need to compute the exact value of $|\alpha^b|$ to compute $|\alpha^*|$.)

Lemma 1. *If $|\alpha^b| > |\alpha^a|$, α^b occurs in the substring B' of B, where $B' = \alpha^a\delta\gamma^a$.*

Proof. By definition of α^b, there are at least two occurrences of α^b in B. Obviously, one occurrence occ_1 of α^b appears as a suffix of α. Since $|\alpha^b| > |\alpha^a|$, occ_1 cannot be included in B'. Let occ_2 denote an occurrence of α^b in B other than occ_1. Let s_2 and e_2 be the starting and the ending positions of occ_2 in B, respectively. Let s_a and e_a be the starting and the ending positions of substring B' in B, respectively.

We show that occ_2 is included in B', i.e., $s_a \leq s_2$ and $e_2 \leq e_a$. We first prove by contradiction that $s_a \leq s_2$. Suppose $s_2 < s_a$. We have two cases according to whether occ_2 is overlapped with δ or not.

- The case when occ_2 is not overlapped with δ. Then, occ_2 is included in α and it means that there are at least two occurrences (occ_1 and occ_2) of α^b in α and also in A. It contradicts with the definition of α^a since $|\alpha^b| > |\alpha^a|$.
- The case when occ_2 is overlapped with δ. Let α' be the suffix of α starting at s_2. Since $s_2 < s_a$, $|\alpha'| > |\alpha^a|$. Since α' is a prefix of occ_2, α' is also a prefix

of occ_1. Hence, there are at least two occurrences of α' in α and also in A. It contradicts with the definition of α^a since $|\alpha'| > |\alpha^a|$.

Similarly, we can prove that $e_2 \leq e_a$ by contradiction with the definition of γ^a.   □

Note that this property also holds for other strings. For example, if $|\alpha^c| > |\alpha^a|$, α^c occurs in the substring $\alpha^a \vartheta \gamma^a$ of C.

Using this property, we can compute $|\alpha^*|$ and $|\gamma^*|$ as follows:

1.1 Compute $|\alpha^a|$ by searching for α^R in the suffix array of A^R and, symmetrically, compute $|\gamma^a|$ by searching for γ in the suffix array of A.
1.2 Compute $\ell^b = \max(|\alpha^a|, |\alpha^b|)$ using the suffix array of $(\alpha^a \delta \gamma^a)^R$ as follows. Let α' be the longest suffix of α occurring in $\alpha^a \delta \gamma^a$. (Note that $|\alpha'| \geq |\alpha^a|$ since α^a occurs in $\alpha^a \delta \gamma^a$.) We can find α' by searching for α^R in the suffix array of $(\alpha^a \delta \gamma^a)^R$. By Lemma 1, if $|\alpha'| > |\alpha^a|$, $\ell^b = |\alpha'|$ and, otherwise, $\ell^b = |\alpha^a|$.
    Symmetrically, compute $\max(|\gamma^a|, |\gamma^b|)$ using the suffix array of $\alpha^a \delta \gamma^a$.
1.3 Similarly, compute $\ell^c = \max(|\alpha^a|, |\alpha^c|)$ using the suffix array of $(\alpha^a \vartheta \gamma^a)^R$. Then, $|\alpha^*| = \max(\ell^b, \ell^c)$.
    Symmetrically, compute $\max(|\gamma^a|, |\gamma^c|)$ and $|\gamma^*|$.

Since the suffix array of a string S can be constructed using $O(|S|)$ time and $O(|S|)$ space [10,12,13], and one can search for a string P using the suffix array of S with some auxiliary information in $O(|P|)$ [1,11], computing $|\alpha^*|$ and $|\gamma^*|$ requires $O(|A| + |\alpha^* \delta \gamma^*| + |\alpha^* \vartheta \gamma^*|)$ time and $O(|A|)$ working space. Note that the suffix array constructed in each substep is needed only in the substep.

In Step 2, we construct the GSA for three strings A, $\alpha^* \delta \gamma^* d$, and $\alpha^* \vartheta \gamma^* d$, where d is the symbol following γ^* in γ. The GSA contains all the a-suffixes of the alignment ρ. (Note that the suffixes of γ in A are the a-suffixes of type 1 of ρ and the suffixes of A longer than $\alpha^* \beta \gamma$ can be *implicitly* converted to the a-suffixes of type 5 of ρ [19].) The reason why $\gamma^* d$ is necessary is as follows. Let $\omega \gamma$ be an a-suffix of type 3 (of B). To determine the order of $\omega \gamma$ among a-suffixes of ρ, we may need a prefix of γ. Since $\gamma^* d$ occurs only once in each string, the order of $\omega \gamma$ is determined by $\omega \gamma^* d$. Obviously, Step 2 requires $O(|A| + |\alpha^* \delta \gamma^*| + |\alpha^* \vartheta \gamma^*|)$ time and space.

In Step 3, we delete suffixes of $\gamma^* d$ in $\alpha^* \delta \gamma^* d$ and $\alpha^* \vartheta \gamma^* d$ because these are redundant with suffixes of A (a-suffixes of type 1). Consider a suffix ω of $\gamma^* d$. In the GSA, there are two ω's derived from $\alpha^* \delta \gamma^* d$ and $\alpha^* \vartheta \gamma^* d$. The two ω's are adjacent in the GSA. We can delete redundant suffixes by scanning the entire GSA, which requires $O(|A| + |\alpha^* \delta \gamma^*| + |\alpha^* \vartheta \gamma^*|)$ time and space.

Theorem 1. *Given an alignment* $\rho = \alpha(\beta/\delta/\vartheta)\gamma$, *the SAA of* ρ *can be constructed using* $O(|A| + |\alpha^* \delta \gamma^*| + |\alpha^* \vartheta \gamma^*|)$ *time and working space.*

2.3   Alignment with Multiple Regions

In this section we consider alignments with multiple non-common regions. Let $A = \alpha_1 \beta_1 \ldots \alpha_k \beta_k \alpha_{k+1}$, $B = \alpha_1 \delta_1 \ldots \alpha_k \delta_k \alpha_{k+1}$, and $C =$

$\alpha_1 \vartheta_1 \ldots \alpha_k \vartheta_k \alpha_{k+1}$. We denote the alignment of the strings by $\rho = \alpha_1(\beta_1/\delta_1/\vartheta_1)\alpha_2(\beta_2/\delta_2/\vartheta_2)\alpha_3 \ldots \alpha_{k+1}$. Without loss of generality, we assume that α_i ($2 \leq i \leq k$) occurs only once in each string. (Otherwise, we merge α_i with adjacent non-common regions, e.g., $\beta_{i-1}\alpha_i\beta_i$ is regarded as one non-common region). For $1 \leq i \leq k$, let α_i^a, α_i^b, and α_i^c be the longest suffixes of α_i occurring at least twice in the strings A, B, and C, respectively. Let α_i^* be the longest of α_i^a, α_i^b, and α_i^c. Then, these are a-suffixes of ρ, which are classified into 5 types ($1 \leq i \leq k$).

1. a suffix of α_{k+1},
2. $\omega_i^a \alpha_{i+1} \ldots \alpha_{k+1}$ where ω_i^a is a (non-empty) suffix of $\alpha_i^* \beta_i$.
3. $\omega_i^b \alpha_{i+1} \ldots \alpha_{k+1}$ where ω_i^b is a (non-empty) suffix of $\alpha_i^* \delta_i$.
4. $\omega_i^c \alpha_{i+1} \ldots \alpha_{k+1}$ where ω_i^c is a (non-empty) suffix of $\alpha_i^* \vartheta_i$.
5. $\alpha_i'(\beta_i/\delta_i/\vartheta_i)\alpha_{i+1} \ldots \alpha_{k+1}$, where α_i' is a suffix of α_i longer than α_i^*.

The SAA of ρ is a lexicographically sorted list of all the a-suffixes of ρ. The SAA requires the space linear to the number of a-suffixes, i.e., $O(|A| + \sum_{i=1}^{k}(2|\alpha_i^*| + |\delta_i| + |\vartheta_i|))$ space.

For $1 \leq i \leq k$, let γ_i^a, γ_i^b, and γ_i^c be the longest prefix of α_{i+1} occurring at least twice in the strings A, B, and C, respectively, and let γ_i^* be the longest of γ_i^a, γ_i^b, and γ_i^c. Let B' be the concatenation of the k strings $\alpha_i^a \delta_i \gamma_i^a \#_i$ ($1 \leq i \leq k$) and C' be the concatenation of the k strings $\alpha_i^a \vartheta_i \gamma_i^a \#_i$ ($1 \leq i \leq k$), where $\#_i$ is a delimiter. That is,

$$B' = \alpha_1^a \delta_1 \gamma_1^a \#_1 \alpha_2^a \delta_2 \gamma_2^a \#_2 \ldots \alpha_k^a \delta_k \gamma_k^a \#_k \text{ and}$$

$$C' = \alpha_1^a \vartheta_1 \gamma_1^a \#_1 \alpha_2^a \vartheta_2 \gamma_2^a \#_2 \ldots \alpha_k^a \vartheta_k \gamma_k^a \#_k.$$

Then, Lemma 1 can be generalized to the following lemma (we omit the proof).

Lemma 2. *For every $i = 1, \ldots, k$, if $|\alpha_i^b| > |\alpha_i^a|$, α_i^b occurs in B'.*

The SAA of ρ can be constructed as follows:

1. Compute $|\alpha_i^*|$ and $|\gamma_i^*|$ ($1 \leq i \leq k$).
   1.1 Compute $|\alpha_i^a|$, for every $i = 1, \ldots, k$, by searching for α_i^R in the suffix array of A^R and, symmetrically, compute $|\gamma_i^a|$ by searching for γ_i in the suffix array of A.
   1.2 Compute $\ell_i^b = \max(|\alpha_i^a|, |\alpha_i^b|)$ using the suffix array of $(B')^R$ as follows. Let α_i' be the longest suffix of α_i occurring in B'. We can find α_i' by searching for $(\alpha_i)^R$ in the suffix array of $(B')^R$. By Lemma 2, if $|\alpha_i'| > |\alpha_i^a|$, $\ell_i^b = |\alpha_i'|$ and, otherwise, $\ell_i^b = |\alpha_i^a|$.
   Symmetrically, compute $\max(|\gamma_i^a|, |\gamma_i^b|)$ using the suffix array of B'.
   1.3 Similarly, compute $\ell_i^c = \max(|\alpha_i^a|, |\alpha_i^c|)$ using the suffix array of $(C')^R$. Then, $|\alpha_i^*| = \max(\ell_i^b, \ell_i^c)$.
   Symmetrically, compute $\max(|\gamma_i^a|, |\gamma_i^c|)$ and $|\gamma_i^*|$.
2. Construct the GSA for $2k + 1$ strings A, $\alpha_i^* \delta_i \gamma_i^* d_i$, and $\alpha_i^* \vartheta_i \gamma_i^* d_i$, where d_i is the symbol following γ_i^* in α_{i+1} ($1 \leq i \leq k$).
3. Delete suffixes of $\gamma_i^* d_i$ derived from $\alpha_i^* \delta_i \gamma_i^* d_i$ and $\alpha_i^* \vartheta_i \gamma_i^* d_i$ ($1 \leq i \leq k$) in the GSA.

2.4   Pattern Search

We can perform pattern search using the SAA in the same way as using classical suffix arrays of strings except for dealing with alignments in a-suffixes. Consider an a-suffix $\omega(\beta_i/\delta_i/\vartheta_i)\ldots\alpha_{k+1}$ where ω does not contain an alignment. We can perform binary search with lcp information like in classical suffix arrays until a prefix of a given pattern P matches ω. If a prefix of P matches ω, we consider only the a-suffix to search for P since ω occurs only once in each string by definition of a-suffixes. Thus, after a prefix of P matches ω, we compare P with β_i, δ_i, and ϑ_i. We can enhance this comparison using the trie of β_i, δ_i, and ϑ_i.

3   Experiments

We show by experiments that our index (the SAA) is an effective data structure for similar data. The SAA requires only about 1/7 of the space required by the GSA to index 11 human genome sequences, which is explained in the following.

3.1   Experimental Data

To measure the space requirement of indexes in practice, we used one reference sequence and 10 individual sequences from 1000 Genomes project website. From the project website, we downloaded pairs of bam and bai files of 10 individual human genomes, where bam files contain *reads* (short segments of length 90-125) of each individual and bai files contain alignment of the reads. We also downloaded their corresponding reference genome, hg19. To convert a set of reads into one sequence, we used samtools[2] (Sequence Alignment/Map tools), by which we obtained 10 individual genome sequences. Since individual genome sequences are aligned to the reference genome sequence, these 11 sequences make a multiple alignment based on the reference genome sequence.

In our experiments, we used chromosome 20 of each genome. The length of the reference sequence is 63,025,520 and the lengths of the individual sequences vary from 62,965,442 to 62,965,512, which are a little shorter than the reference sequence. The sequences consist of five characters $\{A, G, T, C, N\}$, where A, G, T, and C stand for nucleotides Adenine, Guanine, Thymine, and Cytosine, respectively, and N appears in some special cases and is treated exceptionally in general (also in our experiments). In the reference sequence, N's do not appear alone but as chunks of N's. There are six chunks of N's in the reference and their lengths are 60,000, 3,100,000, 150,000, 50,000, 50,000 and 50,000. In the positions where the reference sequence has N's, individual sequences also have N's mostly. In the other positions, most of N's in individual sequences are single N's. The chunks of N's in individual sequences may represent the regions that are not sequenced, the regions that are sequenced but have very low quality, or the regions that are moved to other places. Single N's in individual sequences represent positions where one character from $\{A, G, T, C\}$ cannot be determined

---

[2] http://samtools.sourceforge.net/

Table 1. The number of non-common regions according to length

Length	1	2	3	4	5	6	Total
Number	190,804	3,057	215	47	9	3	194,135

Table 2. The lengths of α^*'s and α^j's ($0 \leq j \leq 10$)

	Total length	Average length		Total length	Average length
α^0's	3,202,864	16.50	α^6's	2,987,607	15.39
α^1's	3,030,406	15.61	α^7's	3,022,359	15.57
α^2's	2,558,396	13.18	α^8's	3,132,487	16.14
α^3's	2,976,231	15.33	α^9's	3,026,544	15.59
α^4's	2,989,375	15.40	α^{10}'s	3,140,456	16.18
α^5's	2,991,517	15.14	α^*'s	3,229,589	16.64

because reads have different characters in $\{A, G, T, C\}$, there are deletions in reads, and/or the quality is low.

3.2 Experimental Results

In this section, we compare the space requirements of the GSA and the SAA for the 11 sequences. For simplicity, we appended N's to the end of each individual sequence so that the length of the individual sequence is the same as that of the reference sequence. Let S_0 be the reference sequence and S_i ($1 \leq i \leq 10$) be each individual sequence. We call an aligned position a *non-common position* if at least two distinct characters in $\{A, G, T, C\}$ appear at this position. Notice that we do not regard a position as a non-common position if N and only one character in $\{A, G, T, C\}$ appear at the position. In our data set, there are 0.3% non-common positions (197,814 among 63,025,520 positions). Consecutive non-common positions become a *non-common region*. There are 194,135 non-common regions whose lengths vary from 1 to 6 (Table 1).

We first compute the lengths of α^*'s in the sequences, which is a main factor for the space requirement of our index. For a common region α_i, we denote by α_i^j the longest suffix of α_i appearing at least twice in sequence S_j ($0 \leq j \leq 10$). Recall that α_i^* is the longest of α_i^0, ..., α_i^{10}. When computing α_i^j, we exclude the part of α_i containing at least 10 consecutive N's since long consecutive N's do not carry any information about $\{A, G, T, C\}$. For each j, the total length and the average length of α_i^j's are shown in Table 2. (We omit the subscript i in α_i^j if not confusing.) For example, in sequence S_1, 3,030,406 characters (4.8%) of the entire 63,025,520 characters are included in α^1's. Since there are 194,135 non-common regions, the average length of α^1's is 15.61.

From the lengths of non-common regions and α^*'s, we calculate the space requirement of the SAA. For a substring $\alpha\beta$ of a sequence S_j where α is a common region and β is a non-common region, we call $\alpha^*\beta$ a *NS-region* (non-shared region) and α' a *S-region* (shared region) where α' is the prefix of α such that $\alpha'\alpha^*$ is α. For a sequence, let n_t be the length of the sequence, n_s be the

Table 3. Distribution of characters in our sequences

	NS-regions (3,427,403 characters)		S-regions (59,598,117 characters)	
	# of $\{A, G, T, C\}$	# of N	# of $\{A, G, T, C\}$	# of N
S_0	3,427,401	2	56,078,119	3,519,998
S_1	3,318,768	108,635	55,263,839	4,334,278
S_2	2,940,879	486,524	49,719,051	9,879,066
S_3	3,272,652	154,751	54,872,472	4,725,645
S_4	3,279,318	148,085	54,969,949	4,628,168
S_5	3,285,414	141,989	54,972,379	4,625,738
S_6	3,275,604	151,799	54,947,026	4,651,091
S_7	3,306,405	120,998	55,010,622	4,587,495
S_8	3,385,161	42,242	55,717,092	3,881,025
S_9	3,311,346	116,057	55,045,612	4,552,505
S_{10}	3,390,329	37,074	55,722,788	3,875,329
Total	36,193,277	1,508,156	602,318,949	53,260,338

total length of S-regions, and n_n be the total length of NS-regions. (Note that n_t, n_s, and n_n are identical in all sequences and $n_t = n_s + n_n$.) Then, the size of the GSA is $11n_t$ and the size of the SAA is $n_s + 11n_n$ ($= n_t + 10n_n$). In our data set, $n_t = 63,025,520$ and $n_n = 3,427,403$, and thus the size of the GSA is 693,280,720 words and the size of the SAA is 97,299,550 words. That is, our index uses only 14.03% space compared to the GSA.

When searching the sequences for a pattern, we may assume that the pattern does not contain N since we do not consider wild-card matches. In this circumstance, we can reduce the space requirement of indexes by eliminating in indexes the suffixes whose first characters are N. To compute the sizes of the two indexes for our data set, we first investigate the distribution of N in our sequences. Table 3 shows the distribution of characters in NS-regions and S-regions for each sequence. For example, in NS-regions of sequence S_1, the number of characters A, G, T, C is 3,318,768 (97%) and the number of character N is 108,635 (3%). In S-regions of sequence S_1, the number of characters A, G, T, C is 55,263,839 (93%) and the number of character N is 4,334,278 (7%).

We compute the sizes of the two indexes when excluding the suffixes whose first characters are N. The size of the GSA is the total number of characters A, G, T, C in NS-regions and S-regions of the 11 sequences, which is 638,512,226 ($36,193,277 + 602,318,949$) words (see Table 3). Next, consider the SAA. For a position in an NS-region, we eliminate the suffix of each sequence starting at this position if the first character of the suffix is N. For a position in an S-region, we eliminate the suffix (a-suffix) starting at this position only if the characters in the position are N in all sequences. In our data set, the total number of A, G, T, C in NS-regions is 36,193,277 and the number of positions in S-regions excluding the positions where characters are N in all sequences is 56,078,133 (see the last row in Table 4). Thus, the size of our index is 92,271,410 words, which is only 14.45% of the size of the GSA.

Table 4. Comparison of the sizes of the GSA and the SAA according to the number of sequences when excluding the suffixes whose first characters are N. Column (C1) is the total number of A, G, T, C in NS-regions and column (C2) is the number of positions in S-regions excluding the positions where characters are N in all sequences. Then, the size of the SAA is (C1) + (C2). The ratio of the size of the SAA to that of the GSA is given in the last column.

	Size of GSA	Size of SAA	(C1)	(C2)	Ratio (%)
$S_0 \sim S_1$	118,088,127	60,455,692	1,914,833	58,540,859	51.20
$S_0 \sim S_2$	170,748,057	62,473,223	4,553,672	57,919,551	36.59
$S_0 \sim S_3$	228,893,181	65,338,136	7,905,004	57,433,132	28.55
$S_0 \sim S_4$	287,142,448	68,758,063	11,706,204	57,051,859	23.95
$S_0 \sim S_5$	345,400,241	72,483,002	15,719,471	56,763,531	20.99
$S_0 \sim S_6$	403,622,871	76,417,395	19,881,933	56,535,462	18.93
$S_0 \sim S_7$	461,939,898	80,305,819	23,928,531	56,377,288	17.38
$S_0 \sim S_8$	521,042,151	84,226,410	27,963,745	56,262,665	16.16
$S_0 \sim S_9$	579,399,109	88,223,461	32,062,878	56,160,583	15.23
$S_0 \sim S_{10}$	638,512,226	92,271,410	36,193,277	56,078,133	14.45

We also compare the space requirements of the GSA and the SAA according to the number of sequences used in indexing (Table 4). Obviously, the space efficiency of our index increases as the number of the sequences increases. The ratio of the space of the SAA to that of the GSA is 51.2% when two sequence are used, and the ratio is 14.45% when the 11 sequences are used.

Acknowledgements. Joong Chae Na was supported by Basic Science Research Program through the National Research Foundation of Korea(NRF) funded by the Ministry of Education, Science and Technology(2012-0003214), and by the IT R&D program of MKE/KEIT [10038768, The Development of Supercomputing System for the Genome Analysis]. Heejin Park was supported by Basic Science Research Program through the National Research Foundation of Korea(NRF) funded by the Ministry of Education, Science and Technology(2012-0006999), by Seoul Creative Human Development Program (HM120006), by the Proteoge-nomics Research Program through the National Research Foundation of Korea funded by the Korean Ministry of Education, Science and Technology, and by the National Research Foundation of Korea(NRF) funded by the Ministry of Science, ICT & Future Planning(2012-054452). Laurent Mouchard was supported by the French Ministry of Foreign Affairs Grant 27828RG (INDIGEN, PHC STAR 2012). Kunsoo Park was supported by National Research Foundation of Korea-Grant funded by the Korean Government(MSIP) (2012K1A3A4A07030483), and by Next-Generation Information Computing Development Program through the National Research Foundation of Korea(NRF) funded by the Ministry of Science, ICT & Future Planning (2011-0029924).

References

1. Abouelhoda, M.I., Kurtz, S., Ohlebusch, E.: Replacing suffix trees with enhanced suffix arrays. Journal of Discrete Algorithms 2(1), 53–86 (2004)
2. Apostolico, A.: The myriad virtues of subword trees. In: Apostolico, A., Galil, Z. (eds.) Combinatorial Algorithms on Words, pp. 85–95. Springer (1985)
3. Burrows, M., Wheeler, D.J.: A block-sorting lossless data compression algorithm. Technical Report 124, Digital Equipment Corporation, Paolo Alto, California (1994)
4. The 1000 Genomes Project Consortium. A map of human genome variation from population-scale sequencing. Nature 467(7319), 1061–1073 (2010)
5. Do, H.H., Jansson, J., Sadakane, K., Sung, W.-K.: Fast relative lempel-ziv self-index for similar sequences. In: Snoeyink, J., Lu, P., Su, K., Wang, L. (eds.) AAIM 2012 and FAW 2012. LNCS, vol. 7285, pp. 291–302. Springer, Heidelberg (2012)
6. Ferragina, P., Manzini, G.: Indexing compressed text. Journal of the ACM 52(4), 552–581 (2005)
7. Grossi, R., Vitter, J.S.: Compressed suffix arrays and suffix trees with applications to text indexing and string matching. SIAM Journal on Computing 35(2), 378–407 (2005)
8. Gusfield, D.: Algorithms on Strings, Tree, and Sequences. Cambridge University Press, Cambridge (1997)
9. Huang, S., Lam, T.W., Sung, W.K., Tam, S.L., Yiu, S.M.: Indexing similar DNA sequences. In: Chen, B. (ed.) AAIM 2010. LNCS, vol. 6124, pp. 180–190. Springer, Heidelberg (2010)
10. Kärkkäinen, J., Sanders, P., Burkhardt, S.: Linear work suffix array construction. Journal of the ACM 53(6), 918–936 (2006)
11. Kim, D.K., Kim, M., Park, H.: Linearized suffix tree: an efficient index data structure with the capabilities of suffix trees and suffix arrays. Algorithmica 52(3), 350–377 (2008)
12. Kim, D.K., Sim, J.S., Park, H., Park, K.: Constructing suffix arrays in linear time. Journal of Discrete Algorithms 3(2-4), 126–142 (2005)
13. Ko, P., Aluru, S.: Space efficient linear time construction of suffix arrays. Journal of Discrete Algorithms 3(2-4), 143–156 (2005)
14. Kreft, S., Navarro, G.: On compressing and indexing repetitive sequences. Theoretical Computer Science 483, 115–133 (2013)
15. Kuruppu, S., Puglisi, S.J., Zobel, J.: Relative lempel-ziv compression of genomes for large-scale storage and retrieval. In: Chavez, E., Lonardi, S. (eds.) SPIRE 2010. LNCS, vol. 6393, pp. 201–206. Springer, Heidelberg (2010)
16. Mäkinen, V., Navarro, G., Sirén, J., Välimäki, N.: Storage and retrieval of individual genomes. In: Batzoglou, S. (ed.) RECOMB 2009. LNCS, vol. 5541, pp. 121–137. Springer, Heidelberg (2009)
17. Mäkinen, V., Navarro, G., Sirén, J., Välimäki, N.: Storage and retrieval of highly repetitive sequence collections. Journal of Computational Biology 17(3), 281–308 (2010)
18. Manber, U., Myers, G.: Suffix arrays: A new method for on-line string searches. SIAM Journal on Computing 22(5), 935–948 (1993)
19. Na, J.C., Crochemore, M., Park, H., Holub, J., Iliopoulos, C.S., Mouchard, L., Park, K.: Suffix tree of alignment: An efficient index for similar data. In: Proceedings of IWOCA 2013 (2013)
20. Navarro, G.: Indexing highly repetitive collections. In: Smyth, B. (ed.) IWOCA 2012. LNCS, vol. 7643, pp. 274–279. Springer, Heidelberg (2012)
21. Ziv, J., Lempel, A.: A universal algorithm for sequential data compression. IEEE Transactions on Information Theory 23(3), 337–343 (1977)

Faster Top-k Document Retrieval in Optimal Space*

Gonzalo Navarro[1] and Sharma V. Thankachan[2]

[1] Department of Computer Science, University of Chile, Chile
gnavarro@dcc.uchile.cl
[2] Department of Computer Science, Louisiana State University, USA
thanks@csc.lsu.edu

Abstract. We consider the problem of retrieving the k documents from a collection of strings where a given pattern P appears most often. We show that, by representing the collection using a Compressed Suffix Array CSA, a data structure using the asymptotically optimal $|\mathsf{CSA}| + o(n)$ bits can answer queries in the time needed by CSA to find the suffix array interval of the pattern plus $O(k \lg^2 k \lg^\epsilon n)$ accesses to suffix array cells, for any constant $\epsilon > 0$. This is $\lg n / \lg k$ times faster than the only previous solution using optimal space, $\lg k$ times slower than the fastest structure that uses twice the space, and $\lg^2 k \lg^\epsilon n$ times the lower-bound cost of obtaining k document identifiers from the CSA. To obtain the result we introduce a tool called the *sampled document array*, which can be of independent interest.

1 Introduction

The problem of *top-k document retrieval* is that of preprocessing a text collection so that, given a search pattern $P[1, m]$ and a threshold k, we retrieve the k documents most "relevant" to P, for some definition of relevance. This is the basic problem of search engines and forms the core of the Information Retrieval (IR) field [5].

The inverted index has been highly successful to solve those top-k queries in many IR scenarios. However, inverted indexes are bound to text collections that can be easily segmented into "words", so that only whole words can be queried, and the distinct words form a reasonably small set. Inverted indexes store, for each word, the list of the documents where it appears, with the associated relevance. Such a structure is not easily applicable in highly synthetic languages like Finnish or German, where long words are built from particles, and even less in languages where word separators are absent and can only be inferred from the meaning, like Chinese, Korean, etc. Out of resorting to complex segmentation heuristics, a simple solution for those cases is to treat the text as an uninterpreted sequence of symbols and look for any substring in those sequences. The model of a collection of documents (strings) where one can find those where a

---

* Funded in part by Fondecyt Grant 1-110066.

O. Kurland, M. Lewenstein, and E. Porat (Eds.): SPIRE 2013, LNCS 8214, pp. 255–262, 2013.

pattern string is relevant is also appealing in other applications like bioinformatics, chemoinformatics, software repositories, multimedia databases, and so on. Supporting document retrieval queries on those general string collections has proved much more challenging.

Sufix trees [24] and suffix arrays [13] are useful tools to search string collections. However, these structures solve the *pattern matching problem*: they can count or list all the occ individual occurrences of P in the collection. Obtaining the k most relevant documents from that set requires time proportional to occ, usually much much larger than k. Only relatively recently [12,8,11,18,22] was this problem solved satisfactorily, finally reaching the optimal time $O(m + k)$. Those solutions, like suffix trees, have the drawback of requiring $O(n \lg n)$ bits of space on a collection of length n, whereas the collection itself would require no more than $n \lg \sigma$ bits, if σ is the alphabet size. In practice these indexes require many times the text size, which renders them impractical on moderate and large text collections.

For the pattern matching problem, the space issue began to be solved in year 2000. Recent Compressed Suffix Arrays (CSAs) efficiently answer queries within space asymptotically equal not only to $n \lg \sigma$ bits, but to the size of the compressed text collection [17]. Moreover, those CSAs can retrieve any substring of any document and hence replace the collection: they can be regarded as compressors that support queries. We call their space |CSA|, which can be tought of as the minimum space in which the text collection can be represented.

A similar result for top-k document retrieval has been more elusive. In their seminal paper, Hon et al. [11] showed that, if the relevance is taken as the number of times P appears in the document (a popular choice in IR), the collection can be represented in $2|\mathsf{CSA}| + o(n)$ bits so that queries are solved in time $O(m \lg \lg \sigma + k \lg^{4+\epsilon} n)$, for any constant $\epsilon > 0$ (this complexity assumes that the CSA searches for P in time $O(m \lg \lg \sigma)$ and computes a cell of the suffix array or its inverse in time $O(\lg^{1+\epsilon} n)$; there exists such a CSA achieving high-order entropy compression of the text [1]). After several time improvements that still used $2|\mathsf{CSA}| + o(n)$ bits [6,3], Hon et al. [10] achieved the best time to date, $O(m \lg \lg \sigma + k \lg k \lg^{1+\epsilon} n)$. Finally, Tsur [23] reduced the space to the asymptotically optimal $|\mathsf{CSA}| + o(n)$ bits, yet with higher time, $O(m \lg \lg \sigma + k \lg k \lg^{2+\epsilon} n)$.

In this paper we (almost) obtain the best from both solutions. We maintain the space in the optimal $|\mathsf{CSA}| + o(n)$ bits, and obtain search time $O(m \lg \lg \sigma + k \lg^2 k \lg^{1+\epsilon} n)$, almost $\lg n$ times faster than the current space-optimal solution and only a $\lg k$ factor away from the fastest one (that uses twice the space). To obtain the result, we introduce a data structure called the *sampled document array*, which may have independent interest.

2   Compressed Top-k Retrieval Indexes

Consider a collection of D strings $\{T_1, T_2, \ldots, T_D\}$ over alphabet $[1, \sigma]$, called *documents*, concatenated into a text $T[1, n] = T_1 \$ T_2 \$ \ldots T_D \$$, where $\$ = 0$ is a special symbol. Consider the suffix tree [24] of T, the suffix array [13] $A[1,n]$ of T, and a Compressed Suffix Array [17] CSA that is able to (1) given a pattern $P[1,m]$, find the area $A[sp,ep]$ of suffixes starting with P in time $t_{\text{search}}(m)$, and (2) given a position i, compute $A[i]$ in time t_{SA}. For example, there is a CSA with $t_{\text{search}}(m) = O(m \lg \lg \sigma)$ and $t_{\text{SA}} = O(\lg^{1+\epsilon} n)$ for any constant $\epsilon > 0$ and using $|\text{CSA}| = nH_h(T)(1+o(1)) + o(n)$ bits of space [1], and another with $t_{\text{search}}(m) = O(m)$ and $t_{\text{SA}} = O(\lg n)$ using $|\text{CSA}| = nH_h(T)(1+o(1))+O(n)$ bits of space, where $H_h(T) \leq \lg \sigma$ is the per-symbol h-th order empirical entropy of T [14] (this is a lower bound on compressibility using any reasonable statistical model). In this paper we focus on the *top-k (most frequent documents) retrieval problem*: given a pattern $P[1,m]$, return the k documents where P appears most often. As explained, this is a reasonable relevance measure, especially when just one pattern is involved.

Each suffix tree leaf (or suffix array cell) can be associated to the document T_d where the corresponding suffix starts. We call $\text{tf}(v,d)$ the number of leaves associated to document d that descend from suffix tree node v (i.e., the number of times the string label of v appears in document d). Then the top-k retrieval problem can be solved by first finding the locus v of pattern P, and then retrieving the k documents d with highest $\text{tf}(v,d)$ values. Note that the problem could be solved by attaching the answer to any suffix tree node, but the space would be $O(kn \lg n)$ bits, and work only up to the chosen k value. Now we describe the solutions we build on to obtain our result.

Hon, Shah and Vitter's Solution. Hon et al.'s [11] structure is built (in principle) for a fixed k value. We choose a grouping factor $b = k \lg^{2+\epsilon} n$ and *mark* every bth leaf in the suffix tree (we use a slightly simplified description of their method [19]). Then we mark the lowest common ancestor (LCA) of every consecutive pair of marked leaves. The tree of marked nodes is called τ_k and has $O(n/b)$ nodes. For every marked suffix tree node v, we store the k pairs $(d, \text{tf}(v,d))$ with highest $\text{tf}(v,d)$. Hon et al. prove that any locus node v contains one maximal marked node u so that there are at most $2b$ leaves covered by v but not by u (we will denote $v \setminus u$ that leaf set). Therefore they traverse those leaves using the CSA, and for each one they (1) compute the corresponding document d, (2) compute the frequency $\text{tf}(v,d)$, (3) add d to the top-k list (or correct its frequency from $\text{tf}(u,d)$ to $\text{tf}(v,d)$ if d was already stored in the precomputed top-k list of u).

To carry out (1) on the ith suffix tree leaf, they first compute $A[i]$ in $O(t_{\text{SA}})$ time, and then convert it into a document number by storing a bitmap $B[1,n]$ that marks with a 1 the document beginnings in T [20]. So the document is $d = \text{rank}(B, A[i])$, where $\text{rank}(B,j)$ counts the number of 1s in $B[1,j]$. Since B has D 1s, it can be represented using $D \lg(n/D) + O(D) + o(n)$ bits, which is $o(n)$ if $D = o(n)$, and answer rank queries in constant time [21]. To carry out (2) they need additional $|\text{CSA}|$ bits (see Sadakane [20]), and time $O(t_{\text{SA}} \lg n)$. The node $u \in \tau_k$ is found using the CSA plus a constant-time LCA on τ_k for the leftmost and rightmost marked leaves in $[sp,ep]$, whereas the leaves covered by v are simply $[sp,ep]$. Thus the total query time is $O(t_{\text{search}}(m) +$

$b\,t_{\mathsf{SA}}\lg n) = O(t_{\mathsf{search}}(m) + k\,t_{\mathsf{SA}}\lg^{3+\epsilon} n)$. On the two CSAs we have mentioned, this is $O(t_{\mathsf{search}}(m) + k\lg^{4+\epsilon} n)$.

As storing the top-k list needs $O(k\lg n)$ bits, the space for τ_k is $O((n/b)k\lg n) = O(n/\lg^{1+\epsilon} n)$ bits. One τ_k tree is stored for each k power of 2, so that at query time we increase k to the next power of 2 and solve the query within the same time complexity. Summed over all the powers of 2, the space becomes $O(n/\lg^\epsilon n) = o(n)$ bits. Therefore the total space is $2|\mathsf{CSA}| + o(n)$ bits.

Several subsequent improvements [6,3,10] reduced the time to $O(t_{\mathsf{search}}(m) + k\,t_{\mathsf{SA}}\lg k\lg^\epsilon n)$, yet still using $2|\mathsf{CSA}| + o(n)$ bits of space, that is, twice the space of an optimal (under the hth order empirical entropy model) representation of the collection. Only this year [23] the space was reduced to the optimal $|\mathsf{CSA}| + o(n)$ bits, yet the time raises to $O(t_{\mathsf{search}}(m) + k\,t_{\mathsf{SA}}\lg k\lg^{1+\epsilon} n)$.

Tsur's Optimal-Space Index. Building on ideas of Belazzougui et al. [3], Tsur [23] managed to reduce the space to the asymptotically optimal $|\mathsf{CSA}| + o(n)$ bits. Let $u' \in \tau_k$ be the parent of u in τ_k, that is, its nearest marked ancestor in the suffix tree. Tsur proved that, from the $O(b)$ leaves of $u' \setminus u$, only $O(\sqrt{bk})$ have a chance to become part of the top-k list for a locus node v between u' and u. Thus, they simply store those *candidate* documents, and their frequency in u, associated to u. When one traverses the $O(b)$ leaves in $v \setminus u$, one (1) computes the document d as before, (2) if it is not stored as a candidate for u one can just ignore it, (3) if it is in the list then one just increases its frequency by 1. At the end one has enough information to answer the top-k query, without the need of the second $|\mathsf{CSA}|$ bits to compute frequencies below v.

If $b = k\ell$, the number of candidates is $t = O(\sqrt{bk}) = O(k\sqrt{\ell})$. One can encode them efficiently by storing, for each candidate d, the position of one leaf corresponding to d in the area covered by $u' \setminus u$. Those leaf positions are sorted and stored differentially: Let $0 < p_1 < p_2 < \ldots < p_t < 2b$ be the ordered positions, then one encodes $x_1, x_2, \ldots, x_t$, where $x_i = p_i - p_{i-1}$ $(p_0 = 0)$, using, say, γ-codes [4], which occupy $\sum 2\lg x_i = O(t\lg(b/t)) = O(k\sqrt{\ell}\lg\ell)$ bits by the log-sum inequality. The frequencies are encoded in $O(k\lg n + k\sqrt{\ell}\lg\ell)$ bits (the method is not relevant here).

Therefore, the space for top-k answers plus candidates is $O(k\lg n + k\sqrt{\ell}\lg\ell)$ bits, and the total space for a fixed k equals $O((n/b)(k\lg n + k\sqrt{\ell}\lg\ell)) = O(n((\lg n)/\ell + (\lg\ell)/\sqrt{\ell}))$ bits. By choosing $\ell = \lg k\lg^{1+\epsilon} n$, and since $\lg k \le \lg n$, this is $O(n/(\lg k\lg^{\epsilon/2} n))$. Added over all the k values that are powers of 2, this is $O(n/\lg^{\epsilon/2} n)\sum_{i=1}^{\lg D} 1/i = O(n\lg\lg D/\lg^{\epsilon/2} n) = o(n)$ bits.

The total time is $O(t_{\mathsf{search}}(m) + b\,t_{\mathsf{SA}}) = O(t_{\mathsf{search}}(m) + k\,t_{\mathsf{SA}}\lg k\lg^{1+\epsilon} n)$. For the two CSAs we have described, this is $O(t_{\mathsf{search}}(m) + k\lg k\lg^{2+\epsilon} n)$.

Hon, Shah, Thankachan and Vitter's Fastest Index. Hon et al. [10] obtained the fastest solution to date using $2|\mathsf{CSA}| + o(n)$ bits of space. For this sake they consider two independent blocking values, $c < b$. For block value b they build the τ_k trees as before. For block value c they build another set of marked trees ρ_k. These trees are finer-grained than the τ_k trees. Now, given the locus node v, there exists a maximal node $w \in \rho_k$ contained in v, and a maximal node $u \in \tau_k$ contained in w. The key idea is to build a list of top-k to top-$2k$ candidates by joining the precomputed results of w and u, and then correct this result by traversing $O(c)$ suffix tree leaves.

Since we have a maximal node $u \in \tau_k$ contained in any node $w \in \rho_k$, we can encode the top-k list of w only for the documents that are not already in the top-k list of u. Note that a document must appear at least once in $w \setminus u$ if it is in the top-k list of w but not in that of u. Thus the additional top-k candidates of w can be encoded using $O(k \lg(b/k))$ bits, by storing as before one of their positions in $w \setminus u$, and encoding the sorted positions differentially. The frequencies do not need to be encoded, since they can be recomputed as for any other candidate.

The space for a τ_k tree is $O((n/b)k \lg n) = O(n/\lg^{1+\epsilon} n)$ bits using $b = k \lg^{2+\epsilon} n$, which added over all the powers of 2 for k gives $O(n/\lg^{\epsilon} n) = o(n)$ bits, as before. For the ρ_k trees they require $O((n/c)k \lg(b/k))$ bits, which using $c = k \lg k \lg^{\epsilon} n$ gives $O(n \lg \lg n/(\lg k \lg^{\epsilon} n))$ bits. Added over the powers of 2 for k this gives $O(n \lg \lg n/\lg^{\epsilon} n) \sum_{i=1}^{\lg D} 1/i = O(n \lg \lg n \lg \lg D/\lg^{\epsilon} n) = o(n)$ bits.

The time is dominated by that of traversing $O(c)$ cells. Using some speedups [3] over the basic technique [11], the time is $O(t_{SA} \lg \lg n)$ per cell, for a total of $O(t_{search}(m) + k\, t_{SA} \lg k \lg^{\epsilon} n)$ for any constant $\epsilon > 0$. Over the two CSAs we have described, this is $O(t_{search}(m) + k \lg k \lg^{1+\epsilon} n)$.

3 A Faster Space-Optimal Representation

We build upon the schemes of Tsur [23] and Hon et al. [10]. We will use the dual marking mechanism of Hon et al., with trees τ_k and ρ_k, and make it work without using the second $|CSA|$ bits. Without this data, the structure gives us the top-k list of the maximal node $w \in \rho_k$ that is below the locus v, but not their frequencies. Similarly, when we traverse the $O(c)$ extra cells to correct the top-k list, we have no way to compute the frequency of the documents d found in $v \setminus w$.

In order to cope with the second problem, we will use the idea of Tsur: there can be only $O(\sqrt{ck})$ candidates that can make it to the top-k list. If $c = k\ell$, this is $O(k\sqrt{\ell})$. Thus we can record their identities by means of their sorted and differentially encoded positions along $O(c)$ leaves, in total space $O(k\sqrt{\ell} \lg \lg n)$ bits. Now we need a mechanism to store the frequencies, both of the top-k elements and of the $O(k\sqrt{\ell})$ candidates. For this sake we introduce a new data structure.

3.1 The Sampled Document Array

The *document array* $E[1, n]$ of T contains at $E[i]$ the document to which $A[i]$ belongs [15]. It is a convenient structure but it requires $n \lg D$ bits of space. We store just a sampled version of it.

Definition 1. *The* sampled document array *is an array $E'[1, n']$ that stores every sth occurrence of each document d in E, for a sampling step s. That is, if*

$\mathrm{rank}_d(E, i)$ *is the number of times d occurs in* $E[1, i]$, *the cell* $E[i]$ *is stored in* E' *iff* $\mathrm{rank}_{E[i]}(E, i)$ *is a multiple of s. Note that* $n' \leq n/s$.

To E' we associate a bitmap $S[1, n]$ that marks the positions in E that are sampled in E'. The following lemma follows easily.

Lemma 1. *Let x be the number of occurrences of a document d in* $E[sp, ep]$, *and let y be the number of occurrences of d in* $E'[\mathrm{rank}(S, sp-1)+1, \mathrm{rank}(S, ep)]$. *Then* $(y-1)s < x < (y+1)s$.

Proof. The area $E[sp, ep]$ includes y sampled occurrences of d. For the last $y-1$, their $s-1$ preceding non-sampled occurrences are also in $E[sp, ep]$. The $s-1$ occurrences preceding the first sampled one could be before sp, and thus $x \geq (y-1)s+1$. Alternatively, all the ys occurrences corresponding to the y sampled ones could be in the range, which could also include up to $s-1$ non-sampled occurrences to their right, yet their sampled successor could be after ep, thus $x \leq ys + (s-1)$. □

To use this lemma we store E' using a representation [7] that requires $n' \lg D + o(n' \lg D)$ and computes $\mathrm{rank}_d(E', i)$ in time $O(\lg \lg D)$. Further, we represent S in compressed form [21] so that it requires $n' \lg(n/n') + O(n') + o(n)$ bits and supports $\mathrm{rank}(S, i)$ in constant time. We use $s = \lg^2 n$, thus $n' = O(n/\lg^2 n)$ and the space for both E' and S is $o(n)$. Using this representation, we can compute y in Lemma 1 as $\mathrm{rank}_d(E', \mathrm{rank}(S, ep)) - \mathrm{rank}_d(E', \mathrm{rank}(S, sp-1))$ in time $O(\lg \lg D)$.

3.2 Completing the Index

To retrieve any $\mathrm{tf}(w, d)$ for a top-k document in node $w \in \rho_k$, we use S and E' to compute the approximation ys in time $O(\lg \lg D)$, and then need to store only $O(\lg s) = O(\lg \lg n)$ bits in w to correct this approximate count. Each node $w \in \rho_k$ stores (the correction of) the frequency information of both its top-k documents that appear in the top-k list of its maximal descendant node $u \in \tau_k$, and those that do not (in fact, we do not need frequency information associated to τ_k nodes). Similarly, we need to compute $\mathrm{tf}(w, d)$ for any of the $O(\sqrt{ck})$ candidates to top-k in w, thus we must also store (the correction of) those $O(\sqrt{ck})$ frequencies, which dominate the total space of $O(k\sqrt{\ell} \lg \lg n)$ bits. With this information we can discard the second $|CSA|$ bits of Hon et al. [10].

We use $\ell = \lg^2 k \lg^\epsilon n$. The space for one ρ_k tree is $O((n/c)k\sqrt{\ell} \lg \lg n) = O(n \lg \lg n/\sqrt{\ell}) = O(n \lg \lg n/(\lg k \lg^{\epsilon/2} n))$ bits. Adding over all the powers of 2 for k yields $O(n \lg \lg n/\lg^{\epsilon/2} n) \sum_{i=1}^{\lg D} 1/i = O(n \lg \lg n \lg \lg D/\lg^{\epsilon/2} n) = o(n)$ bits. Thus the total space is $|CSA| + o(n)$ bits.

At query time we store the top-k documents of w, plus the $O(\sqrt{ck})$ candidates, together with their frequencies in w, in a dictionary using the document identifiers as keys. Then we traverse the $O(c)$ cells of $v \setminus w$, accessing the CSA to determine each document identifier d. If d is not in the dictionary, it can be discarded, otherwise we increment its frequency. At the end, we scan the $O(\sqrt{ck})$ elements

of the dictionary and keep the k largest ones. The cost is dominated by computing the $O(c)$ CSA cells, plus $O(\lg \lg D)$ time per cell to compute $\mathsf{rank}_d(E', i)$ and
$O(1)$ to operate the dictionary[1]. This adds up to $O(k(t_{\mathsf{SA}} + \lg \lg D) \lg^2 k \lg^\epsilon n)$,
and the latter term absorbs the $\lg \lg D$.

Theorem 1. *The top-k most frequent documents problem, on a collection of
length n, for a pattern of length m, can be solved using $|\mathsf{CSA}| + D \lg(n/D) +
O(D) + o(n)$ bits and in $O(t_{\mathsf{search}}(m) + k\,t_{\mathsf{SA}} \lg^2 k \lg^\epsilon n)$ time, for any constant
$\epsilon > 0$. Here CSA is a compressed suffix array over the collection, $t_{\mathsf{search}}(m)$ is
the time CSA takes to find the suffix array interval of the pattern, and t_{SA} is the
time it takes to retrieve any suffix array cell.*

We also give two simplifications using recent CSAs [1,2] whose size is related to
H_h, the per-symbol empirical entropy of the text collection, for any $h \le \alpha \lg_\sigma n$
and any constant $0 < \alpha < 1$. For the second, since it uses $O(n)$ extra bits, we
set a smaller $c = k(\lg k \lg \lg n \lg \lg D)^2$.

Corollary 1. *The top-k most frequent documents problem, when $D = o(n)$, can
be solved using $nH_h(1 + o(1)) + o(n)$ bits and in $O(m \lg \lg \sigma + k \lg^2 k \lg^{1+\epsilon} n)$
time, for any constant $\epsilon > 0$.*

Corollary 2. *The top-k most frequent documents problem can be solved using
$nH_h(1 + o(1)) + O(n)$ bits and in $O(m + k \lg n(\lg k \lg \lg n \lg \lg D)^2)$ time.*

4   Final Remarks

Reaching asymptotic space optimality (under the hth order empirical entropy
model) for top-k document retrieval indexes is a very recent achievement. In this
work we have improved the time of that space-optimal solution [23]. Our time
complexity is a $\lg^2 k \lg^\epsilon n$ factor away from the minimum time required to obtain
k document identifiers using the CSAs, and a $\lg k$ factor away from the fastest
available solution that uses $2|\mathsf{CSA}| + o(n)$ bits [10].

It is natural to ask if those limits can be reached. Especially if the first limit
is matched, this problem could be finally considered closed in the scenario of
using optimal space based on CSAs. We believe, however, that a $\lg k$ factor in
the time is the unavoidable price of allowing k to be specified at query time,
whereas reaching the time of the currently fastest solution [10] seems feasible.

The other natural question is how much space is necessary to obtain the
optimal $O(m + k)$ time. The best current space used to achieve this time is
$O(n(\lg D + \lg \sigma))$ [18]. While it seems that $n \lg D$ bits are unavoidable in this
case, there have been some efforts to use only $|\mathsf{CSA}| + n \lg D + o(n \lg D)$ bits [9].
However the time achieved is not yet the optimal.

---

[1] For example, we can bucket the universe $[1, D]$ in chunks of $\lg^2 D$ elements, and store
a B-tree of arity $\sqrt{\lg D}$ and height $O(1)$ for the elements falling in each chunk. The
bucket structure adds up to $o(D)$ bits, which can be taken as part of the index. The
B-trees are operated in constant time because they store only $O(\lg^\delta D \lg \lg D)$ bits
per internal node. They occupy overall $O(\sqrt{ck} \lg n) = O(k \lg k \lg^{1+\epsilon/2} n)$ bits, which
is the space we use to answer the query. See [16, App. E] for more details.

References

1. Barbay, J., Gagie, T., Navarro, G., Nekrich, Y.: Alphabet partitioning for compressed rank/select and applications. In: Proc. 21st ISAAC, Part II, pp. 315–326 (2010)
2. Belazzougui, D., Navarro, G.: Alphabet-independent compressed text indexing. In: Proc. 19th ESA, pp. 748–759 (2011)
3. Belazzougui, D., Navarro, G., Valenzuela, D.: Improved compressed indexes for full-text document retrieval. J. Discr. Alg. 18, 3–13 (2013)
4. Bell, T., Cleary, J., Witten, I.: Text compression. Prentice-Hall (1990)
5. Büttcher, S., Clarke, C., Cormack, G.: Information Retrieval: Implementing and Evaluating Search Engines. MIT Press (2010)
6. Gagie, T., Kärkkäinen, J., Navarro, G., Puglisi, S.J.: Colored range queries and document retrieval. Theo. Comp. Sci. 483, 36–50 (2013)
7. Golynski, A., Munro, I., Rao, S.: Rank/select operations on large alphabets: a tool for text indexing. In: Proc. 17th SODA, pp. 368–373 (2006)
8. Hon, W.-K., Patil, M., Shah, R., Bin Wu, S.: Efficient index for retrieving top-k most frequent documents. J. Discr. Alg. 8(4), 402–417 (2010)
9. Hon, W.-K., Shah, R., Thankachan, S.V.: Towards an optimal space-and-query-time index for top-k document retrieval. In: Kärkkäinen, J., Stoye, J. (eds.) CPM 2012. LNCS, vol. 7354, pp. 173–184. Springer, Heidelberg (2012)
10. Hon, W.-K., Shah, R., Thankachan, S., Vitter, J.: Faster compressed top-k document retrieval. In: Proc. 23rd DCC, pp. 341–350 (2013)
11. Hon, W.-K., Shah, R., Vitter, J.: Space-efficient framework for top-k string retrieval problems. In: Proc. 50th FOCS, pp. 713–722 (2009)
12. Hon, W.-K., Shah, R., Wu, S.-B.: Efficient index for retrieving top-k most frequent documents. In: Proc. 16th SPIRE, pp. 182–193 (2009)
13. Manber, U., Myers, G.: Suffix arrays: a new method for on-line string searches. SIAM J. Comp. 22(5), 935–948 (1993)
14. Manzini, G.: An analysis of the Burrows-Wheeler transform. J. ACM 48(3), 407–430 (2001)
15. Muthukrishnan, S.: Efficient algorithms for document retrieval problems. In: Proc 13th SODA, pp. 657–666 (2002)
16. Navarro, G.: Spaces, trees and colors: The algorithmic landscape of document retrieval on sequences. CoRR, arXiv:1304.6023v5 (2013)
17. Navarro, G., Mäkinen, V.: Compressed full-text indexes. ACM Comp. Surv. 39(1), art 2 (2007)
18. Navarro, G., Nekrich, Y.: Top-k document retrieval in optimal time and linear space. In: Proc. 23rd SODA, pp. 1066–1078 (2012)
19. Navarro, G., Valenzuela, D.: Space-efficient top-k document retrieval. In: Klasing, R. (ed.) SEA 2012. LNCS, vol. 7276, pp. 307–319. Springer, Heidelberg (2012)
20. Sadakane, K.: Succinct data structures for flexible text retrieval systems. J. Discr. Alg. 5, 12–22 (2007)
21. Raman, R., Raman, V., Srinivasa Rao, S.: Succinct indexable dictionaries with applications to encoding k-ary trees, prefix sums and multisets. ACM Trans. Alg. 3(4), art 43 (2007)
22. Shah, R., Sheng, C., Thankachan, S.V., Vitter, J.S.: Top-k document retrieval in external memory. In: Bodlaender, H.L., Italiano, G.F. (eds.) ESA 2013. LNCS, vol. 8125, pp. 803–814. Springer, Heidelberg (2013)
23. Tsur, D.: Top-k document retrieval in optimal space. Inf. Proc. Lett. 113(12), 440–443 (2013)
24. Weiner, P.: Linear pattern matching algorithm. In: Proc. 14th Annual IEEE Symposium on Switching and Automata Theory, pp. 1–11 (1973)

Faster Range LCP Queries*

Manish Patil, Rahul Shah, and Sharma V. Thankachan

Louisiana State University, USA
{mpatil,rahul,thanks}@csc.lsu.edu

Abstract. Range LCP (longest common prefix) is an extension of the classical LCP problem and is defined as follows: Preprocess a string $S[1...n]$ so that $max_{a,b \in \{i...j\}} \text{LCP}(S_a, S_b)$ can be computed efficiently for the input $i, j \in [1, n]$, where $\text{LCP}(S_a, S_b)$ is the length of the longest common prefix of the suffixes of S starting at locations a and b. In this paper, we describe a linear space data structure with $O((j - i)^{1/2} \log^{\epsilon}(j - i))$ query time, where $\epsilon > 0$ is any constant. This improves the linear space and $O((j-i) \log \log n)$ query time solution by Amir et. al. [ISAAC, 2011].

1   Introduction and Related Work

Let $S[1...n]$ be a given string of length n, and let S_a represent its suffix $S[a...n]$ starting at location a. Then $\text{LCP}(S_a, S_b)$ represents the length of the longest common prefix of S_a and S_b. Being one of the most important tools in Combinatorial Pattern Matching LCP has been studied extensively. LCP computation is required in order to compute the Ziv-Lempel compression [7], and for finding maximal repeats in a genomic sequence in various Bioinformatics algorithms [1]. In [6] authors have established the relation between LCP and a well known string similarity measure "edit distance". It was shown that computing mismatches and LCPs is sufficient for computing the edit distance between a pair of strings. In this paper, we study an extension of this problem called Range LCP which is defined as follows:

Definition 1. *Index a string $S[1...n]$, such that given a range $[i, j]$ with $1 \leq i \leq j \leq n$, output the following:* $\max_{a,b \in \{i...j\}} LCP(S_a, S_b)$.

Cormode and Muthukrishnan [3] introduced a variant of range LCP (which we call pivot range LCP), in which the maximum LCP between a given suffix and all suffixes in a given interval is sought. They provide a solution with query time $O(\log n \log \log n)$. This result was then improved by [5] to $O(\log n)$ query time. In this paper, we first give a solution to this problem where query time is sensitive to the input range instead of string length i.e., n as captured in theorem below.

Theorem 1. *A string $S[1...n]$ can be indexed in $O(n \log n)$ bits, such that given a range $[i, j]$ with $1 \leq i \leq j \leq n$ and a suffix S_a as query input, we can compute* $\max_{b \in \{i...j\}} LCP(S_a, S_b)$ *in $O(\log^{\epsilon}(j - i))$ time, where $\epsilon > 0$ is any constant.*

---

* Work supported by National Science Foundation (NSF) Grants CCF–1017623 (R. Shah and J. S. Vitter) and CCF–1218904 (R. Shah).

O. Kurland, M. Lewenstein, and E. Porat (Eds.): SPIRE 2013, LNCS 8214, pp. 263–270, 2013.

Amir et. al. [1] have recently studied the range LCP problem. They gave an algorithmic solution that returns a pair of suffixes with the longest LCP (not the LCP itself) in $O((j - i) \log(j - i))$ time with $O((j - i) \log(j - i))$-word space utilization. They also investigated the indexing techniques for the same with space-time tradeoffs. They achieve $O(\log \log n)$ query time with index occupying $O(n \log^{1+\epsilon} n)$ words for arbitrary small constant ϵ. A space-efficient index proposed by them is of linear space and answers range LCP queries in $O((j - i) \log \log n)$ time. We improve this result and achieve better query time with linear space as summarized below.

Theorem 2. *A string $S[1...n]$ can be indexed in $O(n \log n)$ bits, such that range LCP query with input range $[i, j]$, $1 \leq i \leq j \leq n$ can be answered in $O((j - i)^{1/2} \log^{\epsilon}(j - i))$ time, where $\epsilon > 0$ is any constant.*

2 Preliminaries

2.1 Suffix Trees

Given a string $S[1...n]$, a substring $S[a...n]$ with $1 \leq i \leq n$ is called a suffix of S and is denoted by S_a. The lexicographic arrangement of all n suffixes of S in a compact trie is known as the *suffix tree* of S [11], where the ith leftmost leaf represents the ith lexicographically smallest suffix. Each edge in the suffix tree is labeled by a character string and for any node u, $path(u)$ is the string formed by concatenating the edge labels from root to u. For any leaf v, $path(v)$ is exactly the suffix corresponding to v. The space requirement of the suffix tree of S is bounded by $O(n \log n)$ bits. Given two leaves v and w, the longest common prefix of two suffixes corresponding to these leaves is given by the least common ancestor (LCA) of v and w in the suffix tree. As suffix tree supports LCA queries in constant time, each node u can be associated with a number representing length of $path(u)$ (maintaining space utilization of $O(n \log n)$ bits) so that given two leaves, length of the longest common prefix can be reported in constant time.

Suffix array of S is an array $SA[1...n]$, such that $SA[j]$ is the starting position of the jth lexicographically smallest suffix of S. Like suffix tree, suffix array of S also takes $O(n \log n)$ bits. We can maintain an additional array $SA^{-1}[1...n]$ in $O(n \log n)$ bits, called inverse suffix array such that $SA^{-1}[i] = j$ if $SA[j] = i$. Then, given a suffix $S_a = S[a...n]$, its position in the suffix array can be located in constant time.

2.2 Orthogonal Range Successor Queries

An orthogonal range reporting query R on a set of 2-dimensional (2d) points P asks for all points $p \in P$ that belong to the query rectangle $R = [x_1, x_2] \times [y_1, y_2]$. The orthogonal range reporting problem, that is, the problem of constructing a data structure that supports such queries has been studied extensively. Lenhof and Smid [8] introduced an optimization query called the orthogonal range successor query (ORS) and subsequently it has been investigated by many researchers [10,12,2,4]. The answer to an ORS query $R = [x_1, x_2] \times [y_1, +\infty]$ is the point with smallest y-coordinate among all points that are in the rectangle R. In this paper, we use the result by Nekrich and Navarro [9] that can answer the ORS queries in $O(\log^\epsilon n)$ time and occupies $O(n)$ word space for a collection of n points[1]. Similar to ORS, we can also define orthogonal range predecessor query (ORP), where given a query $R = [x_1, x_2] \times [-\infty, y_2]$, the goal is to return the point with largest y-coordinate among all points that are in the rectangle R. We note that both ORS and ORP are equivalent up to a simple change of coordinate system and point transformation. Hence, we assume the availability of linear space structure ORPS with $O(\log^\epsilon n)$ query time that can answer both ORP as well ORS queries.

3   Pivot Range LCP

Pivot Range LCP query is a range LCP query where along with $[i, j]$, a pivot suffix S_a is given as input and our goal is to compute $max_{b \in \{i...j\}} LCP(S_a, S_b)$. Here we assume that pivot suffix S_a is made available to the query in terms of its starting position in the text i.e. a. We rely on the following observation [1] to answer the pivot range LCP queries.

Lemma 1. *Given a string S of length n and its suffixes S_x, S_y and S_z such that $1 \leq SA^{-1}[x] \leq SA^{-1}[y] \leq SA^{-1}[z] \leq n$, we have $LCP(S_x, S_z) \leq LCP(S_x, S_y)$ and similarly $LCP(S_x, S_z) \leq LCP(S_y, S_z)$.*

Our first result for the pivot range LCP queries is summarized in the lemma below, which we improve later in the section.

Lemma 2. *A string $S[1...n]$ can be indexed in $O(n \log n)$ bits, such that given a range $[i, j]$ with $1 \leq i \leq j \leq n$ and a suffix S_a as query input, we can compute $max_{b \in \{i...j\}} LCP(S_a, S_b)$ in $O(\log^\epsilon n)$ time, where $\epsilon > 0$ is any constant.*

Proof. The linear space structure consists of the suffix tree of S, and a ORPS structure (Section 2.2) over the suffix array. To obtain the ORPS structure we convert each position m in the string $S[1...n]$ to a 2d point $(m, SA^{-1}[m])$ so that points lie on an $n \times n$ grid. To answer the query, we first find out the position $\bar{y}$ of $S_y = S_a$ in the suffix array. i.e., $\bar{y} = SA^{-1}[y] = SA^{-1}[a]$. Let $\bar{x}$ be the largest position in suffix array such that $\bar{x} < \bar{y}$ and $x \in [i, j]$. If such $\bar{x}$ does not exists, then we assume $\bar{x} = -1$, with S_x being an empty string. Symmetrically, let $\bar{z}$ be the smallest position in suffix array such that $\bar{y} < \bar{z}$ and $z \in [i, j]$. Moreover let S_{b^*} be the suffix such that $b^* \in [i, j]$ and $LCP(S_a, S_{b^*}) = max_{b \in \{i...j\}} LCP(S_a, S_b)$. Then with all suffixes in suffix tree/suffix array being lexicographically sorted, we have either $\bar{b^*} = \bar{x}$ or $\bar{b^*} = \bar{z}$ by Lemma 1. Given $\bar{y}$, both $\bar{x}$ and $\bar{z}$ can be computed in $O(\log^\epsilon n)$ time with constant $\epsilon > 0$. Here $\bar{x}$ can be obtained using a ORP query $R = [i, j] \times [-\infty, \bar{y}]$ and $\bar{z}$ using a ORS query $R = [i, j] \times [\bar{y}, +\infty]$.

---

[1] Assumes that points lie on an $n \times n$ grid.

Then LCP values can be computed for the pairs (S_x, S_y), (S_y, S_z) in $O(1)$ time using suffix tree. Finally we report maximum LCP value as an output. Therefore, total query time can be bounded by $O(\log^\epsilon n)$.   □

With the goal of making the query time sensitive to the input query range we propose the overlapping blocking scheme. We define blocking parameter β (to be fixed later) and without loss of generality assume that it is always rounded to the next highest power of 2. We partition the string $S[1...n]$ into $2(n/\beta) -$ 1 overlapping substrings each of size β (except possibly last two), such that $S_{\beta,t} = S[1 + (t-1)\beta/2...(t+1)\beta/2]$ for $t = 1, 2, 3, ...$. Now we maintain the ORPS structure over the suffix array of each substring $S_{\beta,t}$ separately, occupying $O(n \log \beta)$ bits. Further, such a collection of ORPS structures is maintained for $\beta = n, n^{1/2}, n^{1/4}, ...$. Total space requirement of such a storage can be bounded by $O(n(\log n + \frac{\log n}{2} + \frac{\log n}{4} + ...)) = O(n \log n)$ bits.

To answer the pivot range LCP query $([i, j], S_y)$ we follow the same process as described in Lemma 2 except this time we query the ORPS structure built for substring $S_{\beta,t}$. Here we choose smallest blocking factor β that can encompass a string of length $j - i + 1$ and then choose the substring with this blocking factor that contains both locations i and j. To be precise we choose $\beta = n^{1/2^\alpha}$, such that $n^{1/2^{\alpha+1}} < j - i + 1 \le n^{1/2^\alpha}$ and $t = 1 + \lfloor i/(\beta/2) \rfloor$. We emphasize that due to such careful selection of $S_{\beta,t}$, we have $\log \beta$ at most twice of $\log(j - i + 1)$. Hence, time required to answer the query is bounded by $O(\log^\epsilon \beta) = O(\log^\epsilon(j - i))$. Thus, we achieve the result summarized in Theorem 1 which forms one of the main building blocks of our final solution.

4   An $O(n)$ Bits LCP Matrices

In this section, we begin by describing an $O(n)$ bits LCP matrix LM which along with index in Theorem 1 can answer range LCP queries in $O(\sqrt{n \log n} \log^\epsilon(j-i))$ time. Then we improve it to $O(\sqrt{n \log \log n} \log^\epsilon(j - i))$ with the help of an additional $O(n)$ bits LCP matrix lm. Below we describe the maintenance of these LCP matrices to achieve the required query performance.

Δ-*LCP Matrix:* Let Δ be the blocking parameter to be decided later. We partition the given string $S[1...n]$ into n/Δ disjoint substrings each of size Δ. A blocking boundary $f_{\Delta,t}$ of partitioning based on Δ is called as Δ-boundary and $f_{\Delta,t} = 1 + (t-1)\Delta$. Now we maintain a $(n/\Delta) \times (n/\Delta)$ LCP matrix LM such that $LM[x, y] = max_{a,b \in \{f_{\Delta,x}...f_{\Delta,y}\}} LCP(S_a, S_b)$ along with pair of suffixes that achieve the maximum LCP value. In essence, we pre-compute the answers to the range LCP queries for all ranges $[i, j]$ where both i and j are aligned with some Δ-boundary and maintain them in a matrix. Note that it is sufficient to maintain only one of the suffixes in $LM[x, y]$ instead of actual LCP value and a pair of suffixes, as this information can be obtained quickly in $(O(\log^\epsilon(j - i)))$ time using Theorem 1. Moreover, the suffix is maintained in the matrix LM by its starting position in the text. Therefore, space required for storing the LCP matrix LM is bounded by $O((n/\Delta)^2 \log n)$ bits.

Query algorithm: Given an input range $[i, j]$, we begin by identifying a maximal (largest) range that is aligned with some Δ-boundaries and is completely covered by input range in constant time i.e., a maximal range $I_\Delta = [f_{\Delta,i'}, f_{\Delta,j'}]$ such that it is completely within the input range $[i, j]$. Let suffix pair (S_{a^*}, S_{b^*}) be the output for the given range LCP query, then there are two possibilities regarding the beginning positions of these suffixes in text i.e., a^* and b^*, relative to the range I_Δ. Either both suffixes begin at a position within the range I_Δ ($a^*, b^* \in [f_{\Delta,i'}, f_{\Delta,j'}]$) or either of them begin at a position outside the range I_Δ (either $a^* \notin I_\Delta$ or $b^* \notin I_\Delta$). Relying on this observation we follow a simple process described below to retrieve candidates from which the pair (S_{a^*}, S_{b^*}) along with the LCP value can be obtained.

1. Retrieve the pre-computed answer pair (S_x, S_y) for the range I_Δ by probing $LM[i', j']$ and insert it into candidate set
2. For each suffix S_x, where $x \in [i, f_{\Delta,i'} - 1]$ or $x \in [f_{\Delta,j'} + 1, j]$ (called as fringe suffix), obtain the suffix S_y by issuing a pivot range query ($[i, j], S_x$) and insert the pair (S_x, S_y) into candidate set

Finally, choose the suffix pair with maximum LCP value among the candidates as (S_{a^*}, S_{b^*}) and report it as an output. Since number of fringe suffixes (hence candidates) is bounded by Δ ($f_{\Delta,i'} - i < \Delta$ and $j - f_{\Delta,j'} < \Delta$), Step 2 only needs $O(\Delta \log^\epsilon(j - i))$ time. Moreover probing cell $LM[i', j']$ requires $O(\log^\epsilon(j - i))$ time. Therefore, overall query time can be bounded by $O(\Delta \log^\epsilon(j - i))$.

To restrict the size of LCP matrix LM, we choose $\Delta = \sqrt{n \log n}$, which establishes the result summarized in the following lemma.

Lemma 3. *A string $S[1...n]$ can be indexed in $O(n \log n)$ bits, such that the range LCP query with input range $[i, j]$, $1 \le i \le j \le n$ can be answered in $O(\sqrt{n \log n} \log^\epsilon(j - i))$ time, where $\epsilon > 0$ is any constant.*

It can be observed that query performance in the above lemma can be improved by reducing the number of fringe suffixes evaluated during query time. Below we show how it can be achieved by maintaining an additional LCP matrix lm in $O(n)$ bits.

δ-LCP Matrix: Let δ be the another blocking parameter such that $\delta < \Delta$ [2]. By following the partitioning with parameter δ as described earlier, we intend to keep $(n/\delta) \times (n/\delta)$ LCP matrix lm such that $lm[x, y] = max_{a,b \in \{f_{\delta,x} \cdots f_{\delta,y}\}} LCP(S_a, S_b)$. However, we can not afford to maintain the precomputed answers explicitly by using $\log n$ bits as we did for matrix LM. Rather we exploit the already precomputed answers in LM and restrict each entry in the matrix lm to $1 + \log(\Delta/\delta)$ bits as follows and total space requirement of lm can be bounded by $O((n/\delta)^2 \log(\Delta/\delta))$.

Let $I_\delta = [i, j]$ be the range enclosed by δ-boundaries i.e., $i = f_{\delta,x}, y = f_{\delta,y}$. We obtain a maximal range $I_\Delta = [f_{\Delta,i'}, f_{\Delta,j'}]$ that is completely contained within

---

[2] Assume both Δ and δ are rounded to the next highest power of 2.

I_δ. Moreover, let (S_{a^*}, S_{b^*}) be the output for the range LCP query with input being range I_δ. As observed in the previous query algorithm, either $a^*, b^* \in I_\Delta$ or at least one of the two suffixes do not originate in the range I_Δ i.e. $a^* \notin I_\Delta$ or $b^* \notin I_\Delta$. We record these possibilities using a bit indicator. We set the indicator bit in $lm[x, y]$ to indicate that output of range LCP query with input I_δ is the same as that of range LCP query with input I_Δ (whenever $a^*, b^* \in I_\Delta$). In such a scenario, suffix pair (S_{a^*}, S_{b^*}) can be retrieved from matrix LM directly and hence need not be maintained in matrix lm at all.

For the remaining case, when either $a^* \notin I_\Delta$ or $b^* \notin I_\Delta$ we maintain the approximate location of the suffix. Without loss of generality, let $a^* \notin I_\Delta$, then instead of storing location a^* explicitly (requiring $\log n$ bits), we keep identification of the δ-partition within the Δ-partition to which a^* belongs to. Precisely, we store $1 + \lfloor (a^* - \lfloor a^*/\Delta \rfloor \Delta)/\delta \rfloor$ occupying $\log(\Delta/\delta)$ bits per entry. The downside of such approximate location information of the suffix is that, no longer we can probe cell $lm[x, y]$ in $O(\log^\epsilon(j - i))$ time as before. However retrieving the a^* value from this encoding is bounded by $O(\delta \log^\epsilon(j - 1))$, since we can issue pivot range LCP queries for all suffix locations sharing the same encoding value i.e. δ-partition identified by $lm[x, y]$.

Query algorithm: Given an input range $[i, j]$, we first identify a maximal range $I_\delta = [f_{\delta,i'}, f_{\delta,j'}]$ such that it is completely within $[i, j]$. Then by similar observation as before we accumulate candidate suffix pairs as follows:

1. Retrieve the pre-computed answer pair (S_x, S_y) for the range I_δ by probing $lm[i', j']$ and insert it into candidate set
2. For each suffix S_x where $x \in [i, f_{\delta,i'} - 1]$ or $x \in [f_{\delta,j'} + 1, j]$ (called as fringe suffix) obtain the suffix S_y by issuing a pivot range query $([i, j], S_x)$ and insert the pair (S_x, S_y) into candidate set

Suffix pair with maximum LCP value among the candidates can then be reported as an output. Both the above steps of candidate retrieval need $O(\delta \log^\epsilon(j - i))$ time. In the first step, probing the cell $lm[i', j']$ requires $O(\log^\epsilon(j - i))$ time if indicator bit is set otherwise the approximate suffix location stored in $lm[i', j']$ can be decoded in $O(\delta \log^\epsilon(j - i))$ time as described earlier. Whereas for the second step, the number of fringe suffixes and hence the number of pivot range LCP queries executed is bounded by $O(\delta)$. Thus, total time required for answering the query is bounded by $O(\delta \log^\epsilon(j - i))$.

To restrict the overall size of LCP matrices, we choose $\Delta = \sqrt{n \log n}$ and $\delta = \sqrt{n \log \log n}$ thus establishing following result.

Lemma 4. *A string $S[1...n]$ can be indexed in $O(n \log n)$ bits, such that the range LCP query with input range $[i, j]$, $1 \le i \le j \le n$ can be answered in $O(\sqrt{n \log \log n} \log^\epsilon(j - i))$ time, where $\epsilon > 0$ is any constant.*

5 Adaptive $O((j-i)^{1/2} \log^\epsilon(j-i))$ Time Solution

To reduce the term $\sqrt{n}$ in query time of Lemma 4 to $(j-i)^{1/2}$, we reuse the idea introduced earlier to make query time of pivot range LCP queries sensitive to the input query range. We partition the string $S[1...n]$ using blocking parameter β into $2(n/\beta) - 1$ overlapping substrings such that $S_{\beta,t} = S[1 + (t-1)\beta/2...(t+1)\beta/2]$ for $t = 1, 2, 3, ...$ We then maintain the pair of LCP matrices i.e., matrix LM and lm for each substring $S_{\beta,t}$ separately, occupying $O(n)$ bits. Such a collection of proposed LCP matrices is maintained for $\beta = n, n/2, n/4, ...$ with overall space usage of $O(n \log n)$ bits. Our final data structure consists of three components each of $O(n \log n)$ bits as below:

- Suffix tree of input string S
- ORPS structure for each substring $S_{\beta,t}$ for $\beta = n, n^{1/2}, n^{1/4}, ...$
- Pair of LCP matrices i.e., LM, lm for each substring $S_{\beta,t}$ for $\beta = n, n/2, n/4, ...$

To answer the range LCP query with input range $[i, j]$, we first obtain substrings $S_{\beta,t}$ and $S_{\beta',t'}$ as follows:

- Substrings $S_{\beta,t}$ is selected in exactly the same way we did earlier to obtain the result for pivot range LCP query in Theorem 1. Recall, we choose $\beta = n^{1/2^\alpha}$ such that $n^{1/2^{\alpha+1}} < j - i + 1 \le n^{1/2^\alpha}$ and $t = 1 + \lfloor i/(\beta/2) \rfloor$. Then, any pivot range LCP query (i, j, S_y) where $y \in [i, j]$ can be answered in $O(\log^\epsilon(j-i))$ time by querying the ORPS structure on $S_{\beta,t}$ (Theorem 1).
- For substring $S_{\beta',t'}$, we choose $\beta' = n/2^{\alpha'}$ such that $n/2^{\alpha'+1} < j - i + 1 \le n/2^{\alpha'}$ and $t = 1 + \lfloor i/(\beta'/2) \rfloor$. In essence we choose smallest blocking factor that can encompass a string of length $j - i + 1$ and then choose the substring with this blocking factor that completely contains the range $[i, j]$.

We use the matrix pair for the substring $S_{\beta',t'}$ to answer the query. Let $m \le 2(j - i + 1)$ be the length of the substring $S_{\beta',t'}$. We note that the matrix LM and lm have been built for $S_{\beta',t'}$ using blocking factors $\Delta = \sqrt{m \log m}$ and $\delta = \sqrt{m \log \log m}$. Therefore, by following the query algorithm for Lemma 4, the number of fringe suffixes evaluated during query time are $O(\sqrt{m \log \log m})$ with each such evaluation costing $\log^\epsilon(j - i)$ due to the pivot range LCP query answered using ORPS structure on substring $S_{\beta,t}$. Moreover, probing the cell in the matrix lm will also cost $O(\sqrt{m \log \log m} \log^\epsilon(j-i))$. Time required for answering the given range LCP query is thus bounded by $O(\sqrt{m \log \log m} \log^\epsilon(j - i)) = O((j - i)^{1/2} \log^{2\epsilon}(j - i))$. Therefore, the proposed data structure achieves the result summarized in Theorem 2.

References

1. Amir, A., Apostolico, A., Landau, G.M., Levy, A., Lewenstein, M., Porat, E.: Range LCP. In: Asano, T., Nakano, S.-i., Okamoto, Y., Watanabe, O. (eds.) ISAAC 2011. LNCS, vol. 7074, pp. 683–692. Springer, Heidelberg (2011)

2. Chan, T.M., Larsen, K.G., Patrascu, M.: Orthogonal range searching on the ram, revisited. In: Symposium on Computational Geometry, pp. 1–10 (2011)
3. Cormode, G., Muthukrishnan, S.: Substring compression problems. In: Proceedings of the sixteenth Annual ACM-SIAM Symposium on Discrete Algorithms. Society for Industrial and Applied Mathematics, pp. 321–330 (2005)
4. Crochemore, M., Iliopoulos, C.S., Kubica, M., Rahman, M.S., Tischler, G., Walen, T.: Improved algorithms for the range next value problem and applications. Theor. Comput. Sci. 434, 23–34 (2012)
5. Keller, O., Kopelowitz, T., Landau, S., Lewenstein, M.: Generalized substring compression. In: Kucherov, G., Ukkonen, E. (eds.) CPM 2009 Lille. LNCS, vol. 5577, pp. 26–38. Springer, Heidelberg (2009)
6. Landau, G.M., Vishkin, U.: Fast parallel and serial approximate string matching. Journal of algorithms 10(2), 157–169 (1989)
7. Lempel, A., Ziv, J.: On the complexity of finite sequences. IEEE Transactions on Information Theory 22(1), 75–81 (1976)
8. Lenhof, H.-P., Smid, M.H.M.: Using persistent data structures for adding range restrictions to searching problems. ITA 28(1), 25–49 (1994)
9. Nekrich, Y., Navarro, G.: Sorted range reporting. In: SWAT, pp. 271–282 (2012)
10. Patrascu, M., Thorup, M.: Time-space trade-offs for predecessor search. In: STOC, pp. 232–240 (2006)
11. Weiner, P.: Linear Pattern Matching Algorithms. In: SWAT, pp. 1–11 (1973)
12. Yu, C.-C., Hon, W.-K., Wang, B.-F.: Improved data structures for the orthogonal range successor problem. Comput. Geom. 44(3), 148–159 (2011)

Learning to Schedule Webpage Updates
Using Genetic Programming

Aécio S.R. Santos[1], Nivio Ziviani[1], Jussara Almeida[1], Cristiano R. Carvalho[1],
Edleno Silva de Moura[2], and Altigran Soares da Silva[2]

[1] Universidade Federal de Minas Gerais,
Department of Computer Science, Belo Horizonte, Brazil
[2] Universidade Federal do Amazonas,
Institute of Computing, Manaus, Brazil

Abstract. A key challenge endured when designing a scheduling policy regarding freshness is to estimate the likelihood of a previously crawled webpage being modified on the web. This estimate is used to define the order in which those pages should be visited, and can be explored to reduce the cost of monitoring crawled webpages for keeping updated versions. We here present a novel approach to generate score functions that produce accurate rankings of pages regarding their probability of being modified when compared to their previously crawled versions. We propose a flexible framework that uses genetic programming to evolve score functions to estimate the likelihood that a webpage has been modified. We present a thorough experimental evaluation of the benefits of our framework over five state-of-the-art baselines.

1 Introduction

The quality of a Web search engine depends on several factors, such as the content gathered by the web crawler, the ranking function that produces the document ordering, and the user interface. By its turn, the success of the crawling process of a web search engine depends the coverage of the crawl, the policy used to select pages to collect, and the freshness of the pages. The focus of this work is on freshness, i.e., on the design of policies for scheduling webpage updates.

Web crawlers usually have access to limited bandwidth and their scheduler should periodically sort a large list of known URLs to define the order in which they should be visited. In this scenario, performing a full scan of all priorly crawled webpages to assure database freshness is unfeasible. To avoid that, crawling architectures (e.g., VEUNI [8]) use a score function to assign a weight to each known webpage (URL). Only the top k pages, k being a parameter, are taken to be visited. After crawling the k pages, the scheduler starts a new crawling cycle, using the score function to rank the known pages to be visited.

We here focus on the problem of estimating the likelihood that a webpage has been modified. Prior work has used machine learning techniques to related tasks (e.g., grouping pages with similar change behaviour [11], and predicting a page's change behaviour [10]), but none has applied them to build score functions. We

O. Kurland, M. Lewenstein, and E. Porat (Eds.): SPIRE 2013, LNCS 8214, pp. 271–278, 2013.

investigate the potential of using a genetic programming (GP) framework to learn these score functions. Our experimental evaluation shows that our solution outperforms existing score functions [3,11], being it a viable alternative to solve the addressed problem and opening opportunities for future work.

2 Background and Related Work

Like [2,3,11], we here consider a binary freshness model where the freshness of page p at time t is 1 if the copy of p is identical to the live copy, or 0, otherwise. The *freshness* of a set C of webpages can then be estimated by the average number of fresh pages in C at time t.

Probabilistic models have been proposed to approximate the history and predict webpage changes. For example, Coffman et al. [5] proposed to model the occurrences of changes on each page p by a Poisson process with parameter λ_p changes per time unit. Cho and Garcia-Molina [3] also investigated estimators for the change frequency of elements that are updated autonomously, in various scenarios. They showed that a web crawler can achieve improvements in freshness by setting its refresh policy to visit pages proportionally more often based on their proposed estimator, which is defined in Section 5.1.

Cho and Ntoulas [4] proposed a sampling-based method to detect webpage changes based on the number of pages that changed in a sample downloaded from the web site, which may be too coarse to represent all of its pages. Tan and Mitra [11] proposed to solve this problem by grouping the pages into k clusters with similar change behavior, and then sorting the clusters based on the mean change frequency of a representative cluster's sample. They proposed four strategies to compute the weights associated with a change in each of the downloaded cycles, which are further described in Section 5.1. Our work differs from [11] as our approach is not sampling based, but uses machine learning to build a score function that allows the scheduling of webpage updates. Once the score function has been learned, which is done off-line, it can be applied quickly, thus allowing large scale crawling using the architecture presented in Section 3.

Radinsky and Bennett [10] proposed a webpage change prediction framework that uses content features, the degree and relationship among the prediction page's observed changes, the relatedness to other pages, and the similarity in the kinds of changes they experienced. We here only use features related to whether the page changed or not during each cycle. However, given the flexibility of GP, our approach can be easily extended to include other features in the future.

3 Crawler Architecture

The incremental crawler architecture considered here has four main components: fetcher, URL extractor, uniqueness verifier, and scheduler [8]. Considering cycle i, the *fetcher* receives from the *scheduler* a set of candidate URLs to be crawled, locates them, and returns a set of URLs actually downloaded. The URL extractor parses each downloaded page and obtains a set of new URLs. The uniqueness

Listing 1.1. Genetic Programming for Crawling (GP4C)

```
 1  Let 𝒯 be a training set of pages crawled in a given period;
 2  Let 𝒱 be a validation set of pages crawled in a given period;
 3  Let N_g be the number of generations;
 4  Let N_b be the number of best individuals;
 5  𝒫 ← Initial random population of individuals;
 6  ℬ_t ← ∅;
 7  For each generation g of N_g generations do {
 8      ℱ_t ← ∅;
 9   For each individual i ∈ 𝒫 do
10          ℱ_t ← ℱ_t ∪ {g, i, fitness(i, 𝒯)};
11      ℬ_t ← getBestIndividuals(N_b, ℬ_t ∪ ℱ_t);
12      𝒫 ← applyGeneticOperations(𝒫, ℱ_t, ℬ_t, g);
13  }
14  ℬ_v ← ∅;
15  For each individual i ∈ ℬ_t do
16      ℬ_v ← ℬ_v ∪ {i, fitness(i, 𝒱)};
17  BestIndividual ← applySelectionMethod(ℬ_t, ℬ_v);
```

verifier checks each URL against the repository of unique URLs[1]. The scheduler chooses a new set of URLs to be sent to the fetcher, thus starting a new cycle.

We here focus on the algorithm for scheduling webpage updates, which is driven by two main goals: *coverage*, the fraction of desired pages that the crawler downloads successfully; and *freshness*, the degree to which the downloaded pages remain up-to-date, relative to the current live web copies. Most prior work focuses on only one of them. This work is focused on freshness.

4 Genetic Programming for Incremental Crawling

We here apply GP to the problem of scheduling webpage updates, using it to derive score functions that capture the likelihood that a page has changed. Pages with higher likelihood should receive higher scores, and thus higher priority in the scheduling process. Our method, called *GP4C – Genetic Programming for Crawling*, uses a GP process adapted from [1], and is presented in Listing 1.1.

As shown in Listing 1.1, GP4C is an iterative process with two phases: *training* (lines 5–13) and *validation* (lines 14–16). Our training and validation sets are built as follows: we train with an initial set of pages and validate the results with a distinct set of pages. This scenario is closer to that of large crawling tasks (e.g., crawling to a world wide search engine), where an initial set of pages to build the training set is crawled first, and then a set of validation pages is crawled. Experimental tests apply the resulting function in a third set of pages.

GP4C starts with the creation of an initial random population of N_p individuals (line 5) that evolves generation by generation using genetic operators (line 12) until a maximum number of generations (N_g). We apply the genetic

---

[1] Note that the size of the set of candidate URLs passed to the fetcher is defined by the amount of memory space available to the uniqueness verifier.

operators of reproduction, crossover and (swap/replacement) mutation at pre-defined rates. In particular, for the crossover operation, the selection of the parents is performed randomly among the top best individuals of the current generation. In the training phase, a fitness function is applied to evaluate all individuals of each generation (lines 9–10), so that only the N_b fittest individuals, across all previous generations, are selected to continue evolving (line 11). After the last generation is built, to avoid *over-fitting*, the validation phase is applied: the fitness function is used over the validation set (lines 15–16), and individuals that perform the best are selected as the final scheduling solutions (line 17).

Each *individual* represents a function that assigns a score to each page when composing the scheduling at the training set. Such score combines information useful for estimating the likelihood of a given page being updated in a period of time, exploring, for instance, its behavior in previous crawls. The training is performed in a period of time considered by us, and each individual is evaluated as being the function to create the scheduling in the whole training period.

An individual is represented by a binary tree with a maximum depth d, where terminals are features that help characterizing a page's updating behavior. We here consider three features: (1) n, the number of times that the page was visited; (2) X, the number of times that the page changed in n visits; and t, the number of cycles since the page was last visited. We also use the following *constant values* as terminals: $0.001; 0.01; 0.1; 0.5; 1; 10; 100; 1000$. As inner nodes of the tree, we use the functions addition $(+)$, subtraction $(-)$, multiplication $(*)$, division $(/)$, logarithm (log), exponentiation (pow), and the exponential function (exp).

The *fitness function* measures the quality of the ranking produced using a given individual for the whole training period. To compute the fitness of an individual, we take the score it produces for each page in the training set of each day and generate a schedule for the crawling to be performed on the next day. We here use as fitness function the ChangeRate metric, defined in Section 5.1.

As in [1], we select the *best individuals* in the validation step by running the GP process N times with distinct random seeds, so as to reduce the risk of finding a low performance local best individual. We pick the best individual among those generated by these N runs, referring to this approach as $GP4C_{Best}$. As in [6], we also consider two other strategies that are based on the average Avg_σ and the sum Sum_σ of the performances of each individual in both training and validation sets, minus the standard deviation of such performance when selecting best individuals. The individual with the highest Sum_σ (or Avg_σ) is selected. We refer to GP4C using these selection strategies as $GP4C_{Sum}$ and $GP4C_{Avg}$.

5   Experimental Evaluation

We used a crawl simulation to ensure that all policies are compared under the same conditions. We built a webpage dataset collected from the Brazilian Web (.br domain) using the crawler presented in [8], whose architecture is described in Section 3. Table 1 summarizes the dataset, referred to as BRDC'12[2], which

---

[2] Available at http://homepages.dcc.ufmg.br/~aeciosantos/datasets/ brdc12/

consists of a fixed set of webpages crawled on between September and November 2012. From a repository of around 200 million URLs we selected 3,059,698 webpages, which were then daily monitored. During the monitoring periods, our crawler ran from 0AM to 11PM, recollecting each selected webpage every day, which allowed us to determine when each page was modified.

Table 1. Overview of our BRDC'12 dataset

Monitoring period	Number of webpages	Number of websites	Number of webpages/site		
			Min	Max	Average
57 days	417,048	7,171	1	2,336	58.15

5.1 Baselines and Evaluation Metric

We compare $GP4C_{Best}$, $GP4C_{Sum}$ and $GP4C_{Avg}$ with five baselines, referred to here as CG, NAD, SAD, AAD and GAD. Given n the number of visits and X the number of times that a page p changed in those n visits, the CG baseline [3] estimates the change frequency of p as:

$$CG = -\log(\frac{n - X + 0.5}{n + 0.5}).$$ (1)

The other four baselines were proposed by Tan and Mitra [11]. In order to compute the change frequency of the pages, they assume that each page p follows a Poisson process with parameter λ_p. That is, the probability that a page p will change in the interval $(0, t]$ is given by $1 - e^{\lambda_p t}$. We set t to be the number of cycles since the page was last downloaded and compute λ_p using the change history of the pages:

$$\lambda_p = \sum_{i=1}^{n} w_i \cdot I_i(p),$$

where n is the number of times the page was downloaded so far, w_i is a weight associated with a change occurred in the ith download of the page ($\sum_{i=1}^{n} w_i = 1$), and $I_i(p)$ is either 1 if page p changed in the ith download, or 0 otherwise.

The weights w_i are computed according to one of the following schemes:

- NAD (*Nonadaptive*): all changes are equally important ($w_i = \frac{1}{n}$, $\forall i = 1..n$).
- SAD (*Shortsighted adaptive*): only the last change is important ($w_1 = \cdots = w_{n-1} = 0$, $w_n = 1$).
- AAD (*Arithmetically adaptive*): more recent changes are more important, and weights decrease according to an arithmetic progression ($w_i = \frac{i}{\sum_{i=1}^{n} i}$).
- GAD (*Geometrically adaptive*): as the previous scheme, but weights decrease more quickly, following a geometric progression ($w_i = \frac{2^{i-1}}{\sum_{i=1}^{n} 2^{i-1}}$).

We also consider two simpler approaches to build score functions, referred to as *Rand* and *Age*. In *Rand*, the scores are randomly chosen, whereas in *Age*, they are equal to the time t since the page was last visited (i.e., downloaded).

Our main evaluation metric is the ChangeRate, defined in [7] to assess the ability of a scheduling policy to detect updates. The ChangeRate at cycle i is the fraction of pages that were downloaded during i that had changed. The intuition is that the higher the concentration of changed pages, the better the scheduling. We use ChangeRate both as evaluation metric and fitness function, leaving the use of alternative metrics (e.g., weighted ChangeRate [4]) for the future.

5.2 Experimental Methodology

We adopted a 5-fold cross validation: 4 folds were equally divided into *training set* and *validation set*, and the last fold was used as *test set*. We report average results for the 5 test sets, along with corresponding 95% confidence intervals.

In order to evaluate the score functions and compute fitness values we simulate a crawl using our dataset. Our simulation starts with a warm-up period $W=2$ days, during which collected data is used to build basic statistics about each page. For each day following warm-up, we apply our proposed score function and each baseline to assign scores to each page. The download of the top-k pages with highest scores produced by each method is then simulated by updating statistics of the page such as number of visits (i.e., downloads), number of changes, etc. We set k equal to 5% of the total number of webpages in the dataset. Whenever the actual number of changed pages on a day is smaller than k, no evaluated algorithm can reach a maximum ChangeRate.

Regarding parametrization of the GP framework, we set N_p equal to 300 individuals, created using the ramped half-and-half method [9]. Due to the stability of results, we set N_g equal to 50 generations as termination criterion. We adopted tournament selection of size 2 to select individuals to evolve and set the crossover, reproduction, replacement mutation and swap mutation rates equal to 90%, 15%, 5% and 5%, respectively. We set the maximum tree depth d to 10 and the maximum depth for crossover to 9. During the evolution process we kept the $N_b = 50$ best individuals discovered through all generations to the validation phase. We ran the GP process using $N=5$ random seed values.

5.3 Results

We now discuss the results produced by our GP4C framework and the baselines using the BRDC'12 dataset. We consider only a basic set of terminals - n, X and t (see Section 4) - to show that our solution can derive functions that perform as good or better than the baselines.

Figure 1 shows the average ChangeRate for each day, for $GP4C_{Best}$ and all baselines. We omit the results for the other GP4C variations as they are either statistically tied or inferior to $GP4C_{Best}$. With 95% confidence, $GP4C_{Best}$ is statistically superior to all baselines in most days, being tied to NAD, AAD, GAD and CG, the most competitive baselines, in only a few days. Specifically, $GP4C_{Best}$ is statistically superior to NAD, AAD, GAD and CG in 22, 47, 49 and 50 of the simulated download cycles, respectively, being statistically tied

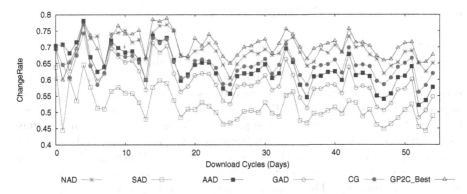

Fig. 1. (Color online) Average ChangeRate on each download cycle

with them in the other days. The only exceptions occur in the three initial days: $GP4C_{Best}$ is statistically inferior to AAD in days 1 and 3 and to GAD in day 1. This result corroborates the flexibility of our framework as it is able to produce results at least as good, if not better, than all five baselines.

Table 2 summarizes these results, showing average ChangeRate along with 95% confidence intervals for all methods, including the Rand and Age baselines (omitted in Figure 1). Once again, our $GP4C$ solutions produce score functions superior to all baselines. Note that the results of Rand and Age are much worse than all other methods. Moreover, even though our $GP4C$ approaches use the exact set of parameters used by the CG baseline [3] (i.e., n, X, t), our methods produce much better results, increasing the average ChangeRate by around 10%. The best baseline is NAD, which uses a different set of parameters that may provide more useful information about a page's updating behavior. Nevertheless, our approaches are still slightly better than NAD and can easily derive other functions if more parameters are given as input.

Table 2. Average ChangeRate for all days along with 95% confidence intervals

Rand	Age	NAD	SAD	AAD	GAD	CG	$GP4C_{Best}$	$GP4C_{Sum}$	$GP4C_{Avg}$
0.1857	0.2130	0.6892	0.5166	0.6344	0.6016	0.6439	0.7058	0.7008	0.7034
±	±	±	±	±	±	±	±	±	±
0.0007	0.0009	0.0056	0.0066	0.0095	0.0059	0.0067	0.0096	0.0176	0.0107

Finally, we note that our GP4C framework can be used for better understanding the scheduling problem. As example, an extremely simple, but also effective, function generated by our method is $t * X$, which yields a final performance superior to most of the baselines, with average ChangeRate above 0.690. It was not the best function found by GP4C, but illustrates how the framework can be applied not only to derive good score functions, but also to give insights about the most important parameters.

6   Conclusions and Future Work

We have presented a GP framework to automatically generate score functions to be used by schedulers of web crawlers to rank webpages according to their likelihood of being modified since they were last crawled. We compared three variations of our framework against seven state-of-the-art baselines, using a webpage dataset collected from the Brazilian Web. Our results show that our best function, $GP4C_{Best}$, is statistically superior to all baselines in most of the simulated download cycles. Moreover, our framework is quite flexible and can derive new score functions by exploiting new features (e.g., Pagerank of the pages, cost for crawling) or alternative fitness functions that balance the objectives of freshness and coverage. This is a direction we intend to pursue in the future.

Acknowledgements. We thank the Brazilian National Institute of Science and Technology for the Web (grant MCT-CNPq 573871/2008-6), Project MinGroup (grant CNPq-CT-Amazônia 575553/2008-1) and authors' grants and scholarships from CNPq.

References

1. Carvalho, A.L., Rossi, C., de Moura, E.S., da Silva, A.S., Fernandes, D.: Lepref: Learn to precompute evidence fusion for efficient query evaluation. Journal of the American Society for Information Science and Technology 63(7), 1383–1397 (2012)
2. Cho, J., Garcia-Molina, H.: Synchronizing a database to improve freshness. In: SIGMOD Record, pp. 117–128 (2000)
3. Cho, J., Garcia-Molina, H.: Estimating frequency of change. ACM Transactions on Internet Technology 3, 256–290 (2003)
4. Cho, J., Ntoulas, A.: Effective change detection using sampling. In: VLDB, pp. 514–525 (2002)
5. Coffman, E.G., Liu, Z., Weber, R.R.: Optimal robot scheduling for web search engines. Journal of Scheduling 1(1) (1998)
6. de Almeida, H.M., Gonçalves, M.A., Cristo, M., Calado, P.: A combined component approach for finding collection-adapted ranking functions based on genetic programming. In: SIGIR, pp. 399–406 (2007)
7. Douglis, F., Feldmann, A., Krishnamurthy, B., Mogul, J.: Rate of change and other metrics: a live study of the world wide web. In: USENIX Symposium on Internet Technologies and Systems, p. 14 (1997)
8. Henrique, W.F., Ziviani, N., Cristo, M.A., de Moura, E.S., da Silva, A.S., Carvalho, C.: A new approach for verifying URL uniqueness in web crawlers. In: Grossi, R., Sebastiani, F., Silvestri, F. (eds.) SPIRE 2011. LNCS, vol. 7024, pp. 237–248. Springer, Heidelberg (2011)
9. Koza, J.R.: Genetic Programming: On the Programming of Computers by Means of Natural Selection. MIT Press (1992)
10. Radinsky, K., Bennett, P.: Predicting content change on the web. In: WSDM (2013)
11. Tan, Q., Mitra, P.: Clustering-based incremental web crawling. ACM Transactions on Information Systems 28, 17:1–17:27 (2010)

Accurate Profiling of Microbial Communities from Massively Parallel Sequencing Using Convex Optimization

Or Zuk[1,2], Amnon Amir[3], Amit Zeisel[3], Ohad Shamir[4], and Noam Shental[5]

[1] Broad Institute of MIT and Harvard
[2] Toyota Technological Institute at Chicago
[3] Department of Physics of Complex Systems, Weizmann Institute of Science
[4] Microsoft Research, New England
[5] Department of Computer Science, The Open University of Israel

Abstract. We describe the Microbial Community Reconstruction (**MCR**) Problem, which is fundamental for microbiome analysis. In this problem, the goal is to reconstruct the identity and frequency of species comprising a microbial community, using short sequence reads from Massively Parallel Sequencing (MPS) data obtained for specified genomic regions. We formulate the problem mathematically as a convex optimization problem and provide sufficient conditions for identifiability, namely the ability to reconstruct species identity and frequency correctly when the data size (number of reads) grows to infinity. We discuss different metrics for assessing the quality of the reconstructed solution, including a novel phylogenetically-aware metric based on the Mahalanobis distance, and give upper-bounds on the reconstruction error for a finite number of reads under different metrics. We propose a scalable divide-and-conquer algorithm for the problem using convex optimization, which enables us to handle large problems (with $\sim 10^6$ species). We show using numerical simulations that for realistic scenarios, where the microbial communities are sparse, our algorithm gives solutions with high accuracy, both in terms of obtaining accurate frequency, and in terms of species phylogenetic resolution.

Keywords: Microbial Community Reconstruction, Massively Parallel Sequencing, Short Reads, Convex Optimization.

1 Introduction

Characterization of the micro-organisms present in a microbial community is of major biological and clinical importance. Since different micro-organisms have different genomes, it is possible to identify species based on their DNA sequences, using either whole-genome sequencing, or sequencing of pre-specified regions. The 16S ribosomal RNA gene (16S rRNA) is of particular interest for identifying microbial communities via sequencing. It has both highly conserved regions, present in almost all microbial species, together with variable regions. The conserved regions allow sequence amplification using universal PCR primers, while

O. Kurland, M. Lewenstein, and E. Porat (Eds.): SPIRE 2013, LNCS 8214, pp. 279–297, 2013.

the variable regions provide information used to distinguish between different species. Large databases [3, 4] with millions of 16S rRNA sequences may enable species identification by querying sequencing results in a database.

Previous methods aiming to characterize microbial communities using microarrays [7] and Sanger sequencing [2] have shown that, in principle, it is possible to identify species present in a sample, yet it is not clear how to get accurate estimation of species frequencies from the analog measurements provided by these technologies. Massively Parallel Sequencing (MPS) [16], also known as Next-Generation Sequencing (NGS), provides high-throughput digital sequence data and can allow a more detailed and accurate picture of the species in the mixture. In this method, one obtains a large number of short sequence reads from the mixture, and the goal is to reconstruct the identities and quantities of the species present. Many studies have used short reads to characterize microbial communities [10], yet they did not demonstrate an ability to identify the specific species present and quantify their abundance in the mixture - reliable recognition was typically achieved only at coarse genus level [12]. The main drawback of MPS is the relatively short read length (typically around 50-400 base-pairs in current technologies), which poses a problem for species reconstruction; short reads do not provide unambiguous evidence in support of the presence of a specific species, as typically the same read may originate from multiple different species, and cannot be uniquely aligned to the reference database.

Recently, more sophisticated methods for quantifying species abundance were developed, for 16S rRNA [6, 17] and whole metagenome shot-gun sequencing data [25]. These methods take into account read-assignment ambiguity and enable increased species resolution, but the question of maximal reconstruction resolution achieved was not systematically studied.

In this paper, we study mathematically the Microbial Community Reconstruction problem (**MCR**) - in which we use MPS data to characterize a microbial community. In a nutshell, the computational and statistical problem we face is as follows: given a large collection of short MPS reads (strings) sampled from a known database of species' sequences (longer strings) according to a certain unknown distribution, our goal is to estimate the sampling frequencies for each species in the database, and specifically recover the support of the distribution, i.e. the list of species with non-zero sampling probabilities. We model the sequencing process statistically, providing a probabilistic generative model for the short read data at hand. We prove conditions for *identifiability* - namely the ability to reconstruct precisely the identity and frequency of species present in the mixture from the short read data as the number of reads is increased. We prove upper-bounds on reconstruction errors for a finite number of reads. We propose a divide-and-conquer algorithm, handling large scale problems with hundreds of thousands of species, which is particularly appealing for sparse microbial communities - that is, realistic scenarios where only hundreds or a few thousands of species are present in the mixture, out of the possible millions of species in the database (see e.g. [5, 18]). We study the reconstruction performance in these realistic settings by simulating reads from the Greengenes 16S rRNA database [4].

Our goal here is to formulate and study the problem mathematically. Practical considerations (e.g. amplification and sequencing biases, restrictions on primers, paired-end reads) together with experimental results for real sequencing data are described in a separate publication [1].

In the spirit of reproducible research, we have implemented all of our algorithms in the Matlab package **COMPASS** (Convex Optimization for Microbial Profiling by Aggregating Short Sequence reads), which is freely available at github: `https://github.com/orzuk/COMPASS`.

2   The MCR Problem Formulation

We describe informally and briefly the biological settings. Our goal is to identify the species present in a given sample. We extract DNA, use 16S rRNA universal primes and amplify the DNA in this region. We then assume that DNA is sheared randomly and sequence it using MPS. We assume that the sequences database contains 16S rRNA sequences for *all* species present in the mixture, and reconstruct the species in the mixture *in silico*. A schematic representation of the **MCR** method is shown in Figure 1.

We denote by N the number of species in the database. The species' 16S rRNA sequences are marked $S_1, .., S_N$, represented as strings over the alphabet $\Upsilon = \{`A', `C', `G', `T'\}$. We assume that the S_i's are *distinct* sequences. The sequences may have *different* lengths $n_1, .., n_N$, with n_i the length of the i-th species' sequence, i.e. $S_i \in \Upsilon^{n_i}$. For the 16S rRNA gene, the lengths n_i are roughly 1500 base-pairs. We define the maximum sequence length as $n_{MAX} \equiv \max_i n_i$. We denote by $s_{i,j}$ the j-th nucleotide in the i-th species' sequence S_i, and by $s_{i,j:k}$ the substring containing nucleotides $j, j+1, .., k$ in the sequence S_i.

We represent the proportion of each species in the mixture using a vector $\mathbf{x}$ of length N, with x_i the frequency of species i. We have $\mathbf{x} \in \Delta_N$, where Δ_N is the N-dimensional simplex, $\Delta_N = \{\mathbf{x} : x_i \geq 0, \sum_{i=1}^{N} x_i = 1\}$. We represent the interior of a set $\mathcal{A}$ as $int(\mathcal{A})$. In particular, $int(\Delta_N)$ is the subset of Δ_N containing vectors with positive entries, $int(\Delta_N) = \{\mathbf{x} : x_i > 0, \sum_{i=1}^{N} x_i = 1\}$.

We observe data in the form of R reads of length L, $r_1, .., r_R \in \Upsilon^L$, with L typically around $\sim 50 - 400$, as in the Illumina and 454 sequencing technologies. We represent the data by a vector of read frequencies, $\mathbf{y} \in \Delta_{4^L}$, with the j-th coordinate given by $y_j = \frac{1}{R} \sum_{i=1}^{R} 1_{\{lex(r_i)=j\}}, \quad \forall j = 1, .., 4^L$. Here $lex(r)$ is the index of r in the lexicographic ordering of all 4^L possible reads (i.e. $lex(`AAA ... A') = 1, .., lex('TTT ... T') = 4^L$). We also define the inverse lexicographic ordering transformation, lex^{-1}, which for a given index j gives the corresponding sequence (e.g. $lex^{-1}(18) = 'AAA ... ATC')$.

In the **MCR** problem, the data vector $\mathbf{y}$ and the database sequences $S_1, .., S_N$ are given as input. Our goal is to reconstruct the species frequencies vector $\mathbf{x}$ from this information. The vector $\mathbf{y}$ is of exponential length (4^L) but very sparse, with only $M \leq R$ non-zero coordinates, where M is the number of *unique* sequence reads. We store and manipulate only the non-zero part of $\mathbf{y}$ - therefore the computational complexity of all of our algorithms will depend on M, and not

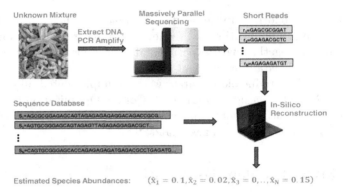

Fig. 1. The steps performed for species reconstruction using the **MCR** method. First, DNA is extracted and amplified using PCR with universal primers matching the 16S rRNA gene. The DNA is then sheared and sequenced using MPS, producing millions of short sequence reads. The sequencing data (reads), together with a database of 16S rRNA sequences, are entered into the computational pipeline providing estimated species abundances as output.

the exponentially large 4^L (see Section 5). In typical MPS experiments with current technologies R may be on the order of $\sim 10^5 - 10^8$.

2.1 Probabilistic Model

We formulate a probabilistic generative model capturing the sequencing process. We assume that the R reads are sampled identically and independently (i.i.d.) from the set of amplified regions in two steps,

1. First, sample a microbial species b from the set of possible species $\{1, .., N\}$, with the probability of species j being sampled proportional to the amount of DNA from this species, $x'_j \equiv Pr(b = j) = \frac{x_j n_j}{\sum_{i=1}^{N} x_i n_i}$.
2. Next, sample a read r from a distribution given by the species b. We represent sampling probabilities using a $4^L \times N$ read-sampling matrix $A = A(S, L)$ whose (i, j)-th entry is the probability to observe read i given that we know it came from species j, $A_{ij} = Pr(r = i|b = j)$.

Remark 1. The vector of sampling probabilities $\mathbf{x}'$ from step 1 is obtained by re-weighting the frequency vector $\mathbf{x}$ according to the sequence lengths. For ease of notation, we disregard this re-weighting, and denote both vectors as $\mathbf{x}$. When all sequences lengths n_j are identical we have indeed $\mathbf{x}' = \mathbf{x}$. More generally, the vectors are different but we can easily convert $\mathbf{x}$ to $\mathbf{x}'$ or $\mathbf{x}'$ to $\mathbf{x}$ using the above relation $x'_j = \frac{x_j n_j}{\sum_{i=1}^{N} x_i n_i}$

The sampling process defines a probability distribution $P_\mathbf{x} = P_\mathbf{x}(\mathbf{y}; A, L)$ on the space of possible frequencies Δ_{4^L},

$$P_\mathbf{x}(\mathbf{y}; A, L) = \begin{cases} \sum_{j=1}^{N} A_{ij} x_j & \mathbf{y} = \mathbf{e}^{(i)} \\ 0 & \text{otherwise} \end{cases} \quad (1)$$

where $\mathbf{e}^{(i)} \in \Delta_{4^L}$ is the i-th vector in the standard basis, $e_i^{(i)} = 1, e_j^{(i)} = 0 \; \forall j \neq i$. The data can be represented as R i.i.d. random variables, $\mathbf{y}^{(1)}, .., \mathbf{y}^{(R)} \sim P_{\mathbf{x}}(\mathbf{y}; A, L)$, with the sample frequency $\mathbf{y}$ represented as, $\mathbf{y} = \frac{1}{R} \sum_{i=1}^{R} \mathbf{y}^{(i)}$. We denote the **MCR** problem with read sampling matrix A by $\mathbf{MCR}(L, A)$. In its simplest form, A can be constructed as follows,

$$A_{ij} = \frac{\sum_{k=1}^{n_j - L + 1} 1_{\{lex^{-1}(i) = s_{j,k:k+L-1}\}}}{n_j - L + 1} \tag{2}$$

This matrix represents *uniform* sampling of *error-free* reads along the sequence of the chosen species j, assuming $L \leq n_j \; \forall j$. A non-zero element A_{ij} means that read i appears in the sequence of species j. If $L > n_i$ we assume that the 'tail' of each read is sampled uniformly from Υ (see Appendix Section A.1).

Remark 2. The above construction of A assumes no read errors and no biases. Incorporating more realistic sequencing models with non-uniform read density due to amplification biases, read errors (substitutions and indels), alignments errors etc. can be done by changing the definition of A from eq. (2). The same database S may thus yield different matrices A, and the statistical and algorithmic properties of a certain **MCR** problem depend on the database S only through the matrix A. The assumption in step 1 is that species DNA fragments are sampled according to their DNA frequencies out of the total DNA present in a sample. The model cannot accommodate deviations from this assumption which may arise from amplification biases and limited library complexity, which may distort the species frequencies - that is, the fraction of reads originating from a certain species may not represent the species' true frequency in the mixture. Accounting and correcting for such biases require analyzing multiple samples together.

In similar to the read frequencies vector $\mathbf{y}$, the matrix A is also huge ($4^L \times N$) but very sparse. In particular, the number of non-zero rows in A, denote K, is much smaller than 4^L, as most of the rows in A are zero and need not be stored. In the simple model above, $K \leq \sum_{j=1}^{N}(n_j - L + 1)$, which is roughly equal to the database size in nucleotides. In more complicated models involving read error, K will be larger, but still much smaller than 4^L. The computational complexity of our algorithms depends on K (see Section 5).

An estimator $\hat{\mathbf{x}}$ of the frequency vector $\mathbf{x}$ is simply a function from the set of all reads and database, to the n-dimensional simplex, $\hat{\mathbf{x}} : \Delta_{4^L} \times S \to \Delta_N$, $\hat{\mathbf{x}} = \hat{\mathbf{x}}(\mathbf{y}, S)$ (here S is the set of all possible sequences databases, i.e. the space of all ordered finite collections of strings over Υ).

We can solve the **MCR** problem by finding an estimator $\hat{\mathbf{x}}$ minimizing an empirical loss function. That is, define $\hat{\mathbf{y}} = A\hat{\mathbf{x}}$, the empirical reads distribution given the estimator $\hat{\mathbf{x}}$. We would like to minimize the loss $l(\hat{\mathbf{y}}, y)$, and define the following estimator,

$$\hat{\mathbf{x}} = argmin_{\mathbf{x} \in \Delta_N} l(A\mathbf{x}, \mathbf{y}) \tag{3}$$

A natural loss function is the Kullback-Leibler divergence $l_{KL}(\mathbf{y}, \hat{\mathbf{y}}) = D(P_{\mathbf{y}} || P_{\hat{\mathbf{y}}})$. This formulation is equivalent to maximizing the likelihood of the data $\mathbf{y}$, according to the probabilistic model in eq. (1). Maximizing the likelihood using the EM algorithm was proposed in [13] for a very similar likelihood formulation - this approach, however, is currently not scalable to a large number of species. We choose instead the l_2 loss $l_2(\mathbf{y}, \hat{\mathbf{y}}) = || \mathbf{y} - \hat{\mathbf{y}} ||_2$, mainly for computational considerations. The l_2 loss leads to a standard optimization problem and many off-the-shelf solvers can be used.

We expect real mixtures to be sparse, with only a few hundreds to a few thousands species present (out of hundreds of thousands). it is therefore appealing to use a sparsity-promoting loss in the cost function in eq. (3), for example by penalizing l_0 norm of $\mathbf{x}$. This is especially important when the number of reads is limited, to avoid over-fitting of the solution to the randomly sampled reads. The l_0 norm is not convex, leading to an intractable computational problem. The most common remedy of replacing the l_0 norm by the convex l_1 norm does not work in our problem since for probability vectors in the simplex $\mathbf{x} \in \Delta_N$ the constraint $||\mathbf{x}||_1 = 1$ trivially holds. Promoting sparsity for probability distributions in the simplex by convex relaxation was recently proposed [20], but the approach does not scale to our problem's size. Instead, we developed a scalable divide-and-conquer thresholding algorithm (see Section 5) which minimizes the l_2 error, while enforcing sparsity implicitly, by a repeated truncation of non-zero frequencies. The resulting solution is guaranteed to be sparse, while still keeping the l_2 error low as desired.

2.2   Evaluating the Solution: Metrics

To evaluate reconstruction accuracy, we need a measure comparing the reconstructed solution $\hat{\mathbf{x}}$ with the correct solution $\mathbf{x}$. Different applications may require different metrics - for example, in some applications we may be interested only in the *identity* of the species, while in other applications one would want to detect changes in *frequencies*. It may be important to identify the particular species or strain, or one may be satisfied with coarser reconstruction at the genus or family level. There are two major groups of performance metrics:

1. Phylogenetically-Unaware criteria: These metrics take into account only the species identities and frequencies. Examples include the l_p norm between the two vectors, recall-precision and Jaccard index. We use the simple l_2 norm as a representative of this group. This metric measures the deviation in species *frequencies* between the true and reconstructed solutions, $\mathcal{D}_{l_2}(\mathbf{x}, \hat{\mathbf{x}}) = \sqrt{\sum_{i=1}^{M}(x_i - \hat{x}_i)^2}$.

2. Phylogenetically-Aware criteria: These metrics take into account the phylogenetic relationship between species. The main intuition here is that identifying a species close to the true species is in fact almost as good as reconstructing the correct species. Examples include unifrac [14], weighted unifrac [15], and DPCoA [19]. We propose a novel Phylogenetically-aware criterion, using a Mahalanobis distance,

$$\mathcal{D}_{MA}(\mathbf{x}, \hat{\mathbf{x}}; D) = \sqrt{(\mathbf{x} - \hat{\mathbf{x}})^\top D(\mathbf{x} - \hat{\mathbf{x}})} = \sqrt{\sum_{i,j} D_{ij}(x_i - \hat{x}_i)(x_j - \hat{x}_j)}.$$

The matrix D is constructed to capture the phylogenetic distance between species (for example, from the species 16S rRNA sequences themselves). High (low) values of D_{ij} correspond to pairs of species (i, j) which are closely-related (remote). For concreteness, we choose specifically $D = A^\top A$, which represent the similarity between species based on their 16S rRNA sequences. The resulting Mahalanobis distance measures the agreement between the true and reconstructed solutions, in terms of both the species identities and their frequencies, while taking into account the similarities between closely related species.

3   Species Identifiability

In this section we study species identifiability - that is, the ability to correctly identify the species and their frequencies as the number of reads, R, goes to infinity.

Definition 1. *We say that the problem* $\mathbf{MCR}(L, A)$ *is identifiable, if for every* $\mathbf{x}^{(1)} \neq \mathbf{x}^{(2)} \in \Delta_N$, *there exists* $\mathbf{y} \in \Delta_{4^L}$ *such that* $P_{\mathbf{x}^{(1)}}(\mathbf{y}; A; L) \neq P_{\mathbf{x}^{(2)}}(\mathbf{y}; A; L)$.

Species identifiability captures fundamental limits of our ability to reconstruct the species frequency vector from the observed reads data. If the problem $\mathbf{MCR}(L, A)$ is identifiable, then *in principle* it is possible to correctly reconstruct the species frequencies vector $\mathbf{x}$, since different vectors will generate different distributions on the observed reads. If the problem is not identifiable, recovering the correct frequencies vector $\mathbf{x}$ may not be possible, regardless of the data size and computational resources available, since other (incorrect) frequency vectors give rise to an identical distribution on the observed reads data.

The identifiability question is not unique to the $\mathbf{MCR}$ problem, and arises more generally when reconstructing the identity of long sequences in a mixture using short reads. For example, conditions for the identification of isoforms from RNA-seq data were given in [11]. The different Isoforms in [11] are analogous to the different species in our problem, yet the precise modeling assumptions and identifiability criteria are different in the two problems. Identifiability is determined by both the similarity between the sequences of different species, and the read length. Longer and more diverse sequenced regions provide more information on the DNA sequence of different species in the mixture, and allow to distinguish between the underlying species more easily. However, even when the sequenced regions are informative enough, short sequenced reads obtained from these region may map to multiple species, thus species identification can be hard when reads are too short. We next formalize this intuition mathematically, showing how identifiability is determined by the input sequence database (and the read length L) through the matrix A, which represents the relation between the unknown vector $\mathbf{x}$ and the observed data $\mathbf{y}$ (see Appendix for proofs of all Propositions),

Proposition 1. *Let $A_{(1)}$ be the matrix constructed from A, concatenated with an all 1's row vector $\mathbf{1}_N$, $A_{(1)} \equiv \begin{pmatrix} A \\ \mathbf{1}_N \end{pmatrix}$. The reconstruction problem $\mathbf{MCR}(L, A)$ is identifiable if and only if $rank(A_{(1)}) = N$.*

As the read length increases, it becomes increasingly easier to distinguish between species,

Proposition 2. *Assume $N > 4$. Suppose that the database S is composed of N distinct sequences such that no sequence is a substring of another sequence, i.e. $s_{i,j:k} \neq s_{i'} \; \forall i \neq i' \in \{1, .., N\}, \forall j, k \in \{1, .., n_i\}$. Let $A^{(u,L)}$ be the sampling matrix obtained by uniform sampling of reads with read length L, according to eq. (2). Then there is a critical read length L_c, $1 < L_c \leq \max_i n_i$ such that the problem $\mathbf{MCR}(L, A^{(u,L)})$ is identifiable if and only if $L \geq L_c$.*

Remark 3. We assume that no sequence in the database is a substring of another database sequence for mathematical convenience. This assumption usually holds in practice provided a long enough region is sequenced, and can be relaxed while still obtaining similar identifiability results. In addition, we demonstrated identifiability for a uniform read sampling distribution, but a similar result can be obtained for other read sampling distributions.

Species identifiability is a worst-case measure, as it requires *all* species to be identified correctly. In practice, we may settle for a weaker notion - for example we would still consider a reconstruction as successful if all species except a small minority were identified correctly. We next define *partial* identifiability, which is a weaker property characterizing our ability to correctly reconstruct identities and frequencies of specific species, while for other species the reconstruction may remain ambiguous.

Definition 2. *We say that the problem $\mathbf{MCR}(L, A)$ is partially identifiable for species j, if for any $\mathbf{x}^{(1)}, \mathbf{x}^{(2)} \in \Delta_N$ such that $P_{\mathbf{x}^{(1)}}(\mathbf{y}; A, L) = P_{\mathbf{x}^{(2)}}(\mathbf{y}; A, L) \forall \mathbf{y} \in \Delta_{4^L}$, we have $x_j^{(1)} = x_j^{(2)}$.*

We can check partial identifiability using the following proposition,

Proposition 3. *The problem $\mathbf{MCR}(L, A)$ is partially identifiable for species j, if and only if the standard basis vector $\mathbf{e}^{(j)} \in \Delta_N$ is orthogonal to the null-space of $A_{(1)}$, that is $A_{(1)}\mathbf{x} = 0 \Rightarrow x_j = 0 \; \forall \mathbf{x} \in \mathbb{R}^N$.*

We present the identifiability properties achieved for real 16S rRNA data in the Appendix (Section A.5).

4   Reconstruction Error

While identifiability ensures that one can *in principle* reconstruct correctly the species vector $\mathbf{x}$, it essentially assumes an unlimited number of reads and computational power. Here we study the reconstruction error in more realistic scenarios, with a finite number of reads. We prove general rigorous upper-bounds on reconstruction error, in terms of the matrix A and the number of reads R. In the Appendix (Section A.8) we examine the actual error achieved in practice using simulations.

The next proposition gives bounds on the approximation error of the true frequency vector $\mathbf{x}^*$ by the estimator $\hat{x}$, which we obtain using the empirically-observed frequencies $\mathbf{y}$,

Proposition 4. *Consider the problem $\mathbf{MCR}(L, A)$ with R sequence reads, and let $\hat{\mathbf{x}}$ be the estimator minimizing the l_2 loss, $\hat{\mathbf{x}} = argmin_{\mathbf{x}\in\Delta_N} l_2(A\mathbf{x}, \mathbf{y})$. Then,*

1. Let $\lambda_{\min}(A^\top A)$ be the smallest eigenvalue of $A^\top A$. The Euclidian l_2 distance satisfies:

$$Pr\left(\mathcal{D}_{l_2}(\hat{\mathbf{x}}, \mathbf{x}^*) \leq \frac{2 + \sqrt{\log(1/\delta)}}{\sqrt{R\lambda_{\min}(A^\top A)}}\right) \geq 1 - \delta, \quad \forall \delta \in (0, 1) \qquad (4)$$

2. The Mahalanobis distance with weight matrix $A^\top A$ satisfies:

$$Pr\left(\mathcal{D}_{MA}(\hat{\mathbf{x}}, \mathbf{x}^*; A^\top A) \leq \frac{2 + \sqrt{\log(1/\delta)}}{\sqrt{R}}\right) \geq 1 - \delta, \quad \forall \delta \in (0, 1). \qquad (5)$$

The bound on the convergence rate of the $\mathcal{D}_{l_2}$ error depends on spectral properties of the matrix $A^\top A$. This is related to the database coherence, or similarity between the sequences S_i, encoded as similarity between the rows of A. In particular, when the problem is non-identifiable, the matrix $A^\top A$ has a zero eigenvalue and the reconstruction error may be arbitrarily large.

In contrast, the Mahalanobis bound does not depend on the matrix A or even the dimension N. Even if the problem is non-identifiable, we still achieve convergence under the Mahalanobis distance - yet the entries in the solution vector will not converge to the corresponding entries in the true frequencies vector $\mathbf{x}$, i.e. the reconstruction may assign (part of) the abundance of a specific species to different, yet highly similar species.

5   Divide-and-Conquer Algorithm

Solving a large scale **MCR** problem with hundreds of thousands of species is computationally challenging. Even computing and storing the matrix A is not trivial, let alone minimizing the loss $l(A\mathbf{x}, \mathbf{y})$ in eq. (3). We developed a scalable divide-and-conquer thresholding approach to cope with large problems. In a nutshell, the algorithm divides the species into distinct blocks, solves a reduced-size problem within each block, setting species with low frequency in the solution for each block to zero, merges solutions from different blocks and iterates to reduce problem size. For the reduced size sub-problems we minimize the l_2 loss, resulting in a convex optimization problem in each block which we solve (exactly) using the CVX convex optimization software package [8, 21]. We describe the algorithm in more details in the Appendix (Section A.7).

We implemented the divide-and-conquer algorithm in the **COMPASS** Matlab package (with some computationally demanding parts implemented in C). For a problem of size $N \sim 5 \times 10^5$, running time is a few hours on a standard PC. The algorithm showed accurate reconstruction performance on simulated and real sequence data (see Section A.8 and [1]).

6   Discussion

We formulated the **MCR** problem mathematically, proposed an algorithm for solving large scale problems, and obtained results on reconstruction performance.

We applied our approach on the 16S rRNA gene. However, the approach is generic and could be applied to other genes or regions. The reconstruction performance is determined by properties of the genomic region used (in our case, 16S rRNA). Different genes or regions will provide different information allowing us to distinguish between different species or strains, for example using clade-specific markers [23].

Extending our method to genome-wide metagenomics sequencing is possible, although computationally challenging. Our approach relies on the presence of a database of reference sequences, and cannot be used as is for *de novo* discovery of new species. Currently there are ~ 3000 whole-genome sequences in the NCBI database [9], compared to $\sim 10^6$ 16S rRNA sequences in the Greengenes database, thus the current utility of the whole-genome approach is limited, although it can be useful as a first filter before the remaining reads can be used for *de novo* discovery (assembly). More importantly, as these database are likely to grow in the near future, it will become increasingly appealing to use whole-genome sequencing, especially for identifying small variations in very close strains, or newly born alleles in present strains (where the 16S rRNA sequences may be identical and not allow identification).

Providing efficient algorithms for the **MCR** problem is important - solving the **MCR** problem directly for N in the order of hundreds of thousands is currently infeasible due to memory and time issues. We used a feasible divide-and-conquer approach to cope with this problem yet there is still room for algorithmic improvements, especially when coping with read errors, which increase the size of the matrix A. Designing faster algorithms for handling larger databases will become crucial in light of the expected growth of microbial databases, in terms of both the number species and the regions (including whole-genomes) covered.

References

1. Amir, A., Zeisel, A., Zuk, O., Elgart, M., Stern, S., Shamir, O., Turnbaugh, P.J., Soen, Y., Shental, N.: High resolution microbial community reconstruction by integrating short reads from multiple 16S rRNA regions. In Revision (2013)
2. Amir, A., Zuk, O.: Bacterial community reconstruction using compressed sensing. Journal of Computational Biology 18(11), 1723–1741 (2011)

3. Cole, J.R., Wang, Q., Cardenas, E., Fish, J., Chai, B., Farris, R.J., Kulam-Syed-Mohideen, A.S., McGarrell, D.M., Marsh, T., Garrity, G.M., et al.: The ribosomal database project: improved alignments and new tools for rrna analysis. Nucleic Acids Research 37(suppl. 1), D141–D145 (2009)
4. DeSantis, T.Z., Hugenholtz, P., Larsen, N., Rojas, M., Brodie, E.L., Keller, K., Huber, T., Dalevi, D., Hu, P., Andersen, G.L.: Greengenes, a chimera-checked 16S rRNA gene database and workbench compatible with arb. Applied and environmental microbiology 72(7), 5069–5072 (2006)
5. Eckburg, P.B., Bik, E.M., Bernstein, C.N., Purdom, E., Dethlefsen, L., Sargent, M., Gill, S.R., Nelson, K.E., Relman, D.A.: Diversity of the human intestinal microbial flora. Science 308(5728), 1635–1638 (2005)
6. Eskin, I., Hormozdiari, F., Conde, L., Riby, J., Skibola, C., Eskin, E., Halperin, E.: eALPS: Estimating abundance levels in pooled sequencing using available genotyping data. In: Deng, M., Jiang, R., Sun, F., Zhang, X. (eds.) RECOMB 2013. LNCS, vol. 7821, pp. 32–44. Springer, Heidelberg (2013)
7. Gentry, T.J., Wickham, G.S., Schadt, C.W., He, Z., Zhou, J.: Microarray applications in microbial ecology research. Microbial Ecology 52(2), 159–175 (2006)
8. Grant, M., Boyd, S.: Graph implementations for nonsmooth convex programs. In: Blondel, V., Boyd, S., Kimura, H. (eds.) Recent Advances in Learning and Control. LNCIS, vol. 371, pp. 95–110. Springer, Heidelberg (2008), http://stanford.edu/~boyd/graph_dcp.html
9. Haft, D.H., Tovchigrechko, A.: High-speed microbial community profiling. Nature Methods 9(8), 793–794 (2012)
10. Hamady, M., Knight, R.: Microbial community profiling for human microbiome projects: Tools, techniques, and challenges. Genome Research 19(7), 1141–1152 (2009)
11. Hiller, D., Jiang, H., Xu, W., Wong, W.H.: Identifiability of isoform deconvolution from junction arrays and rna-seq. Bioinformatics 25(23), 3056–3059 (2009)
12. Huse, S.M., Dethlefsen, L., Huber, J.A., Welch, D.M., Relman, D.A., Sogin, M.L.: Exploring microbial diversity and taxonomy using SSU rRNA hypervariable tag sequencing. PLoS Genetics 4(11), e1000255 (2008)
13. Kessner, D., Turner, T., Novembre, J.: Maximum likelihood estimation of frequencies of known haplotypes from pooled sequence data. Molecular Biology and Evolution 30(5), 1145–1158 (2013)
14. Lozupone, C., Knight, R.: UniFrac: a new phylogenetic method for comparing microbial communities. Applied and Environmental Microbiology 71(12), 8228–8235 (2005)
15. Lozupone, C.A., Hamady, M., Kelley, S.T., Knight, R.: Quantitative and qualitative β diversity measures lead to different insights into factors that structure microbial communities. Applied and Environmental Microbiology 73(5), 1576–1585 (2007)
16. Mardis, E.R.: The impact of next-generation sequencing technology on genetics. Trends in Genetics 24(3), 133–141 (2008)
17. Meinicke, P., Aßhauer, K.P., Lingner, T.: Mixture models for analysis of the taxonomic composition of metagenomes. Bioinformatics 27(12), 1618–1624 (2011)
18. Paster, B.J., Boches, S.K., Galvin, J.L., Ericson, R.E., Lau, C.N., Levanos, V.A., Sahasrabudhe, A., Dewhirst, F.E.: Bacterial diversity in human subgingival plaque. Journal of Bacteriology 183(12), 3770–3783 (2001)
19. Pavoine, S., Dufour, A.B., Chessel, D.: From dissimilarities among species to dissimilarities among communities: a double principal coordinate analysis. Journal of Theoretical Biology 228(4), 523–537 (2004)

20. Pilanci, M., El Ghaoui, L., Chandrasekaran, V.: Recovery of sparse probability measures via convex programming. In: NIPS (2012)
21. CVX Research. CVX: Matlab software for disciplined convex programming, ver. 2.0 (2012), http://cvxr.com/cvx
22. Rockafellar, R.T.: Convex Analysis. Princeton Mathematics Series, vol. 28. Princeton University Press (1970)
23. Segata, N., Waldron, L., Ballarini, A., Narasimhan, V., Jousson, O., Huttenhower, C.: Metagenomic microbial community profiling using unique clade-specific marker genes. Nature Methods 9(8), 811–814 (2012)
24. Shawe-Taylor, J., Cristianini, N.: Kernel methods for pattern analysis. Cambridge University Press (2004)
25. Xia, L.C., Cram, J.A., Chen, T., Fuhrman, J.A., Sun, F.: Accurate genome relative abundance estimation based on shotgun metagenomic reads. PloS One 6(12), e27992 (2011)

Appendix

A.1   Dealing with Sequences Shorter Than the Read Length

In rare cases the read length L might be larger than the sequence length n_j for a particular species j. For completeness, we adopt a convention of a read having it's first n_i nucleotides matching the sequence, and the next $n_i - L$ nucleotides distributed uniformly in Υ^{L-n_i}. In this case eq. (2) generalizes to,

$$A_{ij} = \frac{4^{\min(0, n_j - L)} \sum_{k=1}^{\max(1, n_j - L + 1)} 1_{\{lex^{-1}(i)_{1:\min(n_j, L)} = s_{j, k:k+L-1}\}}}{\max(1, n_j - L + 1)} \qquad (6)$$

where $lex^{-1}(i)_{1:k}$ denotes the first k nucleotides in the i-th read (in lexicographic ordering). One can adopt different conventions for this case, for example obtaining a shorter read (of length n_j), or using a 'joker' symbol for the tail (i.e. for example when sequencing the molecule '$AACGCT$' a read of length 10 will be '$AACGCTNNNN$'). The choice of different conventions does not change our result significantly - we chose the above for mathematical convenience.

A.2   Proof of Proposition 1

Proof. From eq. (1), we have $P_x(e^{(i)}; A, L) = [Ax]_i$, $\forall i = 1, .., 4^L$. Therefore identifiability holds if and only if $Ax^{(1)} = Ax^{(2)} \Rightarrow x^{(1)} = x^{(2)}$, $\forall x^{(1)}, x^{(2)} \in \Delta_N$.

The vector $A_{(1)}x$ is of size $4^L + 1$, obtained as a concatenation of Ax with one additional entry, $[A_{(1)}x]_{4^L+1} = \sum_{j=1}^{N} x_j$. For any $x \in \Delta_N$ the last entry $[A_{(1)}x]_{4^L+1}$ is equal to 1. Therefore $A_{(1)}x^{(1)} = A_{(1)}x^{(2)} \iff Ax^{(1)} = Ax^{(2)}$, $\forall x^{(1)}, x^{(2)} \in \Delta_N$.

If $rank(A_{(1)}) = N$, we have $A_{(1)}x^{(1)} = A_{(1)}x^{(2)} \Rightarrow x^{(1)} = x^{(2)}$, $\forall x^{(1)}, x^{(2)} \in \mathbb{R}^N$. Therefore in particular the relation is true for any $x^{(1)}, x^{(2)} \in \Delta_N \subset \mathbb{R}^N$ and identifiability holds.

Conversely, if $rank(A_{(1)}) < N$ then there exists a non-zero vector $x \in \mathbb{R}^N$, $x \neq 0_N$ in the null-space of $A_{(1)}$. Thus $A_{(1)}x = 0$ and in particular $[A_{(1)}x]_{4^L+1} = \sum_{j=1}^{N} x_j = 0$. Take a vector $x^{(1)} \in int(\Delta_N)$. Then there exists $\epsilon > 0$ such that $x^{(2)} \equiv x^{(1)} + \epsilon x \in \Delta_N$. But $Ax^{(1)} = Ax^{(2)}$ and $x^{(1)} \neq x^{(2)}$, therefore the problem $\mathbf{MCR}(L, S, A)$ is not identifiable.

∎

A.3   Proof of Proposition 2

Proof. Take $L = 1$. Then the vector y simply measures the fraction of 'A's, 'C's, 'G's and 'T's in the sample, and is of length 4. The matrix $A^{(u,1)}$ is of size $4 \times N$, and $rank(A^{(u,1)}) \leq 4$. Therefore, there exists a non-zero vector x in the null-space of $A^{(u,L)}$, $A^{(u,L)}x = 0$. Let $x^{(1)} \in int(\Delta_N)$. Then there exists $\epsilon > 0$ such that $x^{(2)} \equiv x^{(1)} + x \in \Delta_N$. But $P_x(x^{(1)}) = P_x(x^{(2)})$ for $x^{(1)}, x^{(2)} \in \Delta_N$. Hence the problem $\mathbf{MCR}(1, A^{(u,1)})$ is not identifiable.

Take $L = n_{MAX}(= \max_i n_i)$. For each species j define the read $r^{(j)} \equiv [S_j : {}'A'^{(L-n_j)}]$ where ${}'A'^{(k)}$ is a string of k consecutive ${}'A's$, and $[a : b]$ denotes the concatenation of the two strings a and b. The read $r^{(j)}$ contains the sequence S_j, followed by a string of 'A's. Since $S_{j'}$ is not a subsequence of S_j for any $j \neq j'$, the read $r^{(j)}$ cannot appear when sequencing any other sequence $j' \neq j$, so $A_{lex(r^{(j)})j'} = 0 \ \forall j' \neq j$, and the $lex(r^{(j)})$-th row of A is all zeros except for the j-th term. This means that A has N independent rows, indexed by $lex(r^{(1)}), .., lex(r^{(N)})$ and $rank(A) = N$. Therefore $rank(A_{(1)}) = N$ and the problem $\mathbf{MCR}(n_{MAX}, A^{(u,n_{MAX})})$ is identifiable.

Suppose that the problem is $\mathbf{MCR}(L, A^{(u,L)})$ is identifiable, and let $L' > L$. By definition, for every $\mathbf{x}^{(1)} \neq \mathbf{x}^{(2)} \in \Delta_N$, there exists $\mathbf{y} \in \Delta_{4^L}$ such that $P_{\mathbf{x}^{(1)}}(\mathbf{y}; A; L) \neq P_{\mathbf{x}^{(2)}}(\mathbf{y}; A; L)$. But the distribution $P_{\mathbf{x}^{(i)}}(\cdot; A; L)$ is obtained by a projection of the distribution $P_{\mathbf{x}^{(i)}}(\cdot; A; L')$ (for $i = 1, 2$), with $P_{\mathbf{x}^{(i)}}(\cdot; A; L) = \sum_{y', y=y'_{1:L}} P_{\mathbf{x}^{(i)}}(\cdot; A; L')$. Therefore, there must exist $\mathbf{y} \in \Delta_{4^{L'}}$ with $P_{\mathbf{x}^{(1)}}(\mathbf{y}'; A; L') \neq P_{\mathbf{x}^{(2)}}(\mathbf{y}'; A; L')$ and the problem $\mathbf{MCR}(L', A^{(u,L')})$ is also identifiable for L'.

■

A.4   Proof of Proposition 3

Proof. In similar to Proposition 1, since $P_{\mathbf{x}}(\mathbf{e}^{(i)}; A, L) = [A\mathbf{x}]_i, \ \forall i = 1, .., 4^L$ we have partial identifiability if and only if $A\mathbf{x}^{(1)} = A\mathbf{x}^{(2)} \Rightarrow x_j^{(1)} = x_j^{(2)}, \forall \mathbf{x}^{(1)}, \mathbf{x}^{(2)} \in \Delta_N$, which holds if and only if $A_{(1)}\mathbf{x}^{(1)} = A_{(1)}\mathbf{x}^{(2)} \Rightarrow x_j^{(1)} = x_j^{(2)}, \ \forall \mathbf{x}^{(1)}, \mathbf{x}^{(2)} \in \Delta_N$.

Assume that $A_{(1)}\mathbf{x} = 0 \Rightarrow x_j = 0 \ \forall \mathbf{x} \in \mathbb{R}^N$. Then, for any two vectors $\mathbf{x}^{(1)}, \mathbf{x}^{(2)} \in \Delta_N$ take $\mathbf{x} = \mathbf{x}^{(1)} - \mathbf{x}^{(2)}$ to get,

$$A_{(1)}\mathbf{x}^{(1)} = A_{(1)}\mathbf{x}^{(2)} \Rightarrow A_{(1)}(\mathbf{x}^{(1)} - \mathbf{x}^{(2)}) = 0 \Rightarrow [\mathbf{x}^{(1)} - \mathbf{x}^{(2)}]_j = 0 \Rightarrow x_j^{(1)} = x_j^{(2)}. \tag{7}$$

Therefore, $\mathbf{MCR}(L, A)$ is partially identifiable for species j. For the other direction, assume that $\mathbf{MCR}(L, A)$ is partially identifiable for species j. Let $\mathbf{x} \in \mathbb{R}^N$. Take some $\mathbf{x}^{(1)} \in int(\Delta_N)$ and set $\mathbf{x}^{(2)} = \mathbf{x}^{(1)} + \alpha\mathbf{x}$ with $\alpha > 0$ small enough such that $\mathbf{x}^{(2)} \in \Delta_N$. Then,

$$A_{(1)}\mathbf{x} = 0 \Rightarrow A_{(1)}\mathbf{x}^{(1)} = A_{(1)}\mathbf{x}^{(2)} = 0 \Rightarrow x_j^{(1)} = x_j^{(2)} \Rightarrow x_j = 0. \tag{8}$$

■

A.5   Identifiability in the 16S rRNA Database

We checked the ability to identify species based on their 16S rRNA sequences. We downloaded the 16S rRNA Greengenes database from greengenes.lbl.gov [4] (file 'current_prokMSA_unaligned.fasta.gz', version dated 2010). After clustering together species with identical 16S rRNA sequences, we were left with $N =$

455, 055 unique sequences of the 16S rRNA gene, with mean sequence length 1401 - we refer to these N unique sequences as the species. We assume that the entire 16S rRNA gene is available - this can be achieved for example by shot-gun or RNA sequencing (In practice, the choice of primers used when performing targeted DNA sequencing may be restricted due to biochemical considerations. This will affect the region sequenced and therefore all aspects of the reconstruction performance including identifiability - see [1]). Although the sequences are all distinct when considering the entire 16S rRNA sequences, identifiability is not guaranteed since we only observe short reads covering possibly non-unique portions of the 16S rRNA gene, which may cause ambiguities. We plot in Figure 2 the number of uniquely identifiable species as a function of the read length L. Even for very short L, we can identify most species, since the short reads aggregate information from the entire 16S rRNA gene. However, even when L is long ($L = 100$), there is still a small subset of species which are not identifiable.

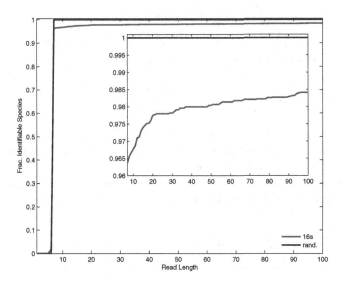

Fig. 2. Partial identifiability as a function of the read length. The red line shows results for a set of $N = 10, 000$ similar species from the Greengenes database. For comparison, the blue line shows results for $N = 10, 000$ sequences of the same length, with uniformly drawn i.i.d. characters. (i.e. $Pr('A') = Pr('C') = Pr('G') = Pr('T') = 0.25$ for each base). The X-axis is read length used. The y-axis shows the fraction of identifiable species. At $L = 7$ we see a big jump in identifiability, as expected, since this is the point at which the number of equations 4^L exceeds the number of species N. For random sequences the problem is identifiable for $L \geq 7$ (i.e., 100% of species are partially identifiable). For the sequences from the 16S rRNA database, the vast majority ($\sim 96.5\%$) of species are partially identifiable for $L = 7$. The number of partially identifiable species then increases slowly with read length (see inset). Even at $L = 100$ the problem is still not identifiable, but $\sim 98.5\%$ of species can be identified. The remaining un-identified species contain groups of species with very close sequences, which can be distinguished only by increasing read length even further.

A.6   Proof of Proposition 4

Proof. Eq. (3) with a l_2 loss implies that Ax is the Euclidean projection of $\mathbf{y}$ on the convex set $A(\Delta_N) \equiv \{\mathbf{z} : \exists \mathbf{x} \in \Delta_N, \mathbf{z} = A\mathbf{x}\}$ (namely, it is the closest point to $\mathbf{y}$ in $A(\Delta_N)$). Similarly, $A\mathbf{x}^*$ is the Euclidean projection of $\mathbf{y}^*$ on $A(\Delta_N)$. Since projections on convex sets can only reduce distances [22], we have,

$$\|A\mathbf{x} - A\mathbf{x}^*\|_2 = \|A\mathbf{x} - \mathbf{y}^*\|_2 \le \|\mathbf{y} - \mathbf{y}^*\|_2 . \tag{9}$$

The left hand side above is equal to the Mahalanobis distance, since

$$\mathcal{D}_{MA}(\mathbf{x}, \mathbf{x}^*; A^\top A) = \sqrt{(\mathbf{x} - \mathbf{x}^*)^\top (A^\top A)(\mathbf{x} - \mathbf{x}^*)} = \|A\mathbf{x} - A\mathbf{x}^*\|_2 . \tag{10}$$

Therefore we get

$$\mathcal{D}_{MA}(\mathbf{x}, \mathbf{x}^*; A^\top A) \le \|\mathbf{y} - \mathbf{y}^*\|_2 . \tag{11}$$

Recall that $\mathbf{y} = \frac{1}{R} \sum_{i=1}^{R} \mathbf{y}^{(i)}$ where the $\mathbf{y}^{(i)}$ are i.i.d. vectors with $E[\mathbf{y}^{(i)}] = \mathbf{y}^*$. Using large-deviation bounds on vectors [24] we get,

$$Pr\left(\|\mathbf{y} - \mathbf{y}^*\|_2 \le \frac{2}{\sqrt{R}} + \sqrt{\frac{\log(1/\delta)}{R}} \right) \ge 1 - \delta, \quad \forall 0 < \delta < 1 \tag{12}$$

Combining eqs. (11,12), we get part 2 of the proposition.

To prove part 1, we need to convert this result to a bound on the Euclidian distance between $\mathbf{x}$ and $\mathbf{x}^*$. The conversion is performed by first writing an eigen-decomposition of $A^\top A$, $A^\top A = U \Lambda U^\top$ where U is an orthogonal matrix and Λ a diagonal matrix with the eigenvalues of $A^\top A$. This gives,

$$\begin{aligned}
\mathcal{D}_{MA}(\mathbf{x}, \mathbf{x}^*; A^\top A)^2 &= (\mathbf{x} - \mathbf{x}^*)^\top (U \Lambda U^\top)(\mathbf{x} - \mathbf{x}^*) \\
&\ge \|U^\top (\mathbf{x} - \mathbf{x}^*)\|_2^2 \lambda_{\min}(A^\top A) \\
&= \|(\mathbf{x} - \mathbf{x}^*)\|_2^2 \lambda_{\min}(A^\top A) \\
&= \mathcal{D}_{l_2}(\mathbf{x}, \mathbf{x}^*)^2 \lambda_{\min}(A^\top A) \tag{13}
\end{aligned}$$

Dividing both sides by $\lambda_{\min}(A^\top A)$, taking the square root and substituting in eq. (5) gives immediately part 1.

∎

A.7   Details of Divide-and-Conqour Algorithm

Box 1: Divide-and-Conquer Reconstruction Algorithm
Input: S - Set of Sequences, y - read measurements, Probabilistic model
Output: x - vector of species frequencies
Parameters: B - block size. τ_B - frequency threshold for each block. $k_{B,j}$ - number of partitions into blocks in j-th iteration, k_F - final number of species allowed

1. Partition to blocks: Set **v** as a binary vector with one entry per species. If this is the first partitioning, set iteration number $j = 1$. Repeat $k_{B,j}$ times:
   (a) Partition species randomly into non-overlapping blocks of size B.
   (b) In each block (B) compute the matrix $A^{(B)}$, (where $^{(B)}$ denotes the restriction of a vector or a matrix to a block B), and solve (exactly) the convex optimization problem (using CVX),

$$\min_{\mathbf{x}^{(B)}} ||A^{(B)}\mathbf{x}^{(B)} - \mathbf{y}||_2 \ s.t., x_i^{(B)} \geq 0 \qquad (14)$$

   (c) Collect all species with frequency above the threshold: if $x_i^{(B)} \geq \tau_B$, set $v_i = 1$. Set $j = j + 1$.
   (d) Collect all linearly dependent species: For each i which is non-identifiable in the block (i.e. species i is orthogonal to the null space of $A^{(B)}$) set $v_i = 1$.
2. Collect results from blocks: Keep only indices i with $v_i = 1$, i.e. species with high enough frequency in at least one block reconstruction.
3. Reduce problem size: Keep only species i with $v_i = 1$. Set $V = \{i, v_i = 1\}$ and set $A = A^{(V)}$, $\mathbf{x} = \mathbf{x}^{(V)}$. If $|V| > k_F$, go back to step 1.
4. Solve for the last time the l_2 minimization problem for the reduced matrix,

$$\min_{\mathbf{x}^{(V)}} ||A^{(V)}\mathbf{x}^{(V)} - \mathbf{y}||_2 \ s.t., x_i^{(V)} \geq 0 \qquad (15)$$

Normalize $\mathbf{x}^{(V)}$ to sum to one, and output the normalized vector as the solution

---

A.8   Simulation Results

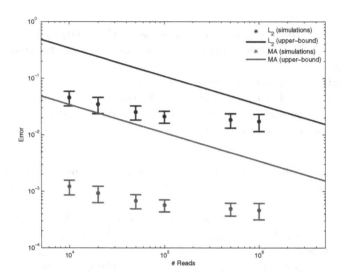

Fig. 3. The curves show the l_2 (blue) and Mahalanobis (red) errors in reconstruction for the example described in the text as function of sample size (number of reads used). Error-bars show mean and 1 standard deviation of error over 100 simulations. Solid curves show the theoretical upper-bounds, taken with $\delta = 1/2$, giving a bound on the median error. For both metrics, the performance achieved in practice is significantly better than the upper bound.

To evaluate the actual reconstruction performance in practice, we have performed a simulation study. In Figure 3 we compare the actual reconstruction performance using simulations to the general rigorous bounds obtained in Section 4.

In our simulations, we studied the performance as a function of the number of reads using the Greengenes 16S rRNA database, with $N = 455,055$ unique 16S rRNA sequences. In each simulation we sampled at random $k = 200$ species out of the total N. We sampled the species frequencies from a power-law distribution with parameter $\alpha = 1$, with frequencies normalized to sum to one. We then sampled sequence read according to the model in eq. (1). Read length was $L = 100$. The number of reads R was varied from 10^4 to 10^6.

We performed reconstruction using Algorithm 1, with the following parameters: block size $B = 1000$, threshold frequency $\tau_B = 10^{-3}$. The parameter $k_{B,j}$ represents a trade-off between time complexity and accuracy, and was initialized to 1 at $j = 1$, then set to 10 when total size $|V|$ was below $150,000$. Then, set to 20 below $20,000$. The final block size used was $k_F = 1000$.

Very low error ($\sim 2\%$) is achieved for $R > 500,000$, showing that accurate reconstruction is possible for a feasible number of reads. The error rate achieved in practice is much lower than the theoretical bounds, indicating that tighter bounds might be achieved. There are many reasons for the gap between our bounds and simulation results: the concentration inequalities we have used may not be tight, the particular frequency distribution chosen may perform better than the worst-case distribution, and most importantly, the small number of species present in the simulated mixture may enable accurate detection with a smaller sample size. Proving improved bounds on reconstruction performance which consider all these issues including the sparsity of the solution is interesting yet challenging. Standard techniques (e.g. from compressed sensing) would need to be modified to achieve improved bounds since they assume incoherence of the matrix A which does not hold in our case, and do not consider the poisson sampling model we use for the reads.

Distributed Query Processing
on Compressed Graphs Using K2-Trees*

Sandra Álvarez-García[1], Nieves R. Brisaboa[1],
Carlos Gómez-Pantoja[2], and Mauricio Marin[3]

[1] Database Laboratory, University of Coruña, Spain
[2] Universidad Andres Bello, Facultad de Ingeniería, Sazié 2325, Santiago, Chile
[3] Yahoo!Research Latin America, Santiago, Chile

Abstract. Compact representation of Web and social graphs can be
made efficiently with the K^2-tree as it achieves compression ratios about
5 bits per link for web graphs and about 20 bits per link for social graphs.
The K^2-tree also enables fast processing of relevant queries such as direct
and reverse neighbours in the compressed graph. These two properties
make the K^2-tree suitable for inclusion in Web search engines where it is
necessary to maintain very large graphs and to process on-line queries on
them. Typically these search engines are deployed on dedicated clusters
of distributed memory processors wherein the data set is partitioned and
replicated to enable low query response time and high query throughput.
In this context a practical strategy is simply to distribute the data on the
processors and build local data structures for efficient retrieval in each
processor. However, the way the data set is distributed on the processors
can have a significant impact in performance. In this paper, we evaluate
a number of data distribution strategies which are suitable for the K^2-
tree and identify the alternative with the best general performance. In
our study we consider different data sets and focus on metrics such as
overall compression ratio and parallel response time for retrieving direct
and reverse neighbours.

1 Introduction

Efficiency of parallel query processing in large Graphs has become a relevant
issue due to emergent applications in the Web and social networks in which
there exists a Graph that must be held in main memory to be queried in real
time. Efficiency has implications in the ever increasing need to (1) reduce service
latency represented by total response time of individual queries of the order of
few milliseconds, (2) design systems capable of processing hundreds of thousands
queries per second using the least amount of hardware resources possible, and

---

* SAG and NB were founded by MICIN (PGE and FEDER) grants TIN2009-14560-
C03-02, TIN2010-21246-C02-01, and CDTI CEN-20091048 and Xunta de Galicia
(co-funded with FEDER) ref. 2010/17. MM was partially funded by research grant
FONDEF IDeA CA12I10314.

O. Kurland, M. Lewenstein, and E. Porat (Eds.): SPIRE 2013, LNCS 8214, pp. 298–310, 2013.

(3) optimize power consumption in data centers hosting the query processing service. To this end, clusters of dedicated processors are deployed in the respective data center in a "one service – one cluster" manner.

To achieve scalable and flexible services, the query processing task is organized as a distributed memory system where processors compute on local data and communication among processors is performed via message passing. Typically this paradigm is applied in a master/slaves fashion where the master (broker) is in charge of sending queries to a set of slaves (processors). The dataset is assumed to be evenly distributed on the processors. Upon the reception of a query from the broker, the processors compute the local top K answers for the query and send the results back to the broker. The broker then merges the local results to compute the global top K results. This scheme has practical advantages related to dynamically handling processor replication to meet query throughput requirements and support fault tolerance.

In this paper we follow the master/slaves approach in the context of serving queries upon a distributed Graph that has been compressed using the K^2-tree method. In this case, the key for achieving efficient performance is to be smart on how to distribute the data across the processors. We propose a number of data distribution alternatives and present an evaluation study using actual datasets executed on a cluster of processors. The experimental results tell us that a strategy we call Latin Square offers the best performance in general.

2   Related Work

Notice that previous works focus on off-line processing whereas we are interested in on-line query processing.

Parallel Boost Graph Library (PBGL) [8] (based on Boost Graph Library [1]), is a generic library written in C++ that implements distributed graph data structures and graph algorithms. To implement a parallel algorithm, it applies existing sequential algorithms to distributed data structures. It supports a rich set of parallel graph implementations and property maps. In contrast with Pregel and HipG expressiveness, PBGL offers a very general model to implement parallel algorithms. Pregel [13] is a scalable infrastructure to mine graphs, where each program is expressed as a sequence of iterations. This infrastructure is inspired by the Bulk Synchronous Parallel (BSP) model [14], which represents a program as a sequence of *supersteps*. Pregel partitions the graph using a hash function applied to the vertex identifier $modN$, where N is the number of partitions, and all its outgoing edges are assigned to the same partition. The partitioning method can be user-defined.

Parallel Combinatorial BLAS [6] is a scalable high-performance library that enables graph analysis and data mining. The authors mention that this library is unique among other libraries, because it combines scalability and distributed memory parallelism. The p processors are logically organized as a two-dimensional grid (to limit the communication), and the partitioning of matrices follows this organization, using a 2D block decomposition. As we will see in the experiments, this partitioning does not reach the best results.

HipG [9] is a distributed framework in which the underlying idea is similar to Pregel: the user has to define pieces of sequential work to be executed in each graph node. HipG partitions the graph nodes into equal-size *chunks*. A chunk is a set of graph nodes and their outgoing edges (edges are co-located with their source nodes). Chunks are assigned to workers, which are the responsible for processing the nodes associated to the chunk. HipG is similar to Pregel in two aspects: the vertex-centered programming and composing the parallel program automatically from user-provided simple sequential-like components. The main difference is the BSP-like global synchronization in each superstep used in Pregel. In contrast, HipG uses asynchronous messages with computation synchronized on the user's request.

GraphLab [12] is a parallel abstraction that exploits the sparse structures and computational patterns of Machine Learning algorithms. The same authors extend this tool to a distributed setting: Distributed GraphLab [15]. Finally, PowerGraph [7] introduces a new approach that exploits the structure of power-law graphs, which are difficult to partition and represent in a distributed environment. PowerGraph exposes greater parallelism, reduces network communication and storage costs associated to the graph processing, and provides a highly effective scheme to distributed graph placement. It also provides fault tolerance.

2.1 K^2-Tree

K^2-tree is a compact data structure to represent binary relationships represented over a conceptual adjacency matrix. Rows and columns of an adjacency matrix M represent the objects in the relationship. A cell $M[i,j]$ would have a 1 value if there were a relationship between the object represented by the row i with the object represented by the column j, and a 0 otherwise.

K^2-tree was originally designed to represent web graphs [10], and it takes advantage of the existence of large areas with a high density of *ones* or *zeros*. It achieves a very compact space (less than 5 bits per link) over very sparse matrices allowing to very large datasets fitting in the main memory. K^2-tree also allows an efficient navigation over the compressed structure [4], providing fast retrieving of direct and reverse neighbours.

The K^2-tree construction begins with the subdivision of the adjacency matrix in K^2 submatrices of equal dimensions. Each one of K^2 submatrices are represented with one bit in the first level of the tree, following a top-down and a left-right order. The bit that represents each submatrix will be 1, if the submatrix contains at least one cell with value 1. Otherwise, the bit will be a 0.

The next level of the tree is created by expanding the 1 elements of the previous levels (that is, the not-empty areas), dividing in the same way the corresponding submatrix in K^2 submatrices. This method continues recursively until the subdivision gets to cell-level. Variations of this structure have been proposed by using different K values depending on the level of the tree or by compressing the last levels through a submatrix vocabulary which is encoded with Direct Access Codes [5].

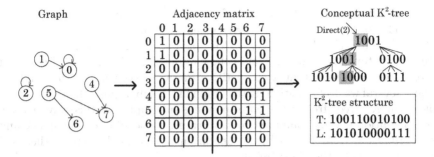

Fig. 1. An example of a binary relationship represented with a k^2-tree

Figure 1 shows an example of this tree creation for $k = 2$. It represents a binary relationship of a set of 8 elements whose graph is shown in the left, with its corresponding adjacency matrix is shown in the middle of the Figure. The K^2-tree structure is represented in the right. The first 1 of the first level means that the up-left 4x4 submatrix has at least a cell with 1 value. The second bit, which is a 0, means that the up-right submatrix does not contain any relation between nodes (that is, all its cells are *zero*) and so on. Therefore, a node with 0 has no children because it represents a submatrix full of *zeros*. Otherwise, each node with a one has 4 children corresponding to the subdivision of the matrix it represents in K^2 submatrices; again, each one will be represented with a *zero* or a *one* depending on whether they have or not at least a cell with a one.

The K^2-tree is only an abstract representation. In fact, it is stored in a very compact way using two bitmaps called T and L. T is a bitmap that stores all the intermediate levels of the K^2-tree, following a level-wise traversal (from left to right) over it. L stores the bits of the last level of the tree, from left to right. It is easy to see in the Figure 1 how T and L store the whole K^2-tree.

Retrieving direct or reverse neighbours is the most common operation performed over an adjacency matrix. It requires obtaining the cells with a 1 value for a given row or column in the adjacency matrix. These operations are solved in a K^2-tree by following a top-down traversal over the tree and they are symmetrical in terms of their computing cost. The example shows the bits of the tree involved in order to obtain the direct neighbours of the element 2 (that is, recovering the *ones* which appear in the second row of the adjacency matrix). This navigation over the K^2-tree is efficiently performed over the bitmaps T and L through an additional structure of counters, created over the bitmap T, which allows to rank operations performing in an efficient way. Note that, given a bit x in T, the children of x are between positions $rank(T, x) * K^2$ and $rank(T, x) * K^2 + K^2 - 1$. More details can be found on [4,10].

3   Our Proposal

In this work we study how K^2-tree can be used as a basis in order to build and query a distributed graph in a parallel environment. Our problem can be summarized in how to partition a graph $G = (N, E)$ (where N denotes the set of the nodes of the graph and E corresponds with the edges that connect the nodes) in a set of $P = \{p_i, i = 1, \ldots, |P|\}$ independent processors.

In this context, the main problem is how to optimize the space and querying response in order to obtain a competitive querying system. To that end we propose several ways of partitioning the graph in $|P|$ subgraphs. Then, processors build local K^2-trees from their corresponding subgraphs. In this way, a basic query operation can be computed by performing, depending on the query and the distribution, from 1 to $|P|$ local operations and a final union of the local answers to compose the global result.

Next, we propose different graph distribution strategies. They map each cell of the global adjacency matrix to only one processor. However, a node of the graph could be implicitly represented in several processors, since its outgoing edges can be stored in different processors.

3.1   Basic Distributions

As explained before, K^2-tree represents the adjacency matrix of a graph. Therefore, some classic matrix partitioning can be applied in order to obtain a distributed graph where each processor stores its adjacency matrix by using a K^2-tree. We propose several distributions where each cell (x, y) of the adjacency matrix is mapped by a simple formula to its corresponding position $p_i(x', y')$, meaning the cell (x', y') is placed at processor p_i. In this way, no additional information has to be stored to perform graph mining over the distributed graph. Figure 2 shows an example of basic distributions for 4 processors.

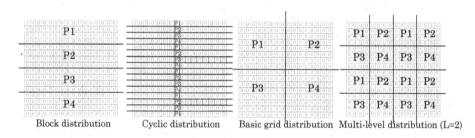

Block distribution   Cyclic distribution   Basic grid distribution   Multi-level distribution (L=2)

Fig. 2. Basic distributions with $|P| = 4$ and a graph with $|N| = 16$ nodes

Block Distribution. We can divide the adjacency matrix in $|P|$ horizontal blocks. Each processor builds a K^2-tree for a subgraph which supports an adjacency matrix with dimensions $(block, |N|)$, where $block = \lceil \frac{|N|}{|P|} \rceil$. The K^2-tree needs to fill out the adjacency matrix with zeros in order to obtain a square matrix, but since big regions of 0 can be compressed by using only a few bits, this asymmetric dimension does not deteriorate the compression. Likewise, a vertical distribution could be used too.

We define a neighbour operation over a node q, $direct(q)$, in terms of the local operations $direct_{p_i}(q')$, meaning the row q' of the processor i is queried. Each obtained local result r is mapped to the global graph through the function $dMap_{p_i}(r)$. The reverse neighbour operation uses the same notation. We can note that a direct neighbour operation only needs one processor to be performed. However, a reverse neighbour operation in answered through the union of the local results of all processors:

- $direct(q) = direct_{p_{1+\lfloor \frac{q}{block} \rfloor}}((q \mod block))$, with $dMap_{p_i}(r) = r$

- $reverse(q) = \bigcup\limits_{i=1}^{|P|} reverse_{p_i}(q)$, with $rMap(r)_{p_i} = r + (p_i - 1) \cdot block$

The main disadvantage of this method is that balance in terms of space strongly relies on the distribution of the adjacency matrix. If it is heterogeneous this distribution will achieve a poor spatial balance.

Cyclic Distribution. This basic distribution tries to minimize the dependency on the distribution by performing a cyclical distribution, where the rows of the global matrix are mapped to processors in a round-robin fashion. As in block distribution it has asymmetrical behaviour for the basic operations:

- $direct(q) = direct(\lfloor \frac{q}{block} \rfloor)_{p_{1+(q \mod |P|)}}$, with $dMap_{p_i}(r) = r$

- $reverse(q) = \bigcup\limits_{i=1}^{|P|} reverse_{p_i}(q)$, with $rMap_{p_i}(r) = p_i + r \cdot block$

The main disadvantage of this distribution is its low compressibility because it breaks the natural clusterization of 1 in the adjacency matrix used for the K^2-tree to save space.

Grid Distribution. As a symmetrical alternative we can distribute the adjacency matrix over $|P|$ square matrices of dimension $sq * sq$, where $sq = \lceil N/\sqrt{|P|} \rceil$, as it is shown in the left-bottom of the Figure 2. Unlike Block and Cyclic distributions direct and neighbour operations are always distributed over sq processors. However it still divides the matrix in big regions, being sensible to the node distribution. This basic grid can be improved by making a recursive L-grid distribution, where L denotes the number of levels of recursion, so we have submatrices with dimensions $sq' * sq'$, where $sq' = \lceil \frac{N}{L\sqrt{|P|}} \rceil$. An example with $L = 2$ is shown on the bottom-right in the Figure 2. Using larger L values, the imbalance produced by an heterogeneous graph distribution is highly minimized. However, with very larger L values the locality of the data can be lost because of the assignment of smaller submatrices to different processors. The effect of the L parameter is discussed in the experimental evaluation. Next we formalise the implementations of the direct and reverse retrieval:

- $direct(q) = \bigcup\limits_{i=1}^{\sqrt{P}} direct_{p_{\sqrt{P}(\lfloor \frac{q}{sq'} \rfloor \mod \sqrt{|P|})+i}}(q \mod sq')$

- $dMap_{p_i}(r) = \sqrt{|P|}sq'\lfloor\frac{r}{sq'}\rfloor + sq'((i-1) \mod \sqrt{|P|}) + (r \mod sq')$
- $reverse(q) = \bigcup\limits_{i=1}^{\sqrt{P}} reverse_{P_{(\lfloor\frac{q}{sq'}\rfloor \mod \sqrt{|P|})+1+(i-1)\sqrt{|P|}}}(q\cdot \mod sq')$
- $rMap_{p_i}(r) = \sqrt{|P|}sq'\lfloor\frac{r}{sq'}\rfloor + sq'\lfloor\frac{i-1}{\sqrt{|P|}}\rfloor + (r \mod sq')$

3.2 Perfect Spatial Balanced Distribution

Basic distributions do not guarantee balance in terms of space, because it strongly depends on the distribution of the edges over the adjacency matrix. Now we are focused on levelling out the final size of the K^2-tree structures that each processor manages, expecting that a spatial balance may bring a well-balanced work load. While the previous section describes typical distributions of an adjacency matrix, this distribution is specific of a final K^2 structure, because it is designed attending to its structural characteristics.

We first consider a global K^2-tree, which stores the full graph (shown in the Figure 3). We propose a distribution of the edges of the graph attending to their position in this global K^2-tree. That is, we allocate the edges to the processors following the order of the last level of the tree. This level corresponds with a $Z - ordering$ over the position of those edges in the adjacency matrix. This distribution will accomplish that if an edge e_i (where i denotes the position of the edge in the last level of the global tree) is allocated on the processor p_x, and another edge e_j is allocated on the processor p_y, and $i <= j$; then $x <= y$. In this way, we expect to avoid representing duplicated elements of the intermediate levels of the tree in multiple processors, since edges with common ancestors are expected to be stored in the same processor. Consequently, this distribution contains a minimum overhead of space in relation to the sequential approach.

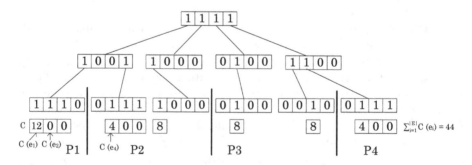

Fig. 3. An example of perfect balanced distribution

If we study how the K^2-tree is built, we can notice that each edge of the tree is represented by using K bits for each level. That is, for a graph with N nodes, the path of an edge from the root to the leaf costs $K^2 log_K N$ bits. However, if we attend to the cumulative costs, not all the edges of the tree cost the same.

An edge placed in a dense region costs less bits than an isolated edge in a sparse region, because the edges in a dense area will share common ancestors that are stored only once. We first define the spatial cost of a graph $G = (N, E)$, denoted as $SC(G)$, as the number of bits that the K^2-tree structure spends in order to represent G. Then, we define the differential cost of an edge e_i, $C(e_i) = SC(G') - SC(G'')$, where $G' = (N, \{e_1, ..., e_i\})$ and $G'' = (N, \{e_1, ..., e_{i-1}\})$, that is the number of bits that the insertion of e_i in G'' costs. We have $SC(G) = \sum_{i=1}^{|E|} C(e_i)$, so we distribute the edges of the graph by storing in each processor a $K^2 - tree$ containing consecutive edges $e_m, ..., e_n$, where $\sum_{i=m}^{n} C(e_i) \approx \frac{SC(G)}{|P|}$. An example of this distribution can be seen in the Figure 3, which shows, for each edge in the last level e_i, its differential cost $C(e_i)$ (in number of bits). Note that the common ancestors for the last edge of each processor i and the first edge of the processor $i+1$ are replicated between these two processors, but this overhead is minimal since the overlapping between two processors is less than $K^2 log_K N$ bits.

3.3   Latin-Square Distribution

We propose another approach focused on workload efficiency. It forces all operations to be distributed over the $|P|$ processors, while it tries to maintain a good balance in terms of space.

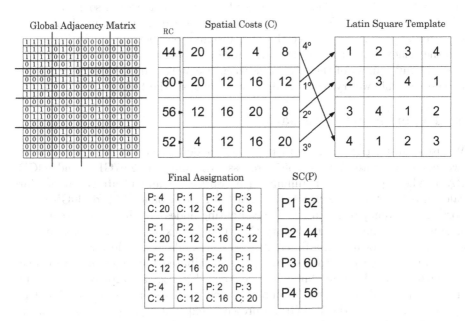

Fig. 4. An example of a Latin Square distribution

We start with the subdivision of the adjacency matrix in a grid of $|P| * |P|$ regions. For each region a K^2-tree is built. The spatial cost of a region (i, j),

$SC(G_{i,j})$ will be the number of bits of the K^2-tree representing this region. Figure 4 shows the grid of spatial costs *(top-center)* for a global adjacency matrix *(top-left)*. We define the cost of a row i, $RC_i = \sum_{j=1}^{|P|} SC(G_{i,j})$, which is also shown in the Figure 4.

We want to distribute this grid of $|P| * |P|$ K^2-trees so that any row and any column of the matrix is distributed over all the processors. Latin Square is an example of matrix that complies with this restriction. It is a two-dimensional matrix of size $n * n$ in which each cell contains a value between $1, \ldots, n$, where each value appears once in each row and column. The normalised Latin Square with dimensions $|P| * |P|$ (shown in the Figure 4) will be our starting template LS, although any other could be used. We could use this template to allocate each K^2-tree $G_{i,j}$ to the processor $LS(i,j)$. In the example, $G_{0,0}$ will be stored in the processor P_1, since $LS(0,0) = 1$, and so on. This distribution complies with our first requirement: any query is distributed over all processors.

However, this procedure is not enough, because we also want to obtain a good spatial balance. In order to improve it, we will permute the rows of our template. We start with the row i that has the maximum spatial cost RC, which is row 2. Then, we assign any available row of the LS template to this row. After that we disable this row, because it cannot be assigned to any other row of the grid. Next, we choose the row with the next maximum cost; in this case, row 3. Then we compute which of the available rows of the template LS obtains a better spatial balance when it is assigned to row 3: it is row 2, so LS_2 will be allocated to the row 3. The same process is performed $|P|$ times, until each row has been assigned. We can see the final distribution (bottom-left in Figure 4) and the achieved spatial costs for each processor (bottom-right). In this way any row and any column of the original adjacency matrix is distributed over all $|P|$ processors while we try to maintain a reasonable spatial balance.

4   Experimental Evaluation

We use a cluster composed by 67 processing nodes. Each node is equipped with two quad-core Intel Xeon E5555 processors running at 2.67GHz, and 24GB RAM. The cluster uses an Infiniband network to communicate the nodes for calculations and I/O purposes, that reaches a peak bandwidth of 40Gb/s per port and a latency of 100 nsec. The computing nodes are allowed to use a message passing communication library (MPI). In the experiments, we ensure that each process is located in a different processing node in the cluster.

We analyse the performance of all our distributed implementations on real graphs from two different contexts. Live Journal is a social graph obtained from [11] that represents the relationships between the user of this community. UK is a Web crawl from the WebGraph project [3][2] that represents links between pages of the Web. Table 1 shows the number of nodes and edges of the graphs, and the size of the K^2-tree structure that stores them.

In this experiment, we distribute the graphs over a system with $|P|$ processors using our strategies. We implement a querying system divided in supersteps. In superstep i, each processor p will perform the following operations:

Table 1. Web and social graphs used in this experimentation

Graph	Domain	Number of nodes	Number of edges	Size (MB)	BPE
LiveJournal	Social Network	4847571	68993773	167.29	20.34
UK-2002	Web	18520486	298113762	149.78	4.21

- It receives Q queries and it computes, depending on the distribution, which processors have to be queried in order to answer each one of them, mapping the global query to several local queries.
- It solves the queries received from the other $|P| - 1$ processors sent in superstep $i - 1$ and it sends the answers to the remiting processor.
- It gathers the answers from the other processors calculated in the superstep $i - 2$, mapping the local results to the global graph to produce the final answer.

We use a value of $Q = 100000$ for each superstep, that is, the querying system processes $100000 * |P|$ queries per superstep. The performance is evaluated from the average time of 10 supersteps. We run experiments with 1 (sequential version), 4, 9, 16 and 25 processors.

We first compare the proposed approaches in terms of space. The total space of the distributed graph is composed by the cost of all local K^2-trees. Depending on the distribution, additional information for the mapping could be required. For instance, the Latin Square distribution needs to store the final assignation of the $|P| * |P|$ K^2-trees replicated in all processors. If we compare the total space regarding to a K^2-tree, the overhead of the distributed approaches is minimum (less than 1%). The only exception is the cyclic distribution, which damages the compression of the K^2-tree (since it loses the locality of the data, getting worse as the number of processors grows). For instance, the cyclic distribution of the graph UK in 25 processors has a space overhead of 44%.

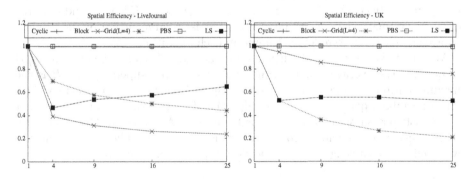

Fig. 5. LiveJournal and UK spatial efficiency comparative

A more interesting study is how well this spatial cost is distributed in the network, because a good balance will allow bigger datasets to fit in the main-memory of each individual processor. Figure 5 shows for LiveJournal (left) and

UK (right), the spatial efficiency achieved with the different distributions. Efficiency is calculated as X/Y where X is the average space occupied by the graph partition in each processor and Y is the maximum space in any processor. As expected, the perfect spatial balanced distribution achieves the best results in both graphs, since it is specifically designed to allocate the same number of bits on each processor. Cyclic distribution also obtains a good spatial efficiency close to 1 because it minimizes the effects caused by a heterogeneous distribution of the matrix (but in return of a significant spatial overhead). Block distribution obtains different results for the Social and the Web graphs, because its efficiency strongly relies on the distribution of the graph. Most of the edges of LiveJournal appear in the first rows and columns of the matrix, so in this case, block distribution achieves a poor spatial efficiency. However, the results achieved in the UK graph for the block distribution are much better since the distribution of this graph favours this partitioning. Grid and Latin Square distributions obtain intermediate results for both graphs.

We also analyse the running time performance of the different distributions. Speed-up is defined as T_s/T_p where T_s is the running time of a sequential algorithm for the problem and T_p the running time of a parallel algorithm. Figure 6 shows the speed-up obtained for LiveJournal (left) and UK (right) executing direct neighbour queries (we omit the results for retrieving reverse neighbours because identical results were obtained). We observe that PSB obtains the worst results. Since the processors which contain elements of a row or column of the adjacency matrix are unknown, the implementation of a direct or reverse operation always queries all processors of the network, deteriorating performance. The results obtained for the block distribution demonstrate its dependency on the distribution, obtaining better results for UK. We also observe that Grid distribution with $L = 4$ achieves good results that can be explained because it only asks to $\sqrt{|P|}$ processors for each query. Finally, Latin Square obtains, in general, very good results. It has a great advantage with respect to other approaches: the original adjacency matrix is represented by $|P| * |P|$ K^2-trees, and only $|P|$ of them are queried in each operation. That is, a large region of the original adjacency matrix is completely ignored in each operation.

We finally study the load efficiency for each distribution calculated as A/B where A is the average running time of processors per superstep and B is the maximum running time in any processor. Figure 7 shows the results achieved for direct neighbour queries in LiveJournal(left) and UK(right). We can observe that cyclic distribution obtains the best load efficiency. As expected, Grid and Latin Square distributions achieve good load efficiency as well, while PSB obtains the worst efficiency causing the poor speed-up reported in Figure 6.

5   Conclusions

In this paper we have proposed alternative methods for distributing compressed Web and Social Graphs on a set of distributed memory processors. The complex features of the K^2-tree compact data structure make the distribution a non-trivial task. For the same reason, at the start of our study, it was not clear to us

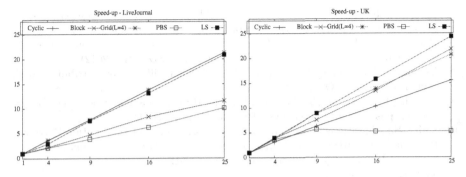

Fig. 6. Speed-up

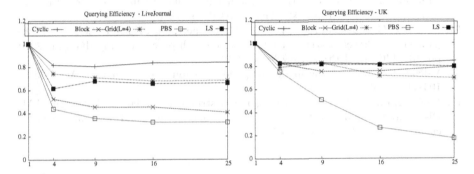

Fig. 7. Querying efficiency

which scheme was most efficient in practice. To answer this question we resorted to actual implementations and experiments executed on a cluster of processors. The results indicate that strategies like Latin Squares are the best choice as it achieves good speed-ups due to a good load balance across processors during query processing. Its $|P| * |P|$ K^2-trees approach, meaning that each processor contains $|P|$ trees, enables an even distribution of the amount of processing per query performed in each processor.

References

1. The boost graph library: user guide and reference manual. Addison-Wesley Longman Publishing Co., Inc., Boston (2002)
2. Boldi, P., Codenotti, B., Santini, M., Vigna, S.: Ubicrawler: A scalable fully distributed web crawler. Software: Practice & Experience 34(8), 711–726 (2004)
3. Boldi, P., Vigna, S.: The WebGraph framework I: Compression techniques. In: WWW, pp. 595–601. ACM Press, Manhattan (2004)
4. Brisaboa, N.R., Ladra, S., Navarro, G.: k2-trees for compact web graph representation. In: SPIRE, pp. 18–30 (2009)
5. Brisaboa, N.R., Ladra, S., Navarro, G.: Dacs: Bringing direct access to variable-length codes. In: SPIRE, pp. 392–404 (2009)

6. Bulu, A., Gilbert, J.R.: The combinatorial blas: design, implementation, and applications. Int. J. High Perform. Comput. Appl. 25(4), 496–509 (2011)
7. Gonzalez, J.E., Low, Y., Gu, H., Bickson, D., Guestrin, C.: Powergraph: distributed graph-parallel computation on natural graphs. In: OSDI 2012 (2012)
8. Gregor, D., Lumsdaine, A.: The parallel bgl: A generic library for distributed graph computations. In: POOSC (2005)
9. Krepska, E., Kielmann, T., Fokkink, W., Bal, H.: Hipg: parallel processing of large-scale graphs. SIGOPS Oper. Syst. Rev. 45(2), 3–13 (2011)
10. Ladra, S.: Algorithms and Compressed Data Structures for Information Retrieval. PhD thesis, Department of Computer Science, University of A Coruña (2011)
11. Leskovec, L.: Snap: Stanford network analysis platform, http://snap.stanford.edu
12. Low, Y., Gonzalez, J., Kyrola, A., Bickson, D., Guestrin, C., Hellerstein, J.M.: Graphlab: A new framework for parallel machine learning. In: Grünwald, P., Spirtes, P. (eds.) UAI, pp. 340–349. AUAI Press (2010)
13. Malewicz, G., Austern, M.H., Bik, A.J.C., Dehnert, J.C., Horn, I., Leiser, N., Czajkowski, G.: Pregel: a system for large-scale graph processing. In: SIGMOD 2010, pp. 135–146. ACM Press, New York (2010)
14. Valiant, L.G.: A bridging model for parallel computation. Commun. ACM 33(8), 103–111 (1990)
15. Yucheng, L., Bickson, D., Gonzalez, J., Guestrin, C., Kyrola, A., Hellerstein, J.M.: Distributed graphlab: a framework for machine learning and data mining in the cloud. VLDB 5(8), 716–727 (2012)

Author Index